Adsorption and Diffusion in Nanoporous Materials

Second Edition

T0315211

Adsorption and Diffusion in Nanoporous Materials

Second Edition

Rolando M.A. Roque-Malherbe

CRC Press
Taylor & Francis Group
Boca Raton London New York

CRC Press is an imprint of the
Taylor & Francis Group, an **informa** business

CRC Press
Taylor & Francis Group
6000 Broken Sound Parkway NW, Suite 300
Boca Raton, FL 33487-2742

First issued in paperback 2020

© 2018 by Taylor & Francis Group, LLC
CRC Press is an imprint of Taylor & Francis Group, an Informa business

No claim to original U.S. Government works

ISBN-13: 978-0-367-57216-7 (pbk)
ISBN-13: 978-1-138-30509-0 (hbk)

Library of Congress Cataloging-in-Publication Data

Names: Roque-Malherbe, Rolando M.A., author.
Title: Adsorption and diffusion in nanoporous materials / Rolando M.A. Roque Malherbe.
Description: Second edition. | Boca Raton : Taylor & Francis, 2018. | "A CRC title, part of the Taylor & Francis imprint, a member of the Taylor & Francis Group, the academic division of T&F Informa plc." | Includes bibliographical references.
Identifiers: LCCN 2017034428 | ISBN 9781138305090 (hardback : alk. paper)
Subjects: LCSH: Porous materials. | Nanostructured materials. | Diffusion. | Adsorption.
Classification: LCC TA418.9.P6 R67 2018 | DDC 620.1/16--dc23
LC record available at https://lccn.loc.gov/2017034428

Visit the Taylor & Francis Web site at
http://www.taylorandfrancis.com

and the CRC Press Web site at
http://www.crcpress.com

Dedication

This book is dedicated to my wife Teresa Fernandez-Mardones, my parents Silvia Malherbe Peña and Rolando Roque-Fernandez, our sons Daniel Dorr-Fernandez, Edelin Roque-Mesa, Rolando Roque-Mesa and Ruben Alcides Perez-Fernandez, our grandsons and all my Pets very specially to Zeolita and Trosia.

Contents

Preface

The increase in the concentration of molecules from a gaseous phase in the neighboring solid surface was recognized in 1777 by Fontana and Scheele, and the term *adsorption* to describe the effect was coined by Kayser in 1881. On the other hand, *diffusion* is a general property of matter that is related with the tendency of systems to occupy all its accessible states. The quantitative study of this phenomenon started in 1850 till 1855 with the works of Adolf Fick and Thomas Graham.

The development of new materials is a basic objective of materials science research. This interest is demanded by the progress in all fields of industry and technology. For example, the evolution of the electronic industry initiated the development of smaller and smaller elements. The dimension of these components is approaching nanometer dimensions, and as this dominion is entered, scientists have found that properties of materials with nanometer dimensions, that is, on the length scale of about 1–100 nm, can differ from those of the bulk material. In these dimensions, adsorption and diffusion are important methods of characterization. Besides, they are processes that determine the governing laws of important fields of application of nanoporous materials.

According to the definition of the International Union of Pure and Applied Chemistry (IUPAC): *Porous materials* are classified into microporous materials, which are materials with pore diameters between 0.3–2 nm; mesoporous materials, which are materials that have pore diameters between 2 and 50 nm; and macroporous materials, which are materials with pores bigger than 50 nm. Between the class of porous materials, nanoporous materials, for example, zeolites and related materials, mesoporous molecular sieves, the majority of silica, and active carbons are the most widely studied and applied materials. In the cases of crystalline and ordered nanoporous materials, such as zeolites and related materials, and mesoporous molecular sieves, its classification as nanoporous materials do not have any discussion. However, in the case of amorphous porous materials, they possess bigger pores together with pores with sizes less than 100 nm. Nonetheless, even in this case, in the majority of instances, the nanoporous component is the most important part of the porosity.

Adsorption and diffusion have a manifold value, as they are not only are powerful means for the characterization of nanoporous materials but are also important industrial operations. The adsorption of a gas can bring information about the microporous volume, the mesoporous area, the volume and size of the pores, and the energetics of adsorption. On the other hand, diffusion controls the molecular transport of gases in porous media and then also brings morphological information, in the case of amorphous materials, and structural information, in the case of crystalline and ordered materials.

Crystalline, ordered, and amorphous microporous and mesoporous materials, such as microporous and mesoporous molecular sieves, amorphous silica, alumina, active carbons, and other materials obtained by different techniques, are the sources of a collection of advanced materials with exceptional properties, and their applications in many fields, such as optics, electronics, ionic conduction, ionic exchange, gas separation, membranes, coatings, catalysts, catalyst supports, sensors, pollution abatement, detergency, and biology.

The present book is a development of some of the author's previous books, book chapters of books, and papers. The author has tried, in the present text, to give a state-of-the-art description of some of the most important aspects of the *theory* and *practice* of adsorption and diffusion, fundamentally, of gases in microporous crystalline, mesoporous ordered, and micro/mesoporous amorphous materials.

In the present monograph, the adsorption process in multicomponent systems will not be discussed with the exception of the final chapter in which adsorption from the liquid phase is analyzed. We are studying here adsorption and diffusion, fundamentally, from the point of view of materials science. That is, we are interested in the methods for the use of single-component adsorption and diffusion in the characterization of the adsorbent surface, pore volume, pore size distribution (PSD), and the study of the parameters characterizing single-component transport processes in porous systems. Besides, in the text, the following areas are studied: (1) adsorption energetic, (2) adsorption thermodynamic, and (3) dynamic adsorption in plug-flow bed reactors. On the other hand, in the text, the structure or morphology and the methods of synthesis and modification of silica, active carbons, zeolites, and related materials and mesoporous molecular sieves are also discussed. Other adsorbents normally used in different applications, such as alumina, titanium dioxide, magnesium oxide, and clays and pillared clays, are not discussed.

From the point of view of the application of dynamic adsorption systems, in the present book, the author will only analyze the use of adsorbents to clean gas or liquid flows by the removal of a low concentrated impurity applying a plug-flow adsorption reactor (PFAR), where the output of the operation of the PFAR is a breakthrough curve.

Finally, in the first place, this book is dedicated, in first place, to my family. It is also devoted to the advisors of my postgraduate studies and the mentors in my postdoctoral fellowships. Concretely, I would recognize Dr. Jürgen Büttner, advisor during my MSc studies, who was the first to explain me the importance of the physics and chemistry of surfaces in materials science. I would also acknowledge my senior PhD tutor, late Prof. Dr. Alekzander A. Zhujovistskii, who in 1934 was the first to recognize the complementary role of the adsorption field and capillary condensation in adsorption in porous materials and was later one of the absolute creators of gas chromatography. He taught me how to *see* inside a scientific data using general principles. In addition, I feel like to recognize my junior PhD tutor, Prof. Dr. Boris S. Bokstein, a very well-known authority in the study of transport phenomena, who gave me a big stimulus in the study of diffusion. I also want to appreciate the mentors of my postdoctoral fellowships, Prof. Dr. Fritz Storbeck, who gave me the opportunity to be in contact with the most advanced methods of surface studies; Prof. Dr. Evgenii D. Shchukin, one of the creators of a new science, physicochemical mechanics, who taught me the importance of surface phenomena in materials science; and the late academician Mijail M. Dubinin and Prof. Dr. A.V. Kiseliov, two of the most important scientists in the field of adsorption science and technology during the past century. Both of them gave me the great opportunity to be in close contact with their philosophy in the study of adsorption systems.

Rolando M.A. Roque-Malherbe
L&C Science and Technology
Hialeah, FL, USA

Author

Prof. Dr. Rolando M.A. Roque-Malherbe was born in 1948 in Güines, Havana, Cuba. He completed his BS in physics from the University of Havana, Havana, Cuba (1970). He specialized (MS equivalent degree) in surface physics from the National Center for Scientific Research–Technical University of Dresden, Germany (1972). He earned his PhD in physics from the Moscow Institute of Steel and Alloys, Moscow, Russia (1978). He completed postdoctoral stays at the Technical University of Dresden, Dresden, Germany; Moscow State University, Moscow, Russia; Technical University of Budapest, Budapest, Hungary; Institute of Physical Chemistry; and Central Research Institute for Chemistry of the Russian and Hungarian Academies of Science (1978–1984). The group lead by him at the National Center for Scientific Research–Higher Pedagogical Institute Varona, Havana, Cuba (1980–1992) was one of the world's leaders in the study and application of natural zeolites. In 1993, after a confrontation with the Cuban regime, he left Cuba with his family as a political refugee. From 1993 to 1999, he worked at the Institute of Chemical Technology, Valencia, Spain; Clark Atlanta University, Atlanta, Georgia; and Barry University, Miami, FL. From 1999 to 2015, he was the dean, director, and full-time professor of the School of Science at Turabo University (TU), Gurabo, Puerto Rico; currently he is an adjunct professor at the L&C Science and Technology, Hialeah, FL, USA. He has published 132 peer-reviewed papers, 7 books, 15 chapters, 16 patents, 42 abstracts, and more than 200 presentations in scientific conferences. He is currently an American citizen.

1 Statistical Physics

1.1 INTRODUCTION

Statistical physics or statistical mechanics named as statistical thermodynamics in the case of equilibrium systems was originated in the work of Maxwell and Boltzmann (1860–1900) on the kinetic theory of gases [1–5]. Later, in his book *Elementary Principles of Statistical Physics*, Gibbs (1902) made the major advance, so far, in the theory and methods of calculation of statistical physics by the introduction of the ensemble conception. More recently, Einstein, Fermi, Bose, Tolman, Langmuir, Landau, Fowler, Guggenheim, Kubo, Hill, Bogoliubov, and others have contributed to the subsequent development and fruitful application of statistical physics [6–10].

1.1.1 THERMODYNAMIC FUNCTIONS AND THEIR RELATIONSHIPS

This component of theoretical physics deals with macroscopic systems comprising a collection of particles, for example, photons, electrons, atoms, or molecules with the composition, structure, and function. To describe these systems, the concepts of quantum state (or Microstate) and thermodynamic state (or Macrostate) are applied in statistical physics to develop a comprehensive methodology for the calculation of the parameters characterizing a thermodynamic state, for example, the functions characterizing a macroscopic system.

To start the discussion, the fundamental equation of thermodynamics for a bulk mixture is applied [1–3]:

$$dU = TdS - PdV + \sum_i \mu_i dn_i$$

where:

$U(S, V, n_i)$ is the internal energy of the system
S is its entropy
V is its volume
T is its temperature
μ_i is the chemical potential
n_i is the number of moles of one of the N components, which form the system

Likewise, using the Legendre transformations (see Appendix 1.1), that is, subtracting TS from U, the Helmholtz free energy $F = U - TS$, a new thermodynamic function depending on T, V, and n, that is, $F(T, V, n)$ is obtained.

Moreover, an additional thermodynamic function is defined, that is, the enthalpy $H = U + PV$, from where the Gibbs function is derived with the help of the Legendre transformation [5–7], that is, $G = H - TS$.

Finally, it is possible to define [10] $\Omega = F - \sum \mu_i n_i$ as the grand potential or Massieu function.

Now, differentiating the thermodynamics functions we get [8,9]

$$dF = -SdT - PdV + \sum_i \mu_i dn_i$$

$$d\Omega = -SdT - PdV + \sum_i \mu_i dn_i$$

$$dH = TdS - VdP + \sum_i \mu_i dn_i$$

$$dG = -SdT + VdP + \sum_i \mu_i dn_i$$

The grand potential, which is generally absent from textbooks on thermodynamics, has a particular meaning in statistical physics, that is, it is the thermodynamic potential for a system with a fixed volume, V; chemical potentials, μ_i; and temperature, T; and, as will be later shown, it is related to the grand canonical partition, which is one of the magnitudes calculated with the help of the methods of statistical thermodynamics.

In Table 1.1, we will report some relations [4].

TABLE 1.1
Thermodynamic Relations

$T = \left(\dfrac{\partial U}{\partial S}\right)_{V, n_i}$	$-P = \left(\dfrac{\partial U}{\partial V}\right)_{S, n_i}$	$\mu_i = \left(\dfrac{\partial U}{\partial n_i}\right)_{S, V, n_i (i \neq j)}$
$-S = \left(\dfrac{\partial F}{\partial T}\right)_{V, n_i}$	$-P = \left(\dfrac{\partial F}{\partial V}\right)_{T, n_i}$	$\mu_i = \left(\dfrac{\partial F}{\partial n_i}\right)_{T, V, n_i (i \neq j)}$
$T = \left(\dfrac{\partial H}{\partial T}\right)_{P, n_i}$	$V = \left(\dfrac{\partial H}{\partial P}\right)_{S, n_i}$	$\mu_i = \left(\dfrac{\partial H}{\partial n_i}\right)_{S, P, n_i (i \neq j)}$
$-S = \left(\dfrac{\partial G}{\partial T}\right)_{P, n_i}$	$V = \left(\dfrac{\partial G}{\partial P}\right)_{T, n_i}$	$\mu_i = \left(\dfrac{\partial G}{\partial n_i}\right)_{T, P, n_i (i \neq j)}$
$-S = \left(\dfrac{\partial \Omega}{\partial T}\right)_{V, \mu_i}$	$-P = \left(\dfrac{\partial \Omega}{\partial P}\right)_{T, \mu_i}$	$-n_i = \left(\dfrac{\partial \Omega}{\partial n_i}\right)_{T, n_i (i \neq j)}$

1.2 DEFINITION OF QUANTUM STATE–MICROSTATE AND THERMODYNAMIC STATE–MACROSTATE

A microstate is a state of the system where all the parameters of the component particles are specified [1,3,13]. In quantum mechanics, for a system in a stationary state, the energy levels along with the state of the particles in terms of quantum numbers are applied to specify the parameters of a microstate, that is, at any given time the system will be in a definite quantum state j, characterized by a certain wave function φ_j, which is a function of a huge number of spatial, spin coordinates, energy E_j, along with a set of quantum numbers [7]. On the other hand, a macrostate is defined as a state of the system, where the distribution of particles over the energy levels is specified. Hence, a macrostate includes different energy levels and particles having particular energies, that is, it contains many microstates [5].

Now, following the principles of thermodynamics [1] for a single-component system, we only need to designate three macroscopic parameters, that is, (P, V, T), (P, V, N), or (E, V, N), where P is the pressure, V is the volume, T is the temperature, and N is the number of particles, in order to specify the thermodynamic state of an equilibrium of a single-component system. In these circumstances, the equation of state for the system relates the three variables to a fourth, for example, for an ideal gas we have [2]

$$PV = nRT = NkT$$

where:
$R = 8.31451$ [JK^{-1}mol^{-1}] is the ideal gas constant
$R = N_A k$
$N_A = 6.02214 \times 10^{23}$ [mol^{-1}] is the Avogadro number
$k = 1.38066$ [JK^{-1}] is the Boltzmann constant

In an ideal gas we assume that the molecules are noninteracting, that is, they do not affect each other's energy levels. Each particle has a certain energy, and at $T > 0$, the system possesses a total energy E. In addition, from quantum mechanics we know that the possible energies, if we consider the particles confined in a cubic box of volume, $V = abc$, (Figure 1.1) are given by [4]

$$E\left(n_1, n_2, n_3\right) = \frac{h^2}{8m}\left(\frac{n_1^2}{a^2} + \frac{n_2^2}{b^2} + \frac{n_3^2}{c^2}\right)$$

For a square box where $a = b = c = L$

$$E\left(n_1, n_2, n_3\right) = \frac{h^2}{8L^2 m}\left(n_1^2 + n_2^2 + n_3^2\right) = \frac{h^2 N^2}{8m}$$

where, $N^2 = (n_1^2 + n_2^2 + n_3^2)$ and n_1, n_2, and n_3 are the quantum numbers, each of which could be any integer number except zero.

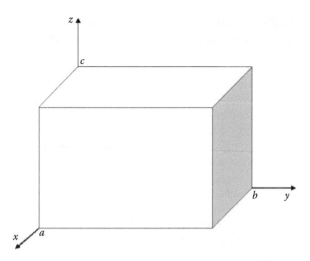

FIGURE 1.1   Box of volume, $V = abc$, where the molecules of the ideal gas are confined.

Then, a macrostate of the ideal gas with energy

$$E = \sum_{n_1} \sum_{n_2} \sum_{n_3} E\left(n_1, n_2, n_3\right)$$

and N molecules are compatible with a huge number of different (n_1, n_2, n_3) quantum numbers corresponding with different microstates. Therefore, a macrostate or thermodynamic state of a system is composed, or is compatible, with a huge number, Ω, of microstates or quantum states [8].

Finally, it is necessary to affirm that the thermodynamic state–macrostate is experimentally observable, whereas the quantum state–microstate is usually not observable.

1.3   DEFINITION OF ENSEMBLE

An ensemble is a hypothetical collection with an extremely high number of noninteracting systems, each of which is in the same macrostate as the system of interest. These systems show a wide variety of microstates, each compatible with the given microstate, that is, an ensemble is an imaginary collection of replications of the system of interest, where N is the number of systems in the ensemble, which is a very large number, that is, $N \rightarrow \infty$ [7].

More concretely, we have the number of systems in the ensemble in a state with a given energy, E_i, denoted by n_i. Then, the total number of systems in the ensemble can be calculated as

$$N = \sum_{i=1}^{\Omega} n_i$$

and the summation is taken over all the Ω accessible E_i energy states that are allowable for the concrete system in study.

Now, it is necessary to make some postulates in order to mathematically deal with the ensemble concept [6,9].

First postulate: The measured time average of a macroscopic property in the system of interest is equal to the average value of that property in the ensemble.

$$\bar{E} = \sum_{i=1}^{\Omega} p_i E_i \tag{1.1}$$

In which p_i is the probability of finding the system in one of the Ω possible states or allowed states in the chosen thermodynamic macroscopic state, and the summation is taken over all the energy states that are allowable for the concrete system in study.

Second postulate: The entropy is defined as [12]

$$S = -k \sum_{i=1}^{\Omega} p_i \ln p_i \tag{1.2}$$

where p_i is the probability of finding the system in one of the Ω possible states or allowed states in the chosen thermodynamic macroscopic state, and k is the Boltzmann constant.

Third postulate: For a thermodynamic system of fixed volume, composition, and temperature, all quantum states, which have equal energy have equal probability of occurrence [7].

Finally, it is necessary to state that in statistical mechanics, for a closed system, the equilibrium state is the state with the maximum entropy, which is one of the statements of the second law of thermodynamics [6].

1.4 THE CANONICAL ENSEMBLE

The canonical ensemble is a system in a heat bath at constant temperature and volume, and with a fixed number of particles N [8], that is, a system, which is in thermal equilibrium with a large bath. Since energy can flow to and from the bath, the system is, as was previously stated, described by the bath temperature T rather than by a fixed energy E [10]. Such a system and the statistical method based on it, are referred to as a canonical ensemble.

We represent the canonical ensemble as a collection of N systems, all in contact with each other and isolated from the rest of the universe [1–3] (Figure 1.2). Consequently, each system in the canonical ensemble is immersed in a bath consisting of the rest of the system replica and is isolated [7].

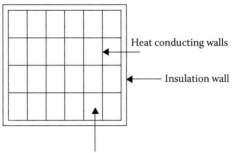

System with thermodynamic parameters V, N, and T,
where the number of particles and temperature are constant

FIGURE 1.2   Representation of the canonical ensemble.

The possible energy states of the systems in the ensemble are $E_j = E_j(V, N)$, since all the systems in the ensemble have the same volume V and number of particles N, then all the systems in the ensemble have the same set of energy states. Thereafter, the number of systems in the energy state E_i is n_i. Therefore, following the third postulate, the probability to select a system in the ensemble with energy E_i is [7]

$$p_i = \frac{n_i}{N} \tag{1.3}$$

where N is the whole number of systems in the canonical ensemble,

$$\sum_{i=1}^{\Omega} p_i = 1$$

and the average energy of the systems in the ensemble is

$$\overline{E} = \sum_{i=1}^{\Omega} p_i E_i$$

where the summations are carried out over all the allowable energy states of the system under study, that is, Ω.

To calculate the probability distribution for the canonical ensemble, we only need to find the conditions for the maximum entropy of the whole canonical ensemble system as expressed in Figure 1.2. In this scheme, the canonical ensemble is represented as a thermodynamic closed system composed of a collection of replicated systems enclosed by a wall, which do not allow the exchange of energy and matter with the rest of the universe. Therefore, based on the second postulate and the additive property of entropy, we need to calculate the maximum of

$$S = -k\mathrm{N} \sum_{i=1}^{\Omega} p_i \ln p_i$$

where the summation is taken over all the states allowable for the system under study. Then the calculation of the maximum is under the following two conditions:

$$\sum_{i=1}^{\Omega} n_i = N$$

and

$$\sum_{i=1}^{\Omega} n_i E_i = E$$

where N is the total number of systems in the ensemble and E is the total energy of the ensemble. Now dividing by N,

$$S = -k \sum_{i=1}^{\Omega} p_i \ln p_i$$

under the following conditions:

$$\sum_{i=1}^{\Omega} p_i = 1 \quad \text{and} \quad \sum_{i=1}^{\Omega} p_i E_i = \overline{E}$$

in which $\overline{E}$ is the average energy in a system of the ensemble. Now, to apply the method of the Lagrange multipliers [13] (see Appendix 1.2) we must define the following auxiliary function [9]:

$$f = -k \sum_{i=1}^{\Omega} p_i \ln p_i + \alpha \sum_{i=1}^{\Omega} p_i + \beta \sum_{i=1}^{\Omega} p_i E_i \tag{1.4}$$

Thereafter, the maximum condition is given by

$$p_i = \frac{\exp[-\beta E_i]}{Z} \tag{1.5}$$

Then, finally, the canonical partition function can be calculated with the following expression:

$$Z = \sum_{i=1}^{\Omega} \exp[-\beta E_i] \tag{1.6}$$

1.5   EVALUATION OF α AND β FOR THE CANONICAL ENSEMBLE

The Helmholtz energy, F, of a concrete system is defined by [8]

$$F = U - TS \tag{1.7}$$

Then [6]

$$\sum_{i=1}^{\Omega} p_i E_i = \overline{E} = U \tag{1.8}$$

Is the system's internal energy in a physical–statistical point of view, together with [9]

$$S = -k \sum_{i=1}^{\Omega} p_i \ln p_i \tag{1.9}$$

is their entropy, whereas T is their temperature. Consequently, within the canonical ensemble, substituting Equation 1.5 into Equations 1.8 and 1.9, and subsequently substituting in Equation 1.7, along with the application of Equation 1.6, it is easy to show that

$$S = -k\beta U + k \ln Z \tag{1.10}$$

$$k\beta U - S = k \ln Z \tag{1.11}$$

Thus, if we take

$$\beta = \frac{1}{kT} \tag{1.12}$$

and

$$F = kT \ln Z \tag{1.13}$$

It follows that Equation 1.10 will be equivalent to Equation 1.7; subsequently, is possible the calculation of all the thermodynamic functions by the use of the partition function, because the Helmholtz free energy F could operate as a starting point (Table 1.2) [8–10]:

TABLE 1.2

Thermodynamic Parameters Calculated with the Help of the Canonical Partition Function

$U = kT^2 \left(\dfrac{\partial \ln Z}{\partial V} \right)$	$P = kT \left(\dfrac{\partial \ln Z}{\partial V} \right)$	$S = k \left[\ln Z + T \left(\dfrac{\partial \ln Z}{\partial T} \right) \right]$
$G = kT \left[-\ln Z + T \dfrac{\partial \ln Z}{\partial V} \right]$	$H = kT \left[T \left(\dfrac{\partial \ln Z}{\partial T} \right) + V \dfrac{\partial \ln Z}{\partial V} \right]$	$\mu_i = -kT \left(\dfrac{\partial \ln Z}{\partial n_i} \right)$

Source: Roque-Malherbe, R., *Mic. Mes. Mat.*, 41, 2000, 227.

1.6 THE GRAND CANONICAL ENSEMBLE

The grand canonical ensemble (GCE) represents a system, which is in a heat bath at a constant temperature and volume, and with a variable number of particles N. That is, a system, which is in thermal equilibrium with a large bath, with which it is possible to exchange particles. Since energy and particles can flow to and from the bath, the system is described by the bath temperature T, and the chemical potential μ [8]. Such a system and the statistical method based on it are referred to as a GCE.

To calculate the probability distribution for the GCE, as was previously stated in the case of the canonical ensemble, we need to find the state of maximum entropy of the whole GCE. In this regard, in Figure 1.3, GCE is represented as a thermodynamic closed system composed of a collection of replicated systems enclosed by a wall, which do not allow the exchange of energy and matter with the rest of the universe [7].

Therefore, based on the second postulate, we need to calculate the maximum of [9].

$$S = -k \sum_{i=1}^{\Omega} \sum_{n=1}^{N} p_{i,n} \ln p_{i,n} \tag{1.14}$$

where the summation is carried out over all the allowable energy states, that is, from 1 to Ω, and for all the possible particle numbers, that is, from 1 to N. Hence, the maximum is calculated under the following three conditions [8]:

$$\sum_{i=1}^{\Omega} \sum_{j=1}^{N} p_{i,j} = 1 \tag{1.15}$$

$$\sum_{i=1}^{\Omega} \sum_{j=1}^{N} E_i(V, j) p_{i,j} = \overline{E} \tag{1.16}$$

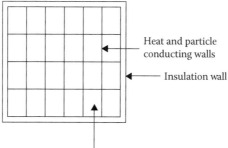

System with thermodynamic parameters V, N, and T, where temperature is constant and the number of particles is variable

FIGURE 1.3 Representation of the grand canonical ensemble.

and

$$\sum_{i=1}^{\Omega}\sum_{j=1}^{N} jp_{i,j} = \overline{N} \tag{1.17}$$

where $\overline{E}$ and $\overline{N}$ are the average energy and number of particles in a system of the ensemble, respectively. At this point, in order to use the method of the Lagrange multipliers to calculate the maximum of the entropy we must define the following auxiliary function [9]:

$$F = -k\sum_{i=1}^{\Omega}\sum_{j=1}^{N} p_{i,j}\ln p_{i,j} + \alpha\sum_{i=1}^{\Omega}\sum_{j=1}^{N} p_{i,j} + \beta\sum_{i=1}^{\Omega}\sum_{j=1}^{N} E_i(V,j)p_{i,j} + \gamma\sum_{i=1}^{\Omega}\sum_{j=1}^{N} np_{i,j} \tag{1.18}$$

Therefore, the maximum condition is given by the following equation:

$$\frac{\partial f}{\partial p_{i,j}} = -k\ln p_{i,j} + 1 - \alpha + \beta E(V,j) + \gamma j = 0 \tag{1.19}$$

Consequently

$$p_{i,j} = \frac{\exp[-\beta E_i(V,j)]\exp[-\gamma j]}{\Theta} \tag{1.20}$$

where:

$$\Theta = \sum_{i=1}^{\Omega}\sum_{j=1}^{N}\exp[-\beta E_i]\exp[-\gamma j] = \sum_{j=1}^{N} Z(j)\exp[-\gamma j] \tag{1.21}$$

is the grand canonical partition function, and $Z(n)$ is the canonical partition function of a system with j particles.

1.7 EVALUATION OF α, β, AND γ FOR THE GRAND CANONICAL ENSEMBLE

Within thermodynamics, the function entropy could be expressed for a concrete single-component system by the following expression [9]:

$$S = \frac{U - \mu n + PV}{T} \tag{1.22}$$

Thereafter, starting with Equation 1.22 it is feasible to show that [1]

$$TdS = dU + PdV - \mu dn \tag{1.23}$$

In which $U = \overline{E}$ is the internal energy of the system, or average energy, T is its temperature, S is its entropy, P is the pressure, V is the volume, μ is the chemical potential together with $n = \overline{N}/N_A$, the average number of moles.

Now, within the context of the GCE differentiating Equation 1.16 is obtained [8]:

$$d\overline{E} = dU = \sum_{i=1}^{\Omega}\sum_{j=1}^{N} E_i(V, j)dp_{i,j} + \sum_{i=1}^{\Omega}\sum_{j=1}^{N} p_{i,j}dE(V, j) \tag{1.24}$$

Thereafter, substituting Equation 1.20 in Equation 1.24 is obtained:

$$d\overline{E} = -\frac{1}{\beta}\left\{\sum_{jN}[\gamma N + \ln p_{jN} + \ln\Theta]dp_{jn}\right\} + \sum_{jN} p_{jN}\frac{\partial E(V, N)_j}{\partial V}dV \tag{1.25}$$

Afterward, using Equation 1.17 it is easy to show that [2]

$$d\overline{N} = \sum_{i,j} jdp_{jn} \tag{1.26}$$

Since Equation 1.25 could be simplified to [8]

$$d\overline{E} + pdV + \frac{\gamma}{\beta}d\overline{N} = -\frac{1}{\beta}d\left\{\sum_{jN}[p_{i,j}\ln p_{i,j}]\right\} \tag{1.27}$$

Hence, matching Equations 1.23 and 1.27 is then feasible to show that

$$\mu = -\frac{\gamma}{\beta} \tag{1.28}$$

Along with

$$TdS = -\frac{1}{\beta}d\left\{\sum_{i,j}[p_{i,j}\ln p_{i,j}]\right\} \tag{1.29}$$

Consequently, if

$$\beta = \frac{1}{kT} \tag{1.30}$$

Together with

$$\gamma = -\frac{\mu}{kT} \tag{1.31}$$

Then, substituting Equations 1.30 and 1.31 in Equation 1.14 and applying Equation 1.22 we get [8]

$$S = \frac{\overline{E}}{T} - \frac{N\mu}{T} + k \ln \Theta = \frac{U}{T} - \frac{N\mu}{T} + \frac{PV}{T} \tag{1.32}$$

Hence,

$$P = kT \left(\frac{\partial \ln \Theta(V,T,\mu)}{\partial V} \right)_{\mu,T} \tag{1.33}$$

and

$$\overline{N} = kT \left(\frac{\partial \ln \Theta(V,T,\mu)}{\partial \mu} \right)_{V,T} \tag{1.34}$$

1.8 CANONICAL PARTITION FUNCTION FOR A SYSTEM OF NONINTERACTING PARTICLES

Once the system's canonical partition function Z has been calculated by the summation of $\exp[-E_i/kT]$ over all the possible, Ω, accessible quantum states of the system, then all the thermodynamic properties of the system are readily established. However, the existence of forces between the molecules composing the system in the study makes Z extremely difficult to be evaluated. Nevertheless, we can relatively easily evaluate Z for a system with no intermolecular forces [1].

Let the Hamiltonian operator, $\overline{H}$, for such a system without interactions between the constituent molecules be the sum of separate terms for individual molecules with no interaction terms between molecules [2]:

$$\overline{H} = \sum_{j=1}^{N} \overline{H}_j \tag{1.35}$$

where $j = 1$ to N in which N is the number of molecules. Then, the energy of the whole system is

$$E_j = \varepsilon_{1,r} + \varepsilon_{2,s} + \ldots\ldots + \varepsilon_{N,w} \tag{1.36}$$

and

$$Z = \sum_j \exp \frac{-[\varepsilon_{1,r} + \varepsilon_{2,s} + \ldots\ldots + \varepsilon_{N,w}]}{kT} \tag{1.37}$$

The allowed energies for the single molecule are [6]

$$\overline{H}_j \phi_{j,r} = \varepsilon_{j,r} \phi_{j,r} \tag{1.38}$$

If the molecules are distinguishable from one another by being confined to different locations in space, for example, the one that occurs in a crystal or in a system of localized adsorption, then the sum over all possible quantum states of this system is [4]

$$Z = \sum_r \sum_s \sum_l ....\sum_w \exp\frac{-[\varepsilon_{1,r}]}{kT}\exp\frac{-[\varepsilon_{2,s}]}{kT}\exp\frac{-[\varepsilon_{3,l}]}{kT}.........\exp\frac{-[\varepsilon_{N,w}]}{kT}$$

Then

$$Z = \sum_r \exp\frac{-[\varepsilon_{1,r}]}{kT}.\sum_s \exp\frac{-[\varepsilon_{2,s}]}{kT}.\sum_l \exp\frac{-[\varepsilon_{3,l}]}{kT}..........\sum_w \exp\frac{-[\varepsilon_{N,w}]}{kT}$$

We can now define the molecular partition function as [8]

$$Z_i = \sum_r \exp\frac{-[\varepsilon_{i,r}]}{kT} \tag{1.39}$$

Consequently, for a system of distinguishable molecules:

$$Z = \prod_i Z_i \tag{1.40}$$

Now, if all the molecules happen to be of the same kind, then $Z_1 = Z_2 = ... = Z_N = q$ [1]. Then

$$Z = q^N \tag{1.41}$$

If the molecules are identical and are restricted to be in a specific zone of space, for example, an ideal gas or an adsorption mobile system, then the molecules are non-localized and there is no way to distinguish one molecule from the other. Therefore, a situation where molecule 1 is in state r and molecule 2 is in state s is the same as molecule 1 is in state s and molecule 2 in state r. The situation previously described could be expressed in terms of quantum mechanics as [2]

$$\phi_r(1)\phi_s(2)\phi_l(3).........\phi_w(N) = \phi_s(1)\phi_r(2)\phi_r(3)......\phi_r(N) \tag{1.42}$$

Consequently, if we permute the molecules, 1,2,3,...,N among all the possible molecular states $r,s,l,.....,w$, we will obtain identical wave functions. Now, as it is shown in Appendix 1.3 the number of permutations of N objects is $N!$ [12]. Consequently, the correct value for Z for a system of N noninteracting indistinguishable molecules is [8].

$$Z = \frac{q^N}{N!} \tag{1.43}$$

1.9   FACTORIZATION OF THE MOLECULAR PARTITION FUNCTION

The energy of a molecule is the sum of different contributions from the motion of its different degrees of freedom, such as translation T, rotation R, vibration V, and the electronic contribution E; therefore [3]

$$\varepsilon_i = \varepsilon_i^T + \varepsilon_i^R + \varepsilon_i^V + \varepsilon_i^E \tag{1.44}$$

Now, given that the molecular energy is a sum of independent contributions, then the molecular or particle partition function could be factorized into a product of different contributions:

$$q = \sum_i \exp\frac{-[\varepsilon_i]}{kT} = \sum_i \frac{-\left[\varepsilon_i^T + \varepsilon_i^R + \varepsilon_i^V + \varepsilon_i^E\right]}{kT}$$

Resulting in

$$q = \sum_i \exp\frac{-\left[\varepsilon_i^T\right]}{kT} \cdot \sum_i \exp\frac{-\left[\varepsilon_i^R\right]}{kT} \cdot \sum_i \exp\frac{-\left[\varepsilon_i^V\right]}{kT} \cdot \sum_i \exp\frac{-\left[\varepsilon_i^E\right]}{kT} = q^T q^R q^V q^E$$

In which [7]

$$q^T = \sum_i \exp\frac{-\left[\varepsilon_i^T\right]}{kT} = \frac{V(2\pi m k T)^{\frac{3}{2}}}{h^3} \tag{1.45}$$

is the translational partition function, h is the Planck constant, m is the mass of the molecule, k is the Boltzmann constant, and T is the absolute temperature. Further [3]:

$$q^R = \sum_i \exp\frac{-\left[\varepsilon_i^R\right]}{kT} = \frac{I k T}{h^2} \tag{1.46}$$

is the rotational partition function for a two-atom homonuclear rotor molecule, where I is the rotor moment of inertia. A two-particle rotor consist of particles of mass m constrained to remain at a fixed distance from each of them [1,2]. The rotational partition function for N atomic molecule with three axis of rotation is [10].

$$q^R = \frac{\pi^{\frac{1}{2}}}{\sigma_i} \prod_i^{3N-6} \left(\frac{8\pi^2 I_i kT}{h^2} \right)^{\frac{1}{2}} \tag{1.47}$$

Valid for $T \gg \theta_i$, where σ is the symmetry number, I_i are the moments of inertia along with $\theta_i = 8\pi^2 I_i k / h^2$; besides [4]

$$q^V = \sum_i \exp \frac{-\left[\varepsilon_i^V\right]}{kT} = \frac{1}{1 - \exp\left[-\theta_V / kT\right]} \tag{1.48}$$

is the vibrational partition function for a molecular vibration mode. Hence, in any polyatomic molecule, each vibrational mode will have their particular partition function. Thereafter, the overall partition function is [6]

$$q^V = q^V(1)q^V(2)q^V(3)......q^V(r)$$

in which $q^V(r)$ is the partition function for r normal vibration mode. Finally, it is necessary to state that [7].

$$\theta_V = \frac{hf}{k}$$

is the characteristic vibration temperature, where h is the Planck constant, k is the Boltzmann constant, and hf is the characteristic frequency of vibration of the concrete mode. Then, finally,

$$q^E = \sum_i \exp \frac{-\left[\varepsilon_i^E\right]}{kT} = 1 \tag{1.49}$$

is the electronic partition function (Table 1.3).

TABLE 1.3

Components of the Molecular Partition Function

$$q^T = \frac{V(2\pi mkT)^{3/2}}{h^3} \qquad\qquad q^R = \frac{IkT}{h^2}$$

$$q^V = \frac{1}{1 - \exp[-\theta_V / kT]} \qquad\qquad q^E = \sum_i \exp \frac{-[\varepsilon_i^E]}{kT} = 1$$

Source: Honerkamp, J., *Statistical Physics. An Advanced Approach with Applications*, Springer, New York, 1998.

1.10 DENSITY FUNCTIONAL THEORY

The methodology discussed in this section [14–18] is based fundamentally on the work of Hohemberg, Kohn, and Sham [14] who created a novel approach for the calculation of the ground-state electron probability density, $\rho(\bar{r})$, in quantum systems such as atoms and molecules. In this regard, Hohemberg and Kohn [15] demonstrated a theorem that states that the ground energy together with all other properties of a ground-state molecule is exclusively determined by the ground-state electron probability density of the system. Thus, E_{gs} is a functional of ρ that is [17]

$$E_{gs} = E_{gs}[\rho(\bar{r})]$$

where the brackets indicate a functional relation [17] (see Appendix 1.4).

The fundamental variable in density functional theory (DFT), at this point applied to an N particle classical system is the single particle density $\rho(\bar{r})$, which is defined for an N particle system, by integrating the N particle distribution function, $P(\bar{r}_1, \bar{r}_2, \ldots \bar{r}_N)$, over $N-1$ variables [16]:

$$\rho(\bar{r}_1) = N \int \ldots \int d\bar{r}_2 . d\bar{r}_3 \ldots d\bar{r}_N P(\bar{r}_1, \bar{r}_2, \ldots \bar{r}_N)$$

locally representing the number of particles per unit volume. Thus, integrating, the total number N is obtained as follows [18]:

$$\int_V \rho(\bar{r}) d\bar{r} = N$$

This concept of single-particle density is valid for, that is, atoms and molecules along with many classical particle systems, for example, a fluid immersed in an external potential, which is the case of interest here. In this regard, for a classical system of N atoms the Hamiltonian is given by [14]

$$H_N = \sum_{i=1}^{N} \frac{p_i^2}{2m} + \Phi(\bar{r}_1, \bar{r}_2 \ldots \bar{r}_N) + \sum_{i=1}^{N} U_{ext}^i(\bar{r}_i)$$

$$H_N = KE + \Phi + U_{ext}$$

where p_i is the momentum of particle, i, $\Phi(\bar{r}_1, \bar{r}_2 \ldots \bar{r}_N)$ is the total interatomic potential energy, and U_{ext}^i is the one-body external potential, whereas KE is the kinetic energy, Φ is the total interatomic potential energy of the system of N particles [17], and

$$U_{ext} = \sum_{i=1}^{N} U_{ext}^i(\bar{r}_i)$$

Now, $P_N(H_N)$, the probability of finding the system in one of the possible states allowed to the classical system included in the GCE is expressed as follows [4]:

$$P_N(H_N) = \frac{\exp[-\beta H_N]\exp[-\gamma N]}{\Theta}$$

where:

$\beta = 1/kT$

$\gamma = -\mu/kT$

Θ is the grand canonical partition function

Within the DFT, $P_N(H_N)$ must be considered as a single functional of $\bar{\rho}(\bar{r})$, the equilibrium single particle density, that is, $P_N(H_N) = P_N[\bar{\rho}(\bar{r})]$. Then, as $P_N(H_N)$ is a functional of the single particle density, in the frame of the DFT, all the functions describing the GCE could be expressed as functionals of $\bar{\rho}(\bar{r})$.

Now, for the reason that all the thermodynamic functions of the system could be obtained with the help of $P_N(H_N)$, they must be dependent on the equilibrium single particle density. In this way, we could define the intrinsic Helmholtz free energy, which is the Helmholtz free energy of the classical system of interests but with the exclusion of the external field interaction as [17]

$$F[\bar{\rho}(\bar{r})] = \overline{KE + \Phi + kT \ln P_N}$$

where $\overline{KE + \Phi + kT \ln P_N}$ means average in the GCE, provided this definition is related with the rest of thermodynamics functions for the GCE as long as from Equation 1.32.

$$\overline{kT \ln P_N(H_N)} = -TS = n\mu - \bar{E} - kT \ln \Theta$$

Together with the GCE definition of entropy, that is, $S = -k\overline{\ln P_N(H_N)}$ (Section 7); consequently, for many particle classical system immersed in an external potential U_{ext}, arising, for instance, from the adsorption field taking place, for example, when a fluid is confined in a pore. In this case, a function, $\Omega[\rho(\bar{r})]$, which reduces to the grand potential, $\Omega = \Omega[\bar{\rho}(\bar{r})]$, when $\rho(\bar{r}) = \bar{\rho}(\bar{r})$, the equilibrium density can be expressed with the help of the following single functional of the density [16–18]:

$$\Omega[\rho(\bar{r})] = F[\rho(\bar{r})] - \int d\bar{r}\rho(\bar{r})[\mu - U_{ext}(\bar{r})] \tag{1.50}$$

which is obtained by means of a Legendre transformation [12] (see Appendix 1.1) of $F[\rho(\bar{r})]$, which as was previously defined is the intrinsic Helmholtz free energy [17] and μ is the chemical potential of the studied system.

Now, to simplify calculations, Equation 1.50 will be now defined as

$$u(\bar{r}) = \mu - U_{ext}(\bar{r}) \tag{1.51}$$

Then

$$\Omega[\rho(\bar{r})] = F[\rho(\bar{r})] - \int d\bar{r}\rho(\bar{r})u(r) \tag{1.52}$$

Thereafter, for a thermodynamic system of N classical particles, the true equilibrium density, $\bar{\rho}(\bar{r})$, is determined with the help of the Euler–Lagrange equation (see Appendix 1.4) [19], that is, we must determine the minimum value of the functional $\Omega[\rho(\bar{r})]$. That is, the equilibrium density is the function making the functional $\Omega[\rho(\bar{r})]$ a minimum so that we have [19].

$$\frac{\delta\Omega[\rho(\bar{r})]}{\delta\rho(\bar{r})} = 0 \qquad (1.53)$$

that is, $\bar{\rho}(\bar{r})$ is the solution of Equation 1.53, where $\delta/\delta\rho(\bar{r})$ indicates a functional derivative (see Appendix 1.4). At this point, combining Equations 1.50 and 1.53 we get [17]

$$u(\bar{r}) = \mu - U_{ext}(\bar{r}) = \frac{\delta F[\rho(\bar{r})]}{\delta\rho(\bar{r})} \qquad (1.54)$$

In addition, in a minimum condition, explicitly the condition, which allows determining the equilibrium density, that is, the solution of Equation 1.54 is $\bar{\rho}(\bar{r})$, the equilibrium density distribution.

For a classical fluid with inhomogenous density distribution, the functional $F[\rho(\bar{r})]$ representing the intrinsic Helmholtz free energy can be expressed as [16]

$$F[\rho(\bar{r})] = F_{id}[\rho(\bar{r})] + F_{ex}[\rho(\bar{r})] \qquad (1.55)$$

where

$$F_{id}[\rho(\bar{r})] = \int d\bar{r}\rho(\bar{r})\{\ln\left(\rho(\bar{r})\Lambda^3\right) - 1\} \qquad (1.56)$$

is the ideal gas free-energy functional, resulting from a system of noninteracting particles, where $\Lambda = (h^2/2\pi mkT)^{1/2}$ is the thermal wavelength [18]. Moreover, $F_{ex}[\rho]$ represents the excess free energy for the classical system and is analogous to the interaction energy functional in the case of electronic quantum systems [2].

Now, solving Equation 1.53 we could define [17]

$$c^1(\bar{r}) = \frac{\delta[\beta F_{ex}(\rho(\bar{r}))]}{\delta\rho(\bar{r})} \qquad (1.57)$$

For the classical system of N particles, the equilibrium density satisfies [16]

$$\rho(\bar{r}) = \Lambda^{-3}[\exp(\beta\mu)][\exp-(\beta U_{ext} + c^1(\bar{r}, \rho(\bar{r})))] \qquad (1.58)$$

where $c^1(\bar{r})$ is also a functional. In particular, this equation for $c^1(\bar{r}) = 0$, that is, in the case of an ideal gas it reduces to the barometric law for the density distribution in the presence of an external field [7].

To proceed further, the excess intrinsic Helmholtz energy is split into contributions from the short-ranged repulsion and the long-term attraction [18]:

$$F_{ex}[\rho(\bar{r})] = F_{rep}[\rho(\bar{r})] + F_{att}[\rho(\bar{r})]$$

Details about this procedure will be given in Chapter 4.

1.11   THERMODYNAMICS OF IRREVERSIBLE PROCESSES

Let us provide now a brief summary of thermodynamics [22–25] and statistical mechanics of irreversible processes [26–32]. To be definite and keeping the exposition in a simple form, the linear terms in all the equations describing the irreversible processes will be only considered.

In an irreversible process, in conformity with the second law of thermodynamics the magnitude, which rules the time dependence of an isolated thermodynamic system, is the entropy S [6]. Merely processes in the systems, which lead to an increase in entropy, are feasible. So, the necessary and sufficient condition for a stable state in an isolated system is that the entropy has attained its maximum value [6]. Consequently, the most probable state is that in which the entropy is maximum [7]. Irreversible processes are driven by generalized forces, X, and are characterized by transport or Onsager phenomenological coefficients, L [22]. The transport coefficients, L_{ij}, are defined by linear relations between the generalized flux densities, J_i, which are the rates of change with the time of state variables, and the corresponding generalized forces X_{ik} [23]:

$$J_k = -\sum_{i=1}^{N} L_{ki} X_i \tag{1.59}$$

where the Onsager reciprocity relations, which are related with the principle of time reversal symmetry in mechanics, $L_{ij} = L_{ji}$, are valid. Besides, it is postulated that the rate of entropy production per unit volume due to internal processes may be expressed as

$$\frac{ds}{dt} = -\sum_{i=1}^{N} J_i X_i \tag{1.60}$$

where s denote the specific value of S per unit volume of the system.

Familiar examples of the relation between generalized fluxes and forces are the Fick's first law of diffusion, the Fourier's law of heat transfer, the Ohm's law of electricity conduction, and the Newton's law of momentum transfer in a viscous flow.

Since diffusion means molecular or, in general, the particle transport caused by a gradient of concentration (in more rigorous terms it is in reality a gradient of chemical potential). Then, the Fick's first law can be expressed as [30]

$$\bar{J} = -D\bar{\nabla}C \tag{1.61}$$

where:

$\overline{J}$ is the matter flux

$\overline{\nabla}C$ is the concentration gradient

D is the Fickean diffusion coefficient or transport diffusion coefficient, which is the proportionality constant

The units of the above-described parameters are D (longitude²/time), C (mol/volume), and J (mol/area·time). In the International System (SI) the unit of D is m²/s for the concentration C can be expressed in mol/m³, in this case the flux is expressed in mol/m²·s.

Now, it is necessary to affirm that the flux, and therefore also the diffusion coefficient, has to be chosen relative to a frame of reference, since the diffusion flux $\overline{J}$ gives the number of species crossing a unit area, fixed relative to the local center of mass, per unit of time [31]. Hence, the Fick's second law:

$$\frac{\partial C}{\partial t} = -D\nabla^2 C \tag{1.62}$$

is an expression of the law of conservation of matter, that is, $\partial C/\partial t = -D\overline{\nabla}\cdot\overline{J}$. Then, as was previously stated, the real driven force of mass transport is the gradient of the chemical potential; in this sense, in the absence of an external Newtonian force similar to those exerted in a charged species by an electric field, it is expressed as

$$X_i = \overline{\nabla}\left(\frac{\mu_i}{T}\right) \tag{1.63}$$

for the transport of the type of particles, i. On the other hand, heat conductivity means the transfer of energy caused by temperature gradients. The Fourier's law of heat transfer is expressed as

$$\overline{Q} = -\kappa\overline{\nabla}T \tag{1.64}$$

where:

$\overline{Q}$ is the energy flux

$\overline{\nabla}T$ is the temperature gradient

κ is the thermal conductance coefficient, which is the proportionality constant

In terms of the thermodynamics of irreversible processes the generalized force for the energy flux is

$$X_q = -\overline{\nabla}\frac{1}{T} = \frac{\nabla T}{T^2} \tag{1.65}$$

Thus, for a problem with interest for us in the present book, that is, single component diffusion in an adsorbent under conditions of nonuniform temperature [30]:

$$J = -L_p \overline{\nabla}\left(\frac{\mu}{T}\right) - L_c\left(\frac{1}{T^2}\right)\nabla T$$

$$Q = -L_c \overline{\nabla}\left(\frac{\mu}{T}\right) - L_e\left(\frac{1}{T^2}\right)\nabla T$$

where L_p and L_e are the transport coefficients related with particle transport and energy transfer, respectively. Besides, L_c is the cross coefficient, which is only 1 because of the reciprocity relations, since normally the energy transfer is faster than diffusion in adsorption systems. Then, the transport process of matter in an adsorbent may be considered as an isothermal process. In this case, for a single-component diffusion under isothermal conditions in a porous adsorbent [31]:

$$J = -L\overline{\nabla}(\mu) \tag{1.66}$$

where $L = L_p/T$; thereafter, this expression reduces to

$$J = -L\frac{\partial \mu}{\partial x} = -D_0\left(\frac{\partial \ln P}{\partial \ln C}\right)\left(\frac{\partial C}{\partial x}\right) \tag{1.67}$$

If substituted, the expression for the chemical potential [1,2] is $\mu = \mu^0 + RT \ln P$, considering unidimensional diffusion.

Another relation between generalized fluxes and forces, which is interesting for the purpose of the present monograph, is the Newton's law of momentum transfer in a viscous flow. In this sense (Figure 1.4), it is possible to express the transfer of momentum caused by a gradient of velocity in a viscous flow by [7]

$$p_{zx} = -\eta\frac{\partial u_x}{\partial z} \tag{1.68}$$

where:

p_{zx} is the tangential force per unit area or stress between two parallel layers of the viscous flow, that is, the index z indicates the orientation of the plane that is perpendicular to the axis z; second index, x, designates the component of the force exerted across this plane

$\partial u_x/\partial z$ is the velocity gradient

η is the dynamic viscosity measured in Pa·s

In Figure 1.4, it is shown a situation where the velocity of the layer of fluid adjacent to the plane located at $z = 0$, perpendicular to the z axis, is $u_x = 0$. Besides, the velocity of the layer of fluid placed at $z = l$ is $u_x = u_0$, that is it reflects the existence of a velocity gradient, $\partial u_x/\partial z$.

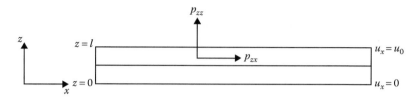

FIGURE 1.4 Viscous fluid flowing in the x direction.

1.12 STATISTICAL MECHANICS OF IRREVERSIBLE PROCESSES

Until the middle of the twentieth century, a typical way of calculating the response of a system to external perturbations consisted of trying the solution of the complete dynamical equations of motion, such as the Liouville equation or the Boltzmann equation, incorporating the perturbation into the system Hamiltonian or evolution matrix. However, about 60 years ago, Green [26], Kubo [27], and others [28–32] introduced simplifications founded in the possibility of the description of the effect of external perturbations in terms of the system's own correlation function [27].

As shown earlier all the needed information to calculate the thermodynamic properties of macroscopic systems is incorporated in the partition function or in the probability distribution function [6–8]. Nevertheless, the probability distribution also contains nonthermodynamic information. In this regard, some macroscopic quantities such as the fluctuations of the thermodynamic magnitudes together with the kinetic coefficients characterizing the linear response of a system to an external perturbation [28,29] as temperature gradient, chemical potential gradient, electric field, and others could be calculated with the help of the correlation functions [30–32].

An external perturbation destroys the state of thermodynamic equilibrium of a system. This fact is expressed in the appearance of, for example, fluxes of energy, matter, electricity, and momentum, or in the time variation of the internal parameters of the system, such as polarization, magnetization, and other parameters [1,7,26]. If the external perturbation is weak enough, then the departure of the system from the equilibrium state will be small, and consequently the system response will be linear, that is, the system response will linearly depend on the perturbation [27].

This linear relation between the system response and the perturbation could be characterized by a set of kinetic coefficients, such as the diffusion coefficient, the thermal and electric conductivities, the electric and magnetic susceptibilities, and other coefficients, which normally depend on temperature, the external parameters, and the frequency of variation of the perturbation [27–32].

1.12.1 Correlation Functions and Generalized Susceptibilities

Here, a classical system is treated that is described in the Γ-space by the phase space coordinates, $q(t)$ and $p(t)$ [6], where

$$q(t) = \{q_1, q_2, \ldots \ldots q_f\}, \text{ and } p(t) = \{p_1, p_2, \ldots \ldots p_f\}$$

provided f, the number of degrees of freedom, is the number of generalized coordinates and momenta needed to describe the system. Hence, if now there are two defined functions of the phase space coordinates [29], then

$$A\{q(t), p(t)\} = A(q, p, t) = A(t)$$

and

$$B\{q(t), p(t)\} = B(q, p, t) = B(t)$$

It is possible to define the time correlation functions $K_{AB}(t)$ by [26]

$$K_{AB}(t) = \langle A(t)B(0)\rangle_\Gamma = \int \ldots \int A(q, p, t)B(q, p, 0)\rho(q, p)dqdp \qquad (1.69)$$

using the ensemble average in the Γ-space with the equilibrium phase space density $\rho(q, p)$, which is the classical equilibrium distribution function in the phase space [6]. The obtained correlation function depends only on the time difference t, since the ensemble average is taken in the equilibrium situation, where the distribution of systems in the ensemble is independent of the absolute value of time [28].

Assuming now $A = B$, it is possible to define the autocorrelation function [5,6]

$$K_{AA}(t) = \langle A(t)A(0)\rangle_\Gamma = \int \ldots \int A(q, p, t)A(q, p, 0)\rho(q, p)dqdp \qquad (1.70)$$

Between the properties of the autocorrelation function is included the symmetry property, which could be expressed as [24]

$$K_{AA}(t) = K_{AA}(-t) \qquad (1.71)$$

One autocorrelation function of interest for the aims of the present book is the case where the function $A(t) = v(t)$, where $v(t)$ is the velocity [24].

$$K_{vv}(t) = \langle v(t)v(0)\rangle = \int \ldots \int v(q, p, t)v(q, p, 0)\rho(q, p)dqdp \qquad (1.72)$$

Since, as will be demonstrated later, it will help to calculate the self-diffusion coefficient and the Einstein relation [34] applying the generalized susceptibilities, which can be expressed as Fourier transforms of the correlation functions [6,7]:

$$\sigma(\omega) = \int_0^\infty \exp(-i\omega t)\left\langle \dot{A}(t)\dot{B}(0)\right\rangle dt \qquad (1.73)$$

For $\dot{A}(t) = \dot{B}(t)$ we will get the Fourier transforms of the autocorrelation functions:

$$\sigma(\omega) = \int_0^\infty \exp(-i\omega t)\left\langle \dot{A}(t)\dot{A}(0)\right\rangle dt \qquad (1.74)$$

For $\omega = 0$:

$$\sigma(0) = \int_0^\infty \left\langle \dot{A}(t)\dot{B}(0) \right\rangle dt$$

in the long-time limit, that is, for $t \rightarrow \infty$ for the autocorrelation function:

$$2t\sigma(0) = \left\langle [A(t) - A(0)]^2 \right\rangle \tag{1.75}$$

1.12.2 CALCULATION OF THE MEAN SQUARE DISPLACEMENT AND THE SELF-DIFFUSION COEFFICIENT

1.12.2.1 Calculation of the Mean Square Displacement with the Help of the Velocity Autocorrelation Function

As an illustration of the utilization of the correlation functions, let us now calculate the mean square displacement (MSD), $\left\langle x^2(t) \right\rangle$, of the particle in time. We will consider here the situation where the system is in equilibrium in the absence of external forces; then, $x(0) = 0$ at $t = 0$; then since, $\dot{x} = v$, one has [31,32]

$$x(t) = \int_0^t v(\tau)d\tau \tag{1.76}$$

Next,

$$\left\langle x^2(t) \right\rangle = \left\langle \int_0^t v(\tau)d\tau \int_0^t v(\tau')d(\tau') \right\rangle = \int_0^t d\tau \int_0^t d(\tau')\langle v(\tau)v(\tau')\rangle \tag{1.77}$$

Now, giving that the velocity autocorrelation function: $\langle v(\tau)v(\tau')\rangle$ depend only on the time difference, $s = \tau' - \tau$, thus,

$$\left\langle v(\tau)v(\tau') \right\rangle = \left\langle v(0)v(s) \right\rangle = K_v(s) \tag{1.78}$$

At this point, changing the integration limits [31]:

$$\left\langle x^2(t) \right\rangle = \int_0^t ds \int_0^{t-s} d\tau K_v(s) + \int_{-t}^0 ds \int_{-s}^t d\tau K_v(s) = \int_0^t ds K_v(s)(t-s) + \int_{-t}^0 ds K_v(s)(t+s)$$

If we now put $s \rightarrow -s$ in the second integral and use the symmetry property of the autocorrelation function: $K_v(s) = K_v(-s)$, subsequently

$$\left\langle x^2(t) \right\rangle = 2\int_0^t ds(t-s)\langle v(0)v(s)\rangle \tag{1.79}$$

Thereafter, for $t \to \infty$, $t \gg s$, and consequently we will get

$$\lim_{t \to \infty} \langle x^2(t) \rangle = 2t \int_0^t ds \langle v(0)v(s) \rangle \tag{1.80}$$

1.12.3 BROWNIAN MOTION MODEL LANGEVIN APPROACH

An adequately small macroscopic particle immersed in a liquid shows a random type of motion called Brownian motion. In general, we designate the movement of a particle arising from thermal agitation by the Brownian motion, that is, the movement of small macroscopic particles, molecules, or atoms. Consequently, there are a diversity of significant physical phenomena that are essentially analogous to the Brownian motion. Therefore, this phenomenon could operate as a sample problem, whose study offers significant insight into the mechanism responsible for the analyzed problem [5]. The first theory explaining the Brownian motion was developed by Einstein in 1905 [32] and later, Langevin created another model to explain this phenomenon [33].

For simplicity, the problem will be treated in one dimension, if considered as a particle of mass m, whose center of mass coordinate is located at $x(t)$ at time t, and whose velocity and acceleration are $v = \dot{x} = dx/dt$ and $a = \dot{v} = dv/dt$, respectively, that is, the particle is immersed in a liquid at temperature T. Therefore, the Langevin model is based on an expression of Newton's second law of motion [5]:

$$m\dot{v} = \Gamma(t) + F_{ext} - \alpha v = F(t) - \alpha v \tag{1.81}$$

where:
F_{ext} are the external forces acting on the moving particle
αv is the term that accounts for drag forces
Function $\Gamma(t)$ is a stochastic noise term, which accounts for random collisions with the liquid at temperature T
$F(t) = \Gamma(t) + F_{ext}$

The function $\Gamma(t)$ is truly a white noise so that [33]

$$\langle \Gamma(t) \rangle = 0 \tag{1.82}$$

and

$$\langle \Gamma(t)\Gamma(\tau + s) \rangle = q\delta(s) \tag{1.83}$$

We will now consider that the system is in the state of equilibrium, where $\langle x \rangle = 0$ because there is no preferred direction in space [7]. Now, we will calculate the MSD, that is, $\langle x^2 \rangle$. Since $v = \dot{x} = dx/dt$ and $a = \dot{v} = dv/dt$, then multiplying Equation 1.81 by x we will get [5]

$$mx\frac{d\dot{x}}{dt} = m\left[\frac{d}{dt}\left(x\dot{x}\right) - \left(\dot{x}\right)^2\right] = -\alpha x\dot{x} + xF(t) \tag{1.84}$$

Consequently, at this time we can calculate the ensemble average of both sides of Equation 1.84:

$$m\left\langle\left[\frac{d}{dt}\left(x\dot{x}\right) - \left(\dot{x}\right)^2\right]\right\rangle = -\alpha\left\langle x\dot{x}\right\rangle + \left\langle xF(t)\right\rangle \tag{1.85}$$

Accordingly, Equation 1.85 becomes

$$m\left\langle\left[\frac{d}{dt}\left(x\dot{x}\right)\right]\right\rangle = m\frac{d}{dt}\left\langle x\dot{x}\right\rangle = kT - \alpha\left\langle x\dot{x}\right\rangle \tag{1.86}$$

Given that $\left\langle xF(t)\right\rangle = \left\langle x\right\rangle\left\langle F(t)\right\rangle = 0$, since $\left\langle x\right\rangle = 0$, $\left\langle F(t)\right\rangle = 0$, and $1/2\,m\left\langle(\dot{x})^2\right\rangle = 1/2\,kT$, on account of the equipartition theorem [6–8]; therefore, solving the differential Equation 1.86, Equation 1.87 is obtained:

$$\left\langle x\dot{x}\right\rangle = C\exp\left(-\frac{\alpha t}{m}\right) + \frac{kT}{\alpha} \tag{1.87}$$

Now, as $x(0) = 0$ at $t = 0$; successively, $0 = C + kT/\alpha$; consequently, Equation 1.87 becomes

$$\left\langle x\dot{x}\right\rangle = \frac{1}{2}\frac{d\left\langle x^2\right\rangle}{dt} = \frac{kT}{\alpha}\left[1 - \exp\left(-\frac{\alpha t}{m}\right)\right] \tag{1.88}$$

In addition, integrating Equation 1.88:

$$\left\langle x^2\right\rangle = 2\frac{kT}{\alpha}\left[t - \frac{m}{\alpha}\left\{1 - \exp\left(-\frac{\alpha t}{m}\right)\right\}\right] \tag{1.89}$$

In addition, evaluating Equation 1.89 for $t \rightarrow \infty$, following equation will be obtained:

$$\left\langle x^2\right\rangle = \frac{2kT}{\alpha}t \tag{1.90}$$

1.12.3.1 Diffusion Equation

Consider now a random walker in one dimension, with probability R of moving to the right, and L for moving to the left. At $t = 0$, we place the walker at $x = 0$, as indicated in Figure 1.5. The walker can jump with the above-stated probabilities, either to the left or to the right for each time step. Hence, giving that every step has length

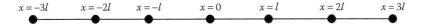

FIGURE 1.5   One-dimensional random walker, which can jump either to the left or to the right.

$\Delta x = l$, along with the fact that we have a jump either to the left or to the right at every time step, assuming that there are equal probabilities for jumping to the left or to the right, that is, $L = R = 1/2$. In this case, the average displacement after N time steps is

$$\left\langle x(N) \right\rangle = \left\langle \sum_{i=1}^{N} \Delta x_i \right\rangle = \sum_{i=1}^{N} \left\langle \Delta x_i \right\rangle = 0 \quad \text{for } \Delta x_i = \pm l$$

At this point, as we have an equal opportunity of jumping either to the left or to the right, the value of $\langle x(N)^2 \rangle$ is

$$\left\langle x(N)^2 \right\rangle = \left\langle \left(\sum_{i=1}^{N} \Delta x_i \right)^2 \right\rangle = \left\langle \sum_{i=1}^{N} \Delta x_i \sum_{j=1}^{N} \Delta x_j \right\rangle = \sum_{i=1}^{N} \left\langle \Delta x_i^2 \right\rangle + \sum_{i \ne j} \left\langle \Delta x_i \Delta x_j \right\rangle = l^2 N$$

Because as a consequence that the steps are not correlated, then $\left\langle \Delta x_i \Delta x_j \right\rangle = \left\langle \Delta x_j \Delta x_i \right\rangle$, which implies that

$$\sum_{i \ne j} \left\langle \Delta x_i \Delta x_j \right\rangle = 0$$

afterward since the time between the jumps is τ, for that reason the jump frequency is $\Gamma = 1/\tau$. Consequently if $\Gamma = $ constant, thenceforth for N steps giving that $N = t/\tau$, then the MSD is given by [7]

$$\left\langle x^2(N) \right\rangle = N l^2 = \left(\frac{l^2}{\tau} \right) t \tag{1.91}$$

Another form to describe diffusion is the Markov process, if this stochastic movement is a process characterized by the fact that the time dependence of the probability, $P(x,t)dx$, that a particle position at time t lies between x and $x + dx$ depends only on the circumstance that $x = x_0$ at $t = t_0$, and nothing on the whole past the history of the particle movement [5,7]. In this regard, the random walk follows the Markov dynamics as described by the Fokker–Planck equation [5,34]:

$$\frac{\partial P(x,t)}{\partial t} = -\frac{\partial}{\partial x} \left[f(x)P(x,t) \right] + \kappa \frac{\partial^2}{\partial x^2} \left[g(x)P(x,t) \right]$$

In particular, for random walk, $f(x) = 0$ and $g(x) = 1$, afterward the diffusion equation for a random walk in one dimension is

$$\frac{\partial P(x,t)}{\partial t} = D\frac{\partial^2 P(x,t)}{\partial x^2} \qquad (1.92)$$

in which, D is the self-diffusion coefficient and $P(x,t)$, where

$$\int_{-\infty}^{\infty} P(x,t)dx = 1$$

is the probability density to find a diffusing particle at the position x, during the time t, if this particle was at $x = 0$ at $t = 0$, provided the solution of this equation for the ensuing initial and boundary conditions, that is, $P(\infty,t) = 0$, $P(-\infty,t) = 0$, and $P(x,0) = \delta(x)$ [35] is

$$P(x,t) = \left(\frac{1}{(4\pi Dt)}\right)^{1/2} \exp\left(-\frac{x^2}{4Dt}\right) \qquad (1.93)$$

For that reason, it is easy to show that the one-dimensional MSD is given by the following equation:

$$\langle x^2 \rangle = \int x^2 P(x,t)dx = 2Dt \qquad (1.94)$$

Afterward, from Equations 1.80 and 1.81:

$$\lim_{t\to\infty}\langle x^2(t)\rangle = 2tD = 2t\int_0^t ds\langle v(0)v(s)\rangle \qquad (1.95)$$

Consequently:

$$D = \int_0^t ds\langle v(0)v(s)\rangle \qquad (1.96)$$

Moreover, after Equation 1.90:

$$\langle x^2 \rangle = \frac{2kT}{\alpha}t = 2Dt \qquad (1.97)$$

Thenceforward:

$$D = \frac{kT}{\alpha} \qquad (1.98)$$

In addition, after Equation 1.91:

$$D = \frac{l^2}{\tau}$$

Accordingly, in a more general form, the self-diffusion coefficient and the Einstein relation for three-dimensional self-diffusion are [10,24]

$$D = \frac{1}{3} \int_0^\infty \langle \bar{v}(t)\bar{v}(0) \rangle dt \qquad (1.99)$$

$$2tD = \frac{1}{3} \langle [\bar{r}(t) - \bar{r}(0)]^2 \rangle \qquad (1.100)$$

APPENDIX 1.1   LEGENDRE TRANSFORMATIONS

The Legendre transformations [13] allow describing a function using a different set of variables. Given a function $f(x, y)$ the total derivative of that function is given as

$$df = \frac{\partial f}{\partial x} dx + \left(\frac{\partial f}{\partial y} \right) dy$$

The coefficients for the partial derivatives are defined as

$$u = \frac{\partial f}{\partial x}, \quad \text{and} \quad v = \frac{\partial f}{\partial y}$$

To change to a new representation, the function $g(u, x)$ is defined as $g(u, x) = f(x, y) - ux$. Implying that

$$dg = df - xdu - udx$$

Using now the total derivative of $f(x, y)$ then

$$dg = -xdu + vdy$$

where

$$x = -\frac{\partial g}{\partial u}, \quad \text{and} \quad v = \frac{\partial g}{\partial y}$$

Consequently, the Legendre transformation construct from a function $f = f(x, y)$, a function $g = g(u, y)$, which by definition depends on u and y.

APPENDIX 1.2   THE LAGRANGE MULTIPLIERS

If we wish to locate the maximum or minimum, for example, in the function $F = F(x,y,z)$, we require that at the maximum or minimum [13].

$$dF(x, y, z) = \left(\frac{\partial F}{\partial x} dx + \frac{\partial F}{\partial y} dy + \frac{\partial F}{\partial z} dz \right) = 0 \qquad (A1.2.1)$$

Since, dx, dy, and dz are linearly independent, because they are independent variables, we have subsequently

$$\frac{\partial F}{\partial x} = \frac{\partial F}{\partial y} = \frac{\partial F}{\partial z} = 0 \qquad (A1.2.2)$$

Now, suppose that we want to locate the maximum or minimum in $F = F(x, y,z)$, not including all the values of x, y, and z, but only those values, which satisfied the condition $G(x, y,z) = C$. In this case, dx, dy, and dz are not linearly independent, since only two of them are linearly independent given that

$$dG(x, y, z) = \left(\frac{\partial G}{\partial x} dx + \frac{\partial G}{\partial y} dy + \frac{\partial G}{\partial z} dz \right) = 0 \qquad (A1.2.3)$$

Lagrange developed a methodology to solve this problem, defining the following auxiliary function:

$$H = F + \alpha G \qquad (A1.2.4)$$

where α is a Lagrange multiplier. Now since $dF = 0$ and $dG = 0$, then $dH = 0$. Therefore:

$$dF(x, y, z) = dF + \alpha dG = 0 \qquad (A1.2.5)$$

and consequently,

$$\left(\frac{\partial F}{\partial x} - \alpha \frac{\partial G}{\partial x} \right) dx + \left(\frac{\partial F}{\partial y} - \alpha \frac{\partial G}{\partial y} \right) dy + \left(\frac{\partial F}{\partial z} - \alpha \frac{\partial G}{\partial z} \right) dz = 0 \qquad (A1.2.6)$$

at (x_0, y_0, z_0). Since α is an arbitrary parameter we can evaluate α at (x_0, y_0, z_0) by applying

$$\frac{\partial F}{\partial z} - \alpha \frac{\partial G}{\partial z} dz = 0 \qquad (A1.2.7)$$

Therefore, the remaining terms are now linearly independent, and consequently

$$\left(\frac{\partial F}{\partial x} - \alpha\frac{\partial G}{\partial x}\right)dx + \left(\frac{\partial F}{\partial y} - \alpha\frac{\partial G}{\partial y}\right)dy = 0 \qquad \text{(A1.2.8)}$$

In which

$$\frac{\partial F}{\partial x} - \alpha\frac{\partial G}{\partial x} = 0 \qquad \text{(A1.2.9)}$$

and

$$\frac{\partial F}{\partial y} - \alpha\frac{\partial G}{\partial y} = 0 \qquad \text{(A1.2.10)}$$

Now, with the help of Equations A1.2.7, A1.2.9, and A1.2.10, we have enough equations to solve how to locate the maximum or minimum.

Finally, it is necessary to affirm that the method could be easily generalized to more than one condition.

APPENDIX 1.3   METHODS OF COUNTING

If one thing can be done in N_1 ways, and after that a second thing can be done in N_2 ways, the two things can be done in succession in that order in $N_1 \times N_2$ ways. This principle, called the *principle of counting*, could be extended to any number of things.

Therefore, following this principle of counting, if we have a set of n things that are lined up in a row or a set of n objects in n boxes or n people and n chairs to be seated. If we ask how many ways we can arrange or permute them to form different permutations, then the answer is [13]

$$P(n,n) = n.(n-1)(n-2)(n-3)\text{.....}1 = n! \qquad \text{(A1.3.1)}$$

ways of arranging them.

On the other hand, if we have n objects but only r boxes or n people and only r chairs then there are

$$P(n,r) = n.(n-1)(n-2)(n-r+1) = \frac{n!}{(n-r)!} \qquad \text{(A1.3.2)}$$

ways of arranging n objects from a population of n objects in $r < n$ boxes. As we have n ways to fill the first box, $n-1$ ways to fill the second box, and finally $(n-r+1)$ ways of filling the r box.

If the objects in the previous case are indistinguishable or equal the number of different ways of arranging from a population of n objects in r boxes is

$$C(n,r) = \frac{P(n,r)}{P(r,r)} = \frac{n!}{(n-r)!r!} \qquad \text{(A1.3.3)}$$

Since the number that we are seeking, $C(n, r)$ is the number of ways for selecting from the total number of objects r objects and then arranging the r objects in r boxes in $P(r, r)$ ways, then applying the principle of counting the total number of ways of selecting n objects to be located in r boxes is $C(n,r)P(r,r) = P(n,r)$.

Similarly, the number of ways of putting N balls in N boxes with N_1 in box 1, N_2 in box 2, and so on is

$$W = \frac{N!}{N_1!N_2!N_3!....N_n!} \tag{A1.3.4}$$

This equation allows us to find, for example, the distribution $\{N_i\}$ at equilibrium, that is, the most probable macrostate distribution of N distinguishable particles over Ω energy levels. The number of ways to assign N_1 particles in the first level is

$$C(N,N_1) = \frac{N!}{(N - N_1)!N_1!}$$

Now, the number of ways to assign N_2 particles in the second level is

$$C(N - N_1,N_2) = \frac{(N - N_1)!}{(N - N_1 - N_2)!N_2!}$$

Consequently,

$$W = C(N,N_1)C(N - N_1 - N_2)......... = \frac{N!}{N_1!N_2!N_3!....N_n!}$$

APPENDIX 1.4　CALCULUS OF VARIATIONS

Suppose, $y(x)$ is defined on the interval $[a, b]$ and so defines a curve on the (x, y) plane. Now, assume that [36]

$$F[y(x)] = \int_a^b f(y(x), y_x(x), x)dx \tag{A1.4.1}$$

where $y_x(x) = dy/dx$. A function such as $F[y(x)]$ is called a functional; this name is used to distinguish $F[y(x)]$ from ordinary real-valued functions, whose domains consist of ordinary variables. The value of this functional will depend on the choice of the function $y(x)$, and the basic problem of the calculus of variations is to find the form of the function, which makes the value of the integral a minimum or maximum, normally a minimum.

In order to derive the extreme value conditions, we will require that the function $f(y(x), y_x(x), x)$ in the integral have continuous partial derivatives of x, y, and y_x. We also require the continuity of the derivatives, because we will need to apply chain rules and the Leibniz rule for differentiation.

Now if $y(x)$ is a curve in $[a, b]$, which minimizes the functional:

$$F[y(x)] = \int_a^b f(y(x), y_x(x), x)dx \qquad \text{(A1.4.2)}$$

Then [36]

$$\delta F[y(x)] = \delta \int_a^b f(y(x), y_x(x), x)dx = \int \left(\frac{\partial f}{\partial x} \delta x + \frac{\partial f}{\partial y_x} \delta y_x \right) = 0$$

Now, it is easy to show that

$$\delta F[y(x)] = \int \left(\frac{\partial f}{\partial x} - \frac{d}{dx} \frac{\partial f}{\partial y_x} \right) \delta y(x)dx = 0$$

Or in terms of the so-called functional derivative

$$\frac{\delta F[y(x)]}{\delta y(x)} = \int \left(\frac{\partial f}{\partial x} - \frac{d}{dx} \frac{\partial f}{\partial y_x} \right) dx = 0$$

Then, the function $y = y(x)$, which minimizes the functional, $F[y(x)]$, must obey the following differential equation:

$$\frac{\partial f}{\partial x} - \frac{d}{dx} \left(\frac{\partial f}{\partial y_x} \right) = 0 \qquad \text{(A1.4.3)}$$

This equation is called the Euler–Lagrange Equation [13].

The functional derivative is a generalization of the usual derivative that arises in the calculus of variations. In a functional derivative, instead of differentiating a function with respect to a variable, one differentiates a functional with respect to a function. The formal definition of the functional derivative for the one variable case is

$$\frac{\delta F[y(x)]}{\delta y(x)} = \lim_{\varepsilon \to 0} \frac{F[y(x) + \varepsilon \delta y(x)] - F(y(x))}{\varepsilon} \qquad \text{(A1.4.4)}$$

REFERENCES

1. P.W. Atkins and J. de Paula, *Physical Chemistry* (9th ed.), W.H. Freeman, and Co., New York, 2017.
2. I.N. Levine, *Physical Chemistry* (6th ed.), McGraw-Hill, New York, 2008.
3. R.H. Fowler and E.A. Guggenheim, *Statistical Thermodynamics* (2nd ed.), Cambridge University Press, Cambridge, UK, 1949.
4. J. Honerkamp, *Statistical Physics: An Advanced Approach with Applications*, Springer, New York, 1998.

5. C. Kittel, *Elementary to Statistical Physics*, John Wiley & Sons and Chapman-Hall, New York, 1958.
6. L. Landau and E.M. Lifshits, *Statistical Physics*, Cambridge University Press, Reading, MS, 1959.
7. R. Reif, *Fundamentals of Statistical and Thermal Physics*, McGraw-Hill, Boston, MA, 1965.
8. T.L. Hill, *Statistical Mechanics*, McGraw-Hill, London-&-New York, 1956.
9. P.T. Lansberg, *Thermodynamics and Statistical Physics*, Dover Books on Physics, London, UK, 1978.
10. R. Roque-Malherbe, *Physical Chemistry of Materials: Energy and Environmental Applications*, CRC Press, Boca Raton, FL, 2009.
11. R. Roque-Malherbe, *Electrochemistry: Energy and Environmental Applications*, Lambert Academic Publishing, Saarbrücken, Germany, 2017.
12. R. Roque-Malherbe, *Mic. Mes. Mat.,* 41 (2000) 227.
13. M.L. Boas, *Mathematical Methods in the Physical Science* (3rd ed.), Wiley-CDA, New York, 2006.
14. P. Hohemberg and W. Kohn, *Phys. Rev.,* 136 (1964) B864.
15. W. Kohn and L.J. Sham, *Phys. Rev.,* 140 (1965) A1441.
16. S.K. Ghosh, *Int. J. Mol. Sci.,* 3 (2002) 260.
17. R. Evans, in *Fundamentals of Inhomogeneous Fluids* (D. Henderson, Ed.), Marcel Dekker, New York, 1992, p. 85.
18. Y. Tang and J. Wu, *J. Chem. Phys.,* 119 (2003) 7388.
19. A.R. Forsyth, *Calculus of Variations*, Dover, New York, 1960.
20. P.I. Ravikovitch and A.V. Neimark, *Phys. Rev. E,* 64 (2000) 011602.
21. M. Thommes, B. Simarsly, M. Groenewelt, P.I. Ravikovitch, and A.V. Neimark, *Langmuir,* 22 (2006) 756.
22. J. Onsager, *Phys. Rev.,* 37 (1931) 405 and 38 (1931) 2265.
23. I. Prigogine, *Introduction to Thermodynamics of Irreversible Process* (3rd ed.), John Wiley & Sons, New York, 1968.
24. R. Kubo, M. Toda, and N. Hashitsume, *Statistical Physics II: Non-Equilibrium Statistical Mechanics*, Springer-Verlag, Berlin, Germany, 1991.
25. S.R. Groot and P. Mazur, *Non-Equilibrium Thermodynamics*, Dover Books on Physics, New York, 2011.
26. M.S. Green, *J. Chem. Phys.,* 22 (1954) 398.
27. R. Kubo, *J. Phys. Soc. Japan,* 12 (1957) 570.
28. Y. Chen, X. Chen, and M.-P. Qiang, *J. Phys. A: Math, & General,* 39 (11) (2006).
29. L. Landau and E.M. Lifshits, *Mechanics*, MIR, Moscow, Russia, 1966.
30. H. Mehrer, *Diffusion in Solids: Fundamentals, Methods, Materials and Diffusion Controlled Processes* (1st ed.), Springer-Verlag, Berlin, Germany, 2007.
31. J. Karger, D.M. Ruthven, and D.N. Theodorou, *Diffusion in Nanoporous Materials*, Vol. 1, Wiley-VCH, New York, 2012.
32. A. Einstein, *Investigations on the Theory of Brownian Movement*, Dower Publications, New York, 1956.
33. P. Langevin, *C. R. Acad. Sci. (Paris),* 146 (1908) 530.
34. H. Risken, *The Fokker-Planck Equation* (2nd ed.), Springer-Verlag, New York, 1996.
35. J. Crank, *The Mathematics of Diffusion* (2nd ed.), Oxford University Press, Oxford, UK, 1975.
36. G.B. Arfken, H.J. Weber, and F.E. Harris, *Mathematical Methods for Physicist* (7th ed.), Wiley-CDA, New York, 2011.

2 Adsorption in Solid Surfaces

2.1 INTRODUCTION

2.1.1 WHAT IS THE MEANING OF THE TERM ADSORPTION

The concept *adsorption* was introduced by Kayser in 1881 to outline the augment in the number of gas molecules over solid surfaces, a phenomenon formerly observed by Fontana and Scheele in 1777 [1].

In this regard, the clean surface of any solid is distinguished by the fact that the atoms constituting it do not have all bonds saturated, in fact producing an adsorption field over it and producing the gathering of molecules adjacent to the surface [2–6]. This phenomenon named as adsorption is a universal tendency, given that, during their manifestation a decrease in the surface tension is experienced by the solid [7–10]. Now, to be definite it is required to state that the concept adsorption is applied for the description of the direct process, whereas for the reverse one the word desorption is utilized [1].

Concluding: Adsorption for gas–solid and liquid–solid interfaces is an increase in concentration of gas molecules in the solid surface or an increase in the concentration of a dissolved substance at the interface of a solid and a liquid phase, a phenomenon caused, in both cases, by surface forces [11,12].

2.1.2 PHASES ALONG WITH COMPONENTS INVOLVED IN THE ADSORPTION PROCESS

The adsorptive or adsorbate is the gas or molecule dissolved in a liquid, which is adsorbed by the solid surface, that is to say, the adsorbent, whenever the gas or liquid is brought in contact with it [1,4]. Moreover, for a gas–solid physical adsorption process, the adsorptive (or adsorbate) is the gas adsorbed by the solid (the adsorbent) every time the gas is brought in contact with the solid [1].

In a thermodynamic sense, during physical adsorption of gases, the *adsorbed phase* is in equilibrium with the *gas phase*. The fact implying that $\mu_a = \mu_g$, in which μ_a and μ_g are, respectively, the chemical potential of the adsorbed and gas phases. Thereafter, applying the *Gibbs phase rule* for adsorption systems [13,18], $P + F = C + 2 + I$, where F is the number of degrees of freedom of the thermodynamic system in equilibrium, I is the number of bidimensional or restricted phases, whereas P, is the number of phases and C is the number of components. Subsequently, for a single component gas–solid adsorption system, in which the number of components is $C = 2$, specifically, gas and solid; whereas the number of restricted phases is $I = 1$, to be precise, the adsorbed phase, been finally the total number of phases, $P = 3$, explicitly: gas, solid, and adsorbed phase, then [14] $F = C + 2 + I - P = 2 + 2 + 1 - 2 = 2$. So, adsorption data are pictured by isotherms [1]: $n_a = F(P,T)$, a relation between the amount adsorbed n_a, the equilibrium adsorption pressure P, and temperature $T =$ constant. Explicitly it is measured by the relationship between the amount adsorbed n_a and the equilibrium pressure P, at a constant temperature T, that is, the adsorption isotherm [2–6]:

$$n_a = F(P)_T \tag{2.1}$$

Gas adsorption is normally considered as a physical adsorption process as the molecular forces implicated are van der Waals type interactions [4,5]. Moreover, physical adsorption might be classified as mobile adsorption, in the case when the adsorbed molecule behave as a gas molecule in the adsorption space or immobile adsorption occupying every time the adsorbed molecule is forced to vibrate around one adsorption site [3].

Now it must be clarified that for open surfaces, adsorption consist of a layer by layer-filling process, where the first layer is filled when $\theta = n_a/N_m = 1$, where θ is defined as the surface recovery and N_m is the monolayer capacity. Consequently, we said then that we have monolayer adsorption, when $\theta = n_a/N_m < 1$, whereas multiplayer adsorption takes place when $\theta = n_a/N_m > 1$ [10]. Meanwhile, physical adsorption of gases in solid surfaces happens when no reaction with exchange of electrons takes place with the formation of chemical bonds during the adsorption process. On the other hand, in the case when during adsorption a reaction with electron exchange between the solid surface and the gas molecules takes place the phenomenon is named chemical adsorption [6].

2.1.3 Porous Materials

First of all, the pore width, D_p, of an adsorbent as equal to the pore diameter in the case of cylindrical pores must be defined, or as the distance between opposite walls in the case of slit-shaped pores as these are the pore types normally found in practice [1]. Likewise, the International Union of Pure and Applied Chemistry (IUPAC) classify them as: (1) microporous, the adsorbents showing pore width within 0.3 nm $< D_p < 2$ nm; (2) mesoporous, the adsorbents those exhibit pore diameters inside 2 nm $< D_p < 50$ nm; and finally, (3) macroporous, the adsorbents those show pore diameters $D_p > 50$ nm [15].

Now, it is necessary to state that the parameters characterizing a porous adsorbent are the specific surface area, labeled S, measured in m²/g, the micropore volume denoted by W^{MP} measured in cm³/g, the total pore volume represented by W, which is the sum of the micropore and mesopore volumes of the adsorbent, measured in cm³/g, together with the pore size distribution (PSD) [8–10], provided the PSD is a graphical representation of $\Delta V_p/\Delta D_p$ versus D_p, where V_p is the pore volume accumulated up to the pore of the width D_p, measured in cc-STP/g Å [5], where the unit, cc-STP, indicates the amount adsorbed measured in cubic centimeters at STP, that is, at standard temperature and pressure, that is, 273.15 K and 760 Torr, that is, 1.01325×10 Pa [6].

Finally, it must be clarified that strictly speaking the surface area is the outer surface, to be exact, the area out of the micropores. On the other hand, if the adsorbent do not have micropores, then the surface area and the outer surface area coincide.

2.2 INTERFACIAL LAYER, GIBBS DIVIDING SURFACE AND GIBBS ADSORPTION

By definition, the *interfacial layer* is a space region between two contacting bulk phases, for instance, solid–gas or solid–liquid, where the properties are significantly different from the properties of the bulk phases [2]. Now, to understand the phenomena occurring in this inhomogeneous region we should enumerate some possible involved properties, such as composition, molecular density, orientation, charge density, pressure tensor, electron density, along with others.

In Figure 2.1 the so-called *Gibbs dividing surface* (GDS) is represented; specifically, the complex profile of interfacial properties occurring in the case of multicomponent systems with coexisting bulk phases where attractive/repulsive molecular interactions involve adsorption or depletion of one or several components [16,17].

Been, the GDS a geometrical surface chosen parallel to the interface, and used to define the volumes of the bulk phases in the calculation of the extent of adsorption, and of other surface excess properties. It is evidenced that in the reference ideal system the concentration remains constant up to the GDS. However, in the real system the concentration changes across the interface of thickness $\gamma = z_\beta - z_\alpha$ from phase α to phase β. Hence, with the help of this representation we define now the *surface excess amount, or Gibbs adsorption* [8] of the ith, component, that is, n_i^σ is defined as the excess of the amount of this component actually present in the system over

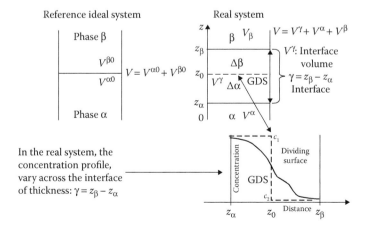

FIGURE 2.1   Gibbs dividing surface (GDS).

that present in a reference ideal system of the same volume as the real system, and in which the bulk concentrations in the two phases remain uniform up to the GDS [16]:

$$n_i^\sigma = n_i - V^{\alpha_0} c_\alpha + V^{\beta_0} c_\beta \qquad (2.2)$$

where:

n_i is the total amount of the component i in the system

c_α and c_β are the concentrations in the two bulk phases, α and β

In addition, V^{α_0} and V^{β_0} are the volumes of the two phases defined by the Gibbs surface.

Figure 2.2 is a schematic representation of a volumetric system for the measurement of adsorption isotherms; then for gas–solid adsorption from Figure 2.1, giving that $c^g = 0$, in the solid phase [18].

$$n^\sigma = n - V^{\alpha_0} c^g = n - c^g (V^g + V^a) \qquad (2.3)$$

In which $V^{\alpha_0} = V^g + V^a$. Now, if A is the adsorbent surface area, and t, is the thickness of the adsorbed layer, then the volume of the adsorbed layer or adsorption space is [8]

$$V^a = At \qquad (2.4)$$

We may define at this point the amount adsorbed as

$$n^a = \int_0^{V^a} c \, dV = A \int_0^t c \, dz \qquad (2.5)$$

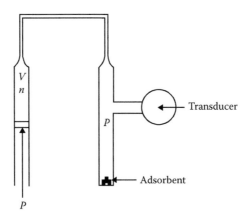

FIGURE 2.2   Schematic representation of a volumetric adsorption experiment.

The total amount of gas molecules in the system is

$$n = n^a + V^g c^g \tag{2.6}$$

therefore,

$$n^a = n - V^g c^g \tag{2.7}$$

Consequently, combining Equations 2.3 and 2.6 we get $n^a = n^\sigma + V^a c^g \approx n^\sigma$, since $V^a c^g \approx 0$. Now we will define more precisely some surface parameters. If the mass of degassed adsorbent is defined as m_s, and measured in g, then the specific surface area is $S = A/m_s$, which is measured in m^2/g. Besides, the specific surface excess amount is defined by $n_a = n^\sigma/m_s \approx n_a/m_s$, where n_a is the amount adsorbed measured in mol/g. The specific surface excess amount is what is usually measured in practice, and is approximately equal to the amount adsorbed, n_a. This magnitude is dependent on the equilibrium adsorption pressure, P, at constant adsorbent temperature, T. Finally, gas adsorption data are, in practice, expressed by adsorption isotherms as follows [1,2]:

$$n_a = \frac{n^\sigma}{m_s} \approx \frac{n^a}{m_s} = F(P)_T \tag{2.8}$$

2.3   GAS–SOLID ADSORPTION THERMODYNAMICS

2.3.1   ADSORPTION INTERACTION FIELDS

As was formerly affirmed the fresh surface of a solid is characterized by the fact that the atoms, which compose it do not have all their bonds saturated. Hence, when a molecule contacts the surface of a solid adsorbent, it becomes subjected to diverse interaction fields characterized by different potentials [1], such as dispersion energy ϕ_D, repulsion energy ϕ_R, polarization energy ϕ_P, field dipole energy $\phi_{E\mu}$, field gradient quadrupole energy, ϕ_{EQ}, sorbate–sorbate interaction energy, ϕ_{AA}, and the acid–base interaction with the active site, ϕ_{AB}, if the surface contains hydroxyl bridge groups [2–6]. Concretely, $\phi_D + \phi_R$ could be represented by the Lennard–Jones 12-6 potential [11,18]: $\phi_D + \phi_R = 4\varepsilon((\sigma/z)^{12} - (\sigma/z)^6)$ where ε, is the potential energy minimum and σ is the gas–solid separation at maximum interaction; appearing, this term in the interaction of any molecule with whichever adsorbent. Moreover, the electrostatic polarization appears when nonpolar molecules interacts with an adsorbent possessing a crystalline electric field, provided this term is expressed as follows [1–3]: $\phi_P = -\alpha E^2/2$, in which α, is the adsorbed molecule or atom and E is the intensity of the adsorbent electric field. Moreover, the field dipole interaction of polar molecules with an adsorbent possessing a crystalline electric field could be written as follows [5]: $\phi_\mu = -\mu E \cos\phi$, in which μ, is the permanent dipole moment of the adsorbed molecule and ϕ, is the angle between the electric field E and the dipole moment [3,7]. Finally, field gradient quadrupole energy is given by [3,6] $\phi_{EQ} = Q(\partial E/\partial z)/2$, where Q is the quadrupole moment of the adsorbed molecule, and $(\partial E/\partial z)$ is the electric field gradient of the adsorbent.

2.3.2 Isosteric and Differential Heats of Adsorption

As was previously affirmed, adsorption is a general tendency of matter since during their occurrence, a decrease in the surface tension is experienced by the solid. Therefore, adsorption is a spontaneous process, where a decrease in the Gibbs free energy, that is, $\Delta G < 0$ takes place [1,18]. In addition, in the course of physical adsorption, molecules from a disordered bulk phase reach a more well-organized adsorbed state, as in this state, molecules are constrained to move in a surface or a pore. Afterward, during adsorption there is a decrease in entropy, that is, $\Delta S < 0$ takes place; hence, as, $\Delta G = \Delta H - T\Delta S$, thenceforth, $\Delta H = \Delta G + T\Delta S < 0$, that is, adsorption is an exothermic process, hence, favored by the decrease in temperature [9].

As it is very well known, the fundamental equation of thermodynamics for a bulk mixture is given by [13] $dU = TdS - PdV + \sum_i \mu_i dn_i$, where U is the internal energy of the system, S is their entropy, V is the volume, T is the temperature, μ_i is the chemical potential along with n_i, which is the number of moles of the components contained in the system. Here, thermodynamic approach is applied that considers the adsorbent plus adsorbed gas or vapor as a solid solution (system aA). Therefore [20]:

$$dU_{aA} = TdS_{aA} - PdV_{aA} + \mu_a dn_a + \mu_A dn_A$$

where U_{aA}, S_{aA}, and V_{aA} are the internal energy, entropy, and volume of the system, respectively; aA, μ_a, μ_A, n_a, and n_A, are the chemical potentials and the number of moles of the adsorbate and the adsorbent in the system aA; defining now, $\Gamma = n_a/n_A$, then $\mu_a = \mu_a(T, P, \Gamma)$ and $\mu_A = \mu_A(T, P)$, therefore [3]: $d\mu_a = \bar{S}_a dT + \bar{V}_a dP + (\partial\mu_a/\partial\Gamma)_{T,P} d\Gamma$, where $\bar{S}_a$ and $\bar{V}_a$ are the partial molar entropy and volume of the adsorbate in the system aA. Now, as in equilibrium the chemical potential of the adsorbate in the aA phase and the gas phase are equal, then $d\mu_a = d\mu_g = \bar{S}_g dT + \bar{V}_g dP$. Subsequently, for $\Gamma = constant$ [16]:

$$\left[\frac{d\ln P}{dT}\right]_\Gamma = \frac{\bar{H}_g - \bar{H}_a}{RT^2} = \frac{q_{iso}}{RT^2} \tag{2.9}$$

where $\bar{H}_g$ and $\bar{H}_a$ are the partial molar enthalpies of the adsorbate in the gas phase and in the system aA. Now with the help of Equation 2.9 it is possible to define the isosteric enthalpy of adsorption [11,20]:

$$\Delta H(n_a) = -(\bar{H}_g - \bar{H}_a) = -q_{iso} \tag{2.10}$$

where q_{iso} is the enthalpy of desorption or isosteric heat of adsorption. The isosteric heat of adsorption is calculated with the help of adsorption isotherms.

One more important adsorption heat is the differential heat of adsorption defined as follows [18]:

$$q_{diff} = \frac{\Delta Q}{\Delta n_a} \tag{2.11}$$

In which ΔQ is the evolved heat during the finite increment measured calorimetrically, whereas Δn_a is the magnitude of adsorbed molecules producing the heat evolution. Moreover, this adsorption heat can be approximately calculated using [5,16].

$$q_{\text{diff}} \approx q_{\text{iso}} - RT \tag{2.12}$$

Nevertheless, it is necessary to recognize that Equation 2.12 is only exactly satisfied in the case of inert adsorbents, but porous adsorbent systems are not generally inert [15].

2.3.3 SOME RELATIONS BETWEEN ADSORPTION MACROSCOPIC AND MICROSCOPIC PARAMETERS

The molar integral change of free energy, at a given temperature T, during adsorption is [2] $\Delta G^{\text{ads}} = \Delta H^{\text{ads}} - T\Delta S^{\text{ads}}$. On the other hand, it is possible to show that the relation between the enthalpy of adsorption (ΔH^{ads}) and the differential heat of adsorption (q_{diff}) for porous systems is [3,20]

$$\Delta H^{\text{ads}} \approx -q_{\text{diff}} - RT + \frac{T}{\Gamma}\left(\frac{\partial \vartheta}{\partial T}\right)_\Gamma \tag{2.13}$$

where $\Gamma = n_a/n_A$, $q_{\text{diff}} \approx q_{\text{iso}} - RT$, n_a, and n_A are the number of moles of the adsorbate and the adsorbent in the system, aA along with ϑ defined by [15] $\vartheta = RT\int_0^P \Gamma d \ln P$. Now assuming that the change in entropy on adsorption is negligible in comparison with the rest of terms in Equation 2.13, and taking into account that the change in the free energy of adsorption (ΔG^{ads}) could be expressed by

$$\Delta G^{\text{ads}} = RT \ln\left(\frac{P}{P_0}\right) \tag{2.14}$$

and that [5,12]

$$-q_{\text{diff}} = U_0 + P_a - \Delta H^{\text{ads}} \tag{2.15}$$

where U_0 and P_a denote the adsorbate–adsorbent and adorbate–adsorbate interactions energies, respectively. Subsequently, from Equations 2.12 through 2.15 it is possible to get

$$RT \ln\left(\frac{P}{P_0}\right) + \left(RT - \frac{T}{\Gamma}\left(\frac{\partial \vartheta}{\partial T}\right)\right) = U_0 + P_a \tag{2.16}$$

Hence, considering that the adsorbed phase is ideal, that is, when $\Gamma = KP$, in this case from the definition of ϑ

$$\vartheta = RT\int_0^P \Gamma d \ln P = RTK\int_0^P \frac{P dP}{P} = RTKP \tag{2.17}$$

Consequently, from Equations 2.16 and 2.17: $RT \ln(P/P_0) + (RT - T/\Gamma(R\Gamma)) = U_0 + P_a$; or:

$$RT \ln\left(\frac{P}{P_0}\right) = U_0 + P_a \tag{2.18}$$

2.4 ADSORPTION IN POROUS MATERIALS

2.4.1 Measurement of Adsorption Isotherms by the Volumetric Method

In Figure 2.3 a volumetric adsorption facility is represented [8].

The procedure to gather an adsorption isotherm using a volumetric facility is as follows: at first, gas from the container is introduced into the dose volume V_1 of the previously evacuated vacuum equipment. Hence, measuring the ensuing pressure P_1; next, stopcock 2 is opened resulting in a pressure P_2, which allows the measurement of the dose volume using the expression: $V_1 = V_c P_2 / P_1 - P_2$, where V_c is a container whose volume was carefully calibrated before attaching it to the adsorption testing facility; afterward, opening stopcock 3, the gas is contacted with the adsorbent at temperature T; it is necessary to state now that the dead volume V_d was yet to be measured as follows: $V_d = V_2 + V_g T_r / T$, with the adsorbent present inside the sample cell with the help of He, a gas not adsorbed normally at the experimental temperature; moreover, for an easier determination of V_d, the volume V_2 is experimentally made approximately zero, that is, $V_2 \approx 0$.

Recapitulating, the application of the volumetric method for the determination of an adsorption isotherm could be reduced to the use of the following procedure: first, adsorbate gas is introduced into the manifold volume V_1, and the amount dosed is measured, generally in cubic centimeters at STP, that is, standard temperature and pressure, that is, 273.15 K and 1.01325×10 Pa.

$$n_{\text{dose}}^{i} = \frac{P_1^{i} V_1}{RT_r} + \frac{P_2^{i-1} V_d}{RT_r} \tag{2.19}$$

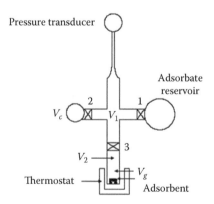

FIGURE 2.3   Volumetric adsorption facility.

In which P_2^{i-1} is the equilibrium pressure of the previous adsorption step (i–1th, step), T_r: ambient temperature, V_1: dose volume, V_d: dead volume, P_1^i initial pressure (ith, step), and n_{dose}^i: initial number of moles during the ith, adsorption step. Thereafter, when equilibrium is attained the quantity of gas not adsorbed is calculated by

$$n_{\text{final}}^i = \frac{P_2^i(V_1 + V_d)}{RT_r} \tag{2.20}$$

where P_2^i: equilibrium pressure of the current measurement (ith, step). Then, the amount adsorbed in the ith isotherm point, $\Delta^i n_a$, is calculated with the help of the following equation:

$$\Delta^i n_a = \frac{n_{\text{dose}}^i - n_{\text{final}}^i}{m_s} \tag{2.21}$$

In which m_s is the mass of the degassed adsorbent. Finally, the isotherm is calculated by summing the different adsorption steps as follows:

$$n_a^i = \sum_{j=1}^{i} \Delta^j n_a^j \tag{2.22}$$

where n_a^i is the magnitude of adsorption or the amount adsorbed up to the ith, adsorption step. Finally, plotting n_a^i versus P_2^i the experimental isotherm is obtained.

2.4.2 CHARACTERIZATION OF POROUS MATERIALS BY ADSORPTION METHODS

Porous materials are of huge practical significance in industrial together with pollution abatement applications. In this regard, microporous materials, such as zeolites and related materials are extensively applied in the petrochemical industry as heterogeneous catalysts, in cracking, and other applications [6], whereas micro/mesoporous materials and mesoporous materials, for instance, silica gels [21], active carbon [22,23], mesoporous molecular sieves [24,25], and other materials [26], are widely used in separation processes, catalysis, and other applications [17]. The successful performance of the adsorption unitary operation in industry and pollution abatement requires a comprehensive characterization of these porous materials with regard to micropore volume, surface area, and PSD [1–9,17–26].

Common porous materials, such as silica [21], active carbons [27], alumina, titania, and porous glasses [3,6] are amorphous. In contrast, zeolites [28], akaganeites [29], Prussian blue analogs (PBAs) [30,31], and metal organic frameworks (MOFs) [32] are crystalline, that is, every atom can be located in a microscopically sized unit cell [17]. Besides, mesoporous molecular sieves such as MCM-41, MCM-48, SBA-15, and others are not crystalline. However, they are ordered [24,25]. Such order is not present in amorphous materials. Consequently, a more complete characterization can be performed in the case of crystalline and ordered materials, whereas for amorphous nanoporous materials a complete and comprehensive characterization is more problematic. Nevertheless, typical properties such as microporous volume, total pore volume, specific surface area, and PSD, and other properties can still be determined.

During the adsorption of vapors in complex porous systems, the adsorption process occurs approximately as follows: initially, micropore filling in which the adsorption behavior is dominated almost entirely by the interactions of the adsorbate and the pore wall [3]. We will consider here that adsorption in the micropores could be considered as a volume filling of the microporous adsorption space, and not as a layer-by-layer surface coverage [7]. Afterward, at higher pressures, external surface coverage consisting of monolayer and multilayer adsorption on the walls of mesopores and open macropores, and capillary condensation taking place in the mesopores [3,6,8].

Vapor adsorption in micropores is the main method for measuring the micropore volume using the Dubinin adsorption isotherm, the *t*-plot method, and other adsorption isotherms [3,7]. Surface coverage is generally described with the Brunauer–Emmett–Teller (BET) adsorption isotherm, which allows the specific surface area of the porous solid to be determined [12]. On the other hand, capillary condensation of vapors is the primary method of assessment of PSD in the range of mesopores [2].

Capillary condensation is associated with a shift of the vapor–liquid coexistence in pores compared to bulk fluid. This means that a confined fluid in a pore condenses at a pressure lower than the saturation pressure at a given temperature, being this a phenomenon in the majority of systems accompanied by hysteresis [5]. In recent years, the standard method for determining the PSD in the mesoporous range with the help of adsorption isotherms was the Barret–Joyner–Hallenda (BJH) method [1,8]. However, this methodology does not estimate the PSD properly. Consequently, a new methodology of adsorption isotherm assessment based on the nonlocal density functional theory (NLDFT), which was originated in the density functional theory (DFT) applied to inhomogeneus fluids has revolutionized the methodology of PSD calculation in porous materials [26,33].

2.5 SOME EXAMPLES OF THE APPLICATION OF THE VOLUMETRIC METHOD

2.5.1 VOLUMETRIC AUTOMATIC SURFACE AREA AND POROSITY MEASUREMENT SYSTEMS

The adsorbed amount as a function of pressure can be obtained by the volumetric and gravimetric methods [1–5], provided the use of volumetric method based fundamentally on nitrogen, carbon dioxide, and argon adsorption isotherms obtained at temperatures of liquid nitrogen (77.35 K), liquid argon (87.27 K), and room temperature, respectively. A model of a volumetric adsorption apparatus is given in the graphical representation shown in Figure 2.4 [5,34], where this volumetric sorption equipment is equipped with pressure transducers in the dosing volume and sample cell sectors. For this reason, the sample cell is isolated throughout the equilibration, which guarantee a very small void volume, and consequently a highly accurate determination of the adsorbed amount [26,35].

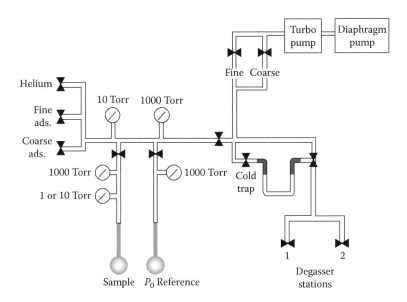

FIGURE 2.4 Standard volumetric adsorption apparatus. (From Reproduced from Yang, R.T. *Adsorbents: Fundamentals, and Applications*, John Wiley & Sons, New York, 2003. With permission.)

The vapor pressure of the adsorbate at the temperature of the adsorption experiment P_0 is measured during the whole analysis by means of a saturation pressure transducer, which allows the vapor pressure to be monitored for each data point. This produces a great accuracy and precision in the determination of the relative pressure: $x = P/P_0$, and thus in the measurement of the PSD [1–6].

Finally, it is necessary to state that the vacuum system of a standard commercial volumetric adsorption apparatus utilize a diaphragm pump as a force-pumping system for the turbomolecular pump, in order to guarantee a complete oil-free environment for the adsorption measurement and the outgassing of the sample, previous to the analysis.

2.5.2 ADSORPTION ISOTHERMS OF NITROGEN AT 77 K IN ZEOLITES

The study of zeolites as adsorbent materials began in firm in 1938 when Prof. R. M. Barrer published a series of papers on the adsorptive properties of zeolites [36]. Moreover, in the past 60 years, both synthetic [37–39] and natural zeolites [40] became one of the most important materials in modern technology; today, the production and application of zeolites for industrial processes have become a multimillion dollar industry.

Zeolites have been shown to be good adsorbents for H_2O, NH_3, H_2S, NO, NO_2, SO_2, and CO_2, linear and branched hydrocarbons, aromatic hydrocarbons, alcohols, ketones, and other molecules, whereas adsorption is not only an industrial application of zeolites, it is also a powerful means of characterizing these materials [1–6,18–26] as the adsorption of a particular molecule gives information about the microporous volume, the mesoporous area, the volume, the size of the pores, the energetic of adsorption, and molecular transport.

In Figure 2.5 the N_2 adsorption isotherm at 77 K of a natural erionite sample AP is shown, where the sample AP contain 85% of erionite, the rest is composed of 15% of montmorillonite (2–10 wt. %), quartz (1–5 wt. %), calcite (1–6 wt. %), feldspars (0–1 wt. %), magnetite (0–1 wt. %), and volcanic glass (3–6 wt. %) [41].

Whereas, in Figures 2.6 and 2.7, the N_2 adsorption isotherms at 77 K of the synthetic zeolites, Na–Y (CBV100, $SiO_2/Al_2O_3 = 5.2$) (Figure 2.6) [3], and Na–Y (SK-40, $SiO_2/Al_2O_3 = 4.8$) are exposed (Figure 2.7) [3].

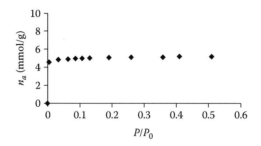

FIGURE 2.5 N_2 adsorption isotherm at 77 K of the natural erionite, sample AP.

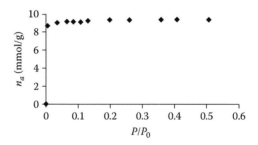

FIGURE 2.6 N_2 adsorption isotherm at 77 K of the synthetic zeolite Na–Y, sample CBV-100.

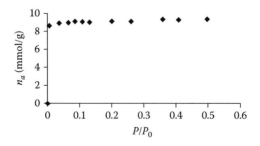

FIGURE 2.7 N_2 adsorption isotherm at 77 K of the synthetic zeolite Na–Y, sample SK-40.

The N_2 adsorption isotherm at 77 K were measured with an accelerated surface area and a Porosimetry System ASAP 2000 from Micromeritics [3,32], which is a volumetric automatic equipment similar to those reported in Figure 2.4.

2.5.3   CALORIMETRY OF ADSORPTION OF NH₃ IN AlPO₄-5 AND FAPO-5 MOLECULAR SIEVES

AlPO₄-5 and FAPO-5 molecular sieves are synthesized using procedures previously described [38,39], and were used for the study of the calorimetry of adsorption of NH_3. The occluded amine in both aluminophosphates was triethylamine (TEA). The quantity of Fe included in the FAPO-5 molecular sieve was determined by X-ray fluorescence using a Camberra spectrometer equipped with a Si–Li detector, whereas the X-ray diffractograms of AlPO₄-5 and FAPO-5 were obtained in a Carl Zeiss TUR-M62 equipment [42]. The crystallinity of the as-synthesized FAPO-5 and AlPO₄-5 molecular sieves was approximately 100%, whereas the weight per cent of metal included in the framework of the FAPO molecular sieve was 1%.

The heat of adsorption of NH_3 in AlPO₄-5 and FAPO-5 molecular sieves was measured in a heat flow calorimeter using the equation [41]:

$$\Delta Q = \kappa \int \Delta T dt$$

where:
   ΔQ is the heat evolved during the finite increment
   Δn_a is the magnitude of adsorption
   κ is the calibration constant
   ΔT is the difference between thermostat temperature, and the sample temperature during adsorption
   t is the time

The calorimeter was a high-vacuum line for adsorption measurements applying the volumetric method, used in a Pyrex glass, homemade vacuum system, including a sample holder, a dead volume, a dose volume, a U-tube manometer, and a thermostat (see Section 4.1). In the sample holder the adsorbent—thermostated with 0.1% of temperature fluctuation—is in contact with a chromel–alumel thermocouple included in an amplifier circuit (amplification factor: 10), and connected with an x–y plotter.

The calibration of the calorimeter was performed using the reported data for the adsorption of NH_3 at 300 K in Na–X zeolite [38]. The differential heat of adsorption was calculated with the help of Equation 2.11 in an increment form: $q_{diff} = \Delta Q / \Delta n_a$.

In Figure 2.8 the results obtained during the measurement of the differential heats of adsorption of NH_3 at 300 K in AlPO₄-5 and FAPO-5 molecular sieves are presented. The error in q_{diff}, is ± 2 kJ/mol, $\theta = n_a / N_a$, n_a is the magnitude of adsorption, and N_a is the maximum magnitude of adsorption in the zeolite.

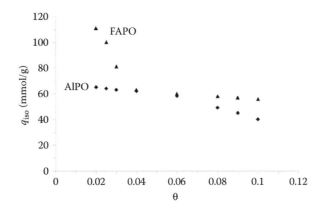

FIGURE 2.8 Differential heat of adsorption of NH_3 at 300 K versus θ in $AlPO_4$-5 and FAPO-5.

REFERENCES

1. K.S.W. Sing, D.H. Everett, R.A.W. Haul, L. Moscou, R.A. Pirotti, J. Rouquerol, and T. Siemieniewska, *Pure App. Chem.*, 57 (1985) 603.
2. F. Rouquerol, J. Rouquerol, K. Sing, P. Llewellyn, and G. Maurin, *Adsorption by Powders and Porous Solids* (2nd ed.), Academic Press, New York, 2013.
3. R. Roque-Malherbe, *Physical Chemistry of Materials: Energy and Environmental Applications*, CRC Press, Boca Raton, FL, 2009.
4. J.U. Keller and R. Staudt, *Gas Adsorption Equilibria: Experimental Methods and Adsorptive Isotherms*, Springer-Verlag, New York, 2004.
5. R.T. Yang, *Adsorbents: Fundamentals, and Applications*, John Wiley & Sons, New York, 2003.
6. R.M. Barrer, *Zeolites and Clay Minerals as Sorbents and Molecular Sieves*, Academic Press, London, UK, 1978.
7. S. Ross and J.P. Olivier, *On Physical Adsorption*, John Wiley & Sons, New York, 1964.
8. D.H. Everet and R.H. Ottewill, *Surface are Determination*, *International Union of Pure and Applied Chemistry*, Butterworth, London, UK, 2013.
9. S. Lowell, J.E. Shields, M.A. Thomas, and M. Thommes, *Characterization of Porous Solids and Powders: Surface Area, Pore Size and Density*, Kluwer Academic Press, Dordrecht, the Netherlands, 2004.
10. J.B. Condon, *Surface Area and Porosity Determination by Physisorption*, Elsevier, Amsterdam, the Netherlands, 2006.
11. M.M. Dubinin, *ACS Symposium Series*, 40 (1977) 1.
12. B.P. Bering and V.V. Serpinskii, *Ixv. Akad. Nauk, SSSR, Ser. Xim.*, 2427 (1974).
13. P.I. Ravikovitch and A.V. Neimark, *Colloids & Surf. A*, 187–188 (2001) 11.
14. R. Roque-Malherbe, *Mic. Mes. Mat.*, 41 (2000) 227.
15. M. Thommes, R. Kohn, and M. Froba, *J. Phys. Chem. B.*, 104 (2000) 7932.
16. F. Marquez-Linares and R. Roque-Malherbe, *J. Nanosci. & Nanotech.*, 6 (2006) 1114.
17. Quantachrome, AUTOSORB-1, Manual, 2003.
18. Micromeritics, ASAP 2000, Description, 1992.
19. G.D. Halsey, *J. Chem. Phys.*, 16 (1948) 931.
20. D.M. Young and A.D. Crowell, *Physical Adsorption of Gases*, Butterworth, London, UK, 1962.

21. T.L. Hill, *An Introduction to Statistical Thermodynamics*, Dover Publications, New York, 1986.
22. V.A. Bakaev, *Dokl. Akad. Nauk SSSR*, 167 (1966) 369.
23. M. Dupont-Pavlovskii, J. Barriol, and J. Bastick, Colloques Internes du CNRS, No. 201 (Termochemie), 1972.
24. R. Roque-Malherbe, *KINAM*, 6 (1984) 35.
25. R. Roque-Malherbe, L. Lemes, L. López-Colado, and A. Montes, in *Zeolites'93 Full Papers Volume* (D. Ming and F.A. Mumpton, Eds.), International Committee on Natural Zeolites Press, Brockport, New York, 1995, p. 299.
26. R. Roque-Malherbe, in *Handbook of Surfaces and Interfaces of Materials*, Vol. 5, (H.S. Nalwa, Ed.), Academic Press, New York, Chapter 12, 2001, p. 495.
27. M.M. Dubinin, *Prog. Surf. Memb. Sci.*, 9 (1975) 1.
28. M.M. Dubinin, E.F. Zhukovskaya, V.M. Lukianovich, K.O. Murrdmaia, E.F. Polstiakov, and E.E. Senderov, *Izv. Akad. Nauk SSSR*, (1965) 1500.
29. P.B. Balbuena and K.E. Gubbins, in *Characterization of Porous Solids* (I.J. Rouquerol, P. Rodriguez-Reynoso, K.S.W. Sing and K.K. Unger, Eds.), Elsevier, Amsterdam, the Netherlands, 1994, p. 41.
30. S.J. Gregg and K.S.W. Sing, *Adsorption Surface Area and Porosity*, Academic Press, London, UK, 1991.
31. D.W. Ruthven, *Principles of Adsorption and Adsorption Processes*, John Wiley & Sons, New York, 1984.
32. W. Rudzinskii, W.A. Steele, and G. Zgrablich, *Equilibria, and Dynamic of Gas Adsorption on Heterogeneus Solid Surfaces*, Elsevier, Amsterdam, the Netherlands, 1996.
33. J.B. Loos, *Modeling of Adsorption, and Diffusion of Vapors in Zeolites*, Coronet Books, Philadelphia, PA, 1997.
34. J.P. Fraissard (Ed.), *Physical Adsortion: Experiment, Theory and Applications*, Kluwer Academic Publishers, the Netherlands, 1997.
35. G. Horvath and K. Kawazoe, *J. Chem. Eng. Japan*, 16 (1983) 470.
36. S.U. Rege and R.T. Yang, in *Adsorption: Theory, Modeling and Analysis* (J. Toth, Ed.), Marcel Dekker, New York, 2002, p. 175.
37. A. Saito and H.C. Foley, *A.I.Ch. E. J.*, 37 (1991) 429.
38. R. Roque-Malherbe, C. de las Pozas, and G. Rodriguez, *Rev. Cub. de Física*, 5 (1985) 107 (Chemical Abstracts 103 No. 221398h).
39. R. Roque-Malherbe, C. de las Pozas, and G. Rodriguez, *Rev. Cub. de Fisica*, 4 (1984) 143 (Chemical Abstracts 103 No. 221398).
40. R. Roque-Malherbe, A. Costa, C. Rivera, F. Lugo, and R. Polanco, *J. Mat. Sci. Eng. A*, 3 (2013) 263.
41. C.H. Baerlocher, W.M. Meier, and D.M. Olson, *Atlas of Zeolite Framework Types* (5th ed.), Elsevier, Amsterdam, the Netherlands, 2001.
42. A.A. Zagorodni, *Ion Exchange Materials: Properties and Applications*, Elzevier, Amsterdam, the Nertherlands 2007.
43. F. Marquez-Linares and R. Roque-Malherbe, *Facets-IUMRS J.*, 3 (2004) 8.
44. R. Roque-Malherbe, R. López-Cordero, J.A. González-Morales, J. Onate, and M. Carreras, *Zeolites*, 13 (1993) 481.
45. E.M. Flanigen, R. Lyle-Patton, and S.T. Wilson, *Stud. Surf. Sci. & Catal.*, 37 (1988) 13.
46. H. Hattori and Y. Ono, *Solid Acid Catalysts: From Fundamentals to Applications*, CRC Press, Boca Raton, FL, 2015.
47. E.M. Flanigen, R.W. Broach, and S.T. Wilson, in *Zeolites in Industrial Separations and Catalysis* (S. Kulpathipanja Ed.), Wiley-WCH Verlag, Weinheim, Germany, 2010, p. 1.
48. T. Ishihara and H. Takinta, in *Catalysis*, Vol. 12, (J.J. Spivey Ed.), Royal Society of Chemistry, London, UK, 1996.

49. A. Suleiman, C. Cabrera, R. Polanco, and R. Roque-Malherbe, *RSC Advances*, 5 (2015) 7637.
50. R. Roque-Malherbe, F. Lugo, and R. Polanco, *App. Surf. Sci.*, 385 (2016) 360.
51. R. Roque-Malherbe, E. Carballo, R. Polanco, F. Lugo, and C. Lozano, *J. Phys. Chem. Solids*, 86 (2015) 65.
52. R. Roque-Malherbe, F. Lugo, C. Rivera, R. Polanco, P. Fierro, and O.N.C. Uwakweh, *Current App. Phys.*, 15 (2015) 571.
53. A. Rios, C. Rivera, G. Garcia, C. Lozano, P. Fierro, L. Fuentes-Cobas, and R. Roque-Malherbe, *J. Mat. Sci. Eng. A*, 2 (2012) 284.
54. R. Roque-Malherbe, O.N.C. Uwakweh, C. Lozano, R. Polanco, A. Hernandez-Maldonado, P. Fierro, F. Lugo, and J.N. Primera-Pedrozo, *J. Phys. Chem. C*, 115 (2011) 15555.
55. R. Roque-Malherbe, C. Lozano, R. Polanco, F. Marquez, F. Lugo, A. Hernandez-Maldonado, and J. Primera-Pedroso, *J. Solid State Chem.*, 184 (2011) 1236.
56. F. Marquez-Linares, O. Uwakweh, N. Lopez, E. Chavez, R. Polanco, C. Morant, J.M. Sanz, E. Elizalde, C. Neira, S. Nieto, and R. Roque-Malherbe, *J. Solid State Chem.*, 184 (2011) 655.
57. R. Roque-Malherbe, R. Polanco, and F. Marquez-Linares, *J. Phys. Chem. C*, 114 (2010) 17773 v.
58. V.A. Bakaev, *Dokl. Akad. Nauk SSSR*, 167 (1966) 369.
59. D.M. Ruthven, *Nature*, *Phys Sci.*, 232 (1971) 70.
60. W. Schirmer, K. Fiedler, and H. Stach, *ACS*, *Symposium Series*, 40 (1977) 305.
61. J. de la Cruz, C. Rodriguez, and R. Roque-Malherbe, *Surface Sci.*, 209 (1989) 215.
62. B.C. Lippens and J.H. de Boer, *J. Catalysis*, 4 (1965) 319.
63. D.W. Breck, *Zeolite Molecular Sieves*, John Wiley & Sons, New York, 1974.
64. S. Brunauer, P.H. Emmett, and E. Teller, *J. Amer. Chem. Soc.*, 60 (1938) 309.
65. A. Galarneau, D. Desplantier, R. Dutartre, and F. Di Renzo, *Mic. Mes. Mat.*, 27 (1999) 297.
66. J.R. Sams, G. Contabaris, and G.D. Halsey, *J. Phys. Chem.*, 64 (1960) 1689.
67. D.H. Everett and J.C. Powl, *J. Chem. Soc. Faraday Trans.*, 72 (1976) 619.
68. M.A. Parent and J.B. Moffat, *Langmuir*, 11 (1996) 4474.
69. R.J. Dombrowski, C.H.M. Lastoskie, and D.R. Hyduke, *Colloids & Surf. A*, 187–188 (2001) 23.

3 Assessment of Adsorption in Porous Adsorbents

3.1 INTRODUCTION

Gas adsorption measurements are widely used for the characterization of the surface area along with the porosity of porous materials [1–19]. This methodology is specifically applied for the calculation of the surface area, pore volume, and pore size distribution (PSD) of porous materials [9,14]. During adsorption in complex porous systems, as was previously stated, adsorption takes place as follows [3]: micropore filling, in which adsorption is dominated almost entirely by the interactions of the adsorbate, and the pore wall; next, at higher pressures, external surface coverage consisting of monolayer and multilayer adsorption on the walls of mesopores, and open macropores, then finally, capillary condensation, which takes place in the mesopores.

In this section, how adsorption in micropores is used as a method for the measurement of the micropore volume using the Dubinin, the t-plot method, and other isotherms [1–28] is explained. Moreover, it will be studied in which form surface coverage is portrayed, using the Brunauer–Emmett–Teller (BET) adsorption isotherm that also permits the determination of the specific surface area of the porous solid [29–34]. Finally, it will be explained in the Horvath–Kawazoe and Saito–Foley methods of assessment of the PSD in the range of micropores [35–37].

3.2 DUBININ ADSORPTION ISOTHERM

As was defined in Chapter 1 the adsorption process is characterized by n_a, the magnitude of adsorption, in this case within the micropore volume, expressed in mmol adsorbed/mass of dehydrated adsorbent. Moreover, the maximum adsorption magnitude, that is, the amount adsorbed that saturates the micropore volume of the adsorbent, is described by the parameter N_m, which is also expressed in moles adsorbed/mass of dehydrated adsorbent.

One of the best examples of microporous adsorbents are zeolites [38–40] and related materials [41–57]. These compounds belong to the group of molecular sieves; that is, three-dimensional microporous crystalline structure-type built from tetrahedral TO_4 groups linked in the corners sharing all oxygen atoms [3]. Particularly, in the case of aluminosilicate zeolites, their framework is built from Al and Si tetrahedrally coordinated atoms, that is, AlO_4 and SiO_4 linked in the corners sharing all oxygen atoms [6]. Within these frameworks the presence of tetracoordinated

Al, generates negative charge that must be balanced by extra-framework cations (one per Al). Consequently, the chemical composition of the aluminosilicate zeolites can be expressed as $M_{x/n}[(AlO_2)_x (SiO_2)_y] \cdot zH_2O$, where M are the balancing cations (charge $+n$) compensating the charge from the T(III) atom in tetrahedral coordination, and z is the water contained in the voids of the zeolite [41], provided the balancing cation on one inorganic or organic species, as ammonium, which can be exchanged by giving the zeolite its ion-exchange property [42].

Other molecular sieves have also been produced by incorporation into the concrete microporous material framework elements, such as P, Pd, Ge, Ga, Fe, B, Be, Cr, V, Zn, Zr, Co, Mn, and others in tetrahedral coordination (TO_4) to produce AlPO, SAPO, and MeAPO molecular sieves [43–85], where molecular sieves is a term coined by McBain in 1932 that also include clays [6], carbon black [49], Prussian blue analogs (PBAs) [50,51], akaganeites [52], metal organic frameworks (MOFs) [53], nitroprussides (NPs) [54,55], single-walled carbon nanotubes [56], silica [57], and other adsorbents. Therefore, to simplify the nomenclature when we refer here to zeolites we are also including the related materials.

For the calculation of the parameters characterizing microporous adsorbents the Dubinin adsorption isotherm equation is one of the best tools. It can be deduced using the theory of volume filling and the Polanyi's adsorption potential [27]. Polanyi in 1914 developed the first consistent adsorption theory. Hence, Dubinin, Polanyi's disciple, took this theory that proposed the existence of a relation between the volume of the adsorption space V_i, and the potential energy of the adsorption field, that is, $\varepsilon_i = F(V_i)$ (Figure 3.1) known as the Polanyi characteristic function, which is considered independent of temperature, where the energy of the adsorption field is given by the following equation: $\varepsilon_i = RT \ln(P_0/P_i)$ [20].

Where P_0 is the vapor pressure of the adsorptive at the temperature T of the adsorption experiment, and P_i is the equilibrium adsorption pressure (ε could also be designed as the differential work of adsorption). Then, applying the Gurvich rule, the relation obtained was $V_i = V^L n_a$ between the volume of the adsorption space V_i and the amount adsorbed, where V^L is the molar volume of the phase conforming the adsorbed phase. Similarly, combining the equation $\varepsilon_i = RT \ln(P_0/P_i)$ with the characteristic function together with the relation between the volume of the adsorption space

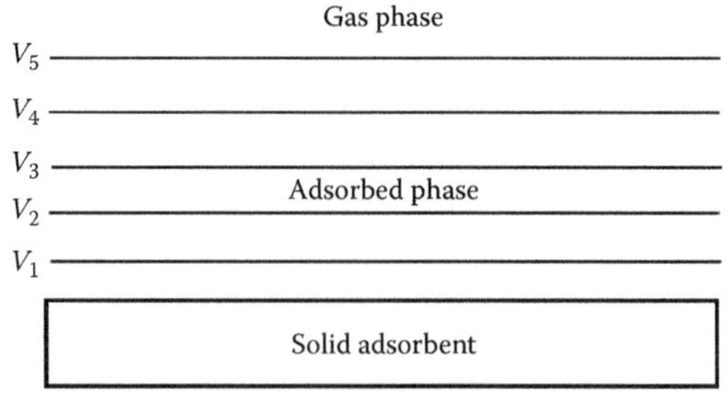

FIGURE 3.1 Polanyi adsorption model.

along with the amount adsorbed we will get [28] $F(V_i) = f(n_a) = \varepsilon_i = RT \ln(P_0/P_i)$; thereafter, using the Weibull distribution function, that is,

$$F(E) = \frac{\beta}{\eta}\left(\frac{E - \gamma}{\eta}\right)^{\beta-1} \exp\left(-\frac{E - \gamma}{\eta}\right)^{\beta} \tag{3.1}$$

Together with the relation between the amount adsorbed n_a, and the differential work of adsorption $\varepsilon_i = RT \ln(P_0/P_i)$ is obtained by the following relation [11]:

$$n_a = N_m \exp\left(-\frac{\varepsilon}{E}\right)^n \tag{3.2}$$

where:
E is a parameter named the characteristic energy of adsorption
N_m is the maximum amount adsorbed in the volume of the micropore
n ($1 < n < 5$) is an empirical parameter

Finally, merging Equations 3.1 and 3.2 is obtained the Dubinin adsorption isotherm equation:

$$n_a = N_a \exp\left(-\frac{RT}{E}\ln\left[\frac{P_0}{P}\right]\right)^n \tag{3.3}$$

which is expressed in linear form as follows:

$$\ln(n_a) = \ln(N_a) - \left(\frac{RT}{E}\right)^n \ln\left(\frac{P_0}{P}\right)^n \tag{3.4}$$

It is a very powerful tool for the description of the experimental data of adsorption in microporous material. In this regard, in Figure 3.2, the Dubinin plot of the adsorption isotherm in the range $0.001 < P/P_0 < 0.03$ is shown, describing the adsorption

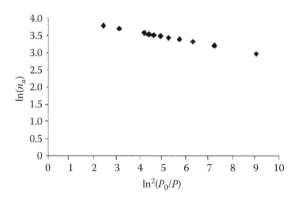

FIGURE 3.2   Dubinin plot sample CBV-720 N_2 at 77 K.

of N_2 at 77 K in a high silica commercial HY zeolite, specifically, sample CBV-720 provided by PQ corporation, provided the isotherm obtained in an Autosorb-1 gas adsorption system [25].

The Dubinin plot shown in Figure 3.2 was carried out as a linear plot:

$$y = \ln(n_a) = \ln(N_a) - \left(\frac{RT}{E}\right)^n \ln\left(\frac{P_0}{P}\right)^n = b - mx$$

where:

$y = \ln(n_a)$
$b = \ln(N_a)$
$m = (RT/E)^n$
$x = \ln(P_0/P)^n$

However, the fitting process of the Dubinin equation can be also carried out with the help of a nonlinear regression method. Now, it will be shown that with the help of the Dubinin equation, it is possible to make an estimation of the isosteric heat of adsorption, q_{iso}, with one experimental isotherm. As $(d(\ln P)/dT)_\Gamma = \bar{H}_g - \bar{H}_a/RT^2 = q_{iso}/RT^2$, making now the following approximation: $\Gamma =$ constant, which is equivalent to $n_a =$ constant, it is possible to calculate the isosteric heat of adsorption employing the expression $q_{iso} = RT^2(d(\ln P)/dT)_\Gamma$. Hence, substituting the Dubinin adsorption equation into the previous equation, it is possible to show that

$$q_{iso} \approx L + E\left[\ln\left(\frac{N_a}{n_a}\right)\right]^{1/n} \tag{3.5}$$

In which, L (kJ/mol) is the liquefaction heat of the adsorbate at the temperature of the adsorption experiment, and E and n are the parameters of the Dubinin adsorption equation.

3.3　OSMOTIC ADSORPTION ISOTHERM

Within the framework of the osmotic theory of adsorption the adsorption phenomenon in a microporous adsorbent is contemplated as the *osmotic* equilibrium between two solutions (vacancy plus molecules) of different concentrations, where one of these solutions is created in the micropores, and the other in the gas phase, and the role of the solvent in the present model is carried out by the vacancies, that is, by vacuum [11]. It is given that the solutions can be only in equilibrium, when one of the solutions is immersed in an external field. Hence, the key assumption of the model is that the potential field can be virtually represented by an osmotic pressure Π, that is, to say, the effect of the adsorption field present in the adsorption field can be formally represented by the difference in pressures between the gas and the adsorbed phase. Consequently, considering that the adsorption space is an inert volume, the adsorption effect is caused by a virtual pressure that is applied to compress the adsorbed phase in this volume [12]. Specifically, we could delineate a mental experiment, where the micropore is dispossessed of their adsorption field. Then, in this situation, we will have a simple volume, to be precise, an empty adsorption space, where the role of the adsorption field is taken by an external pressure Π.

Following the hypothesizes of the osmotic theory of adsorption developed by Dubinin–Bering and Serpinskii [11,12], it is possible to state that the volume occupied by the adsorbate V_a, and the vacancies V_x, or free volume is given by $V_a + V_x = V$. Hence, considering that the volume occupied by an adsorbed molecule, b, and a vacancy is the same, then

$$\frac{V_a}{b} + \frac{V_x}{b} = n_a + N^x = \frac{V}{b} = N_a \tag{3.6}$$

Multiplying this expression by $1/N_a$, we will get

$$\frac{V_a}{N_a b} + \frac{V_x}{N_a b} = \frac{n_a}{N_a} + \frac{N^x}{N_a} = \frac{V}{N_a b} = X_a + X^x = 1 \tag{3.7}$$

where X_a and X^x are the molar fractions of the adsorbed molecules and vacancies, respectively. Thus, at this point, taking into account that adsorption in a micropore can be described as an osmotic process in which vacuum, that is, the vacancies are the solvent, whereas the adsorbed molecules are the solute. Thenceforth, applying the methods of osmosis thermodynamics to the above-described model, it is possible to obtain the following adsorption isotherm equation [12]:

$$n_a = \frac{N_a K_0 P^B}{1 + K_0 P^B} \tag{3.8}$$

called the osmotic isotherm of adsorption. Equation 3.4 reduces for $B = 1$ to a Langmuir-type (LT) isotherm equation describing a volume filling

$$n_a = \frac{N_a K_0 P}{1 + K_0 P} \tag{3.9}$$

Equation 3.8 known in literature as the Sips or Bradley's isotherm equation, fairly well, describes the experimental adsorption data in micoroporous materials [28]. Hence, the linear form of the osmotic equation could be expressed as follows:

$$y = P^B = N_a \left(\frac{P^B}{n_a} \right) + \frac{1}{K} = mx + b \tag{3.10}$$

where:
$y = P^B$
$x = P^B/n_a$
$m = N_a$ is the slope
$b = 1/K$ is the intercept

In Figure 3.3 the linear osmotic ($B = 0.5$) plot of the NH_3 at 300 K in the homoionic magnesium natural zeolite, sample Mg–CMT is shown, where the CMT natural

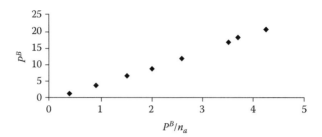

FIGURE 3.3   Osmotic plot ($B = 0.5$) for NH_3 at 300 K in Mg–CMT zeolite adsorption.

zeolite is a mixture of clinoptilolite (42 wt. %) and mordenite (39 wt. %) and other phases (15 wt. %), where others are montmorillonite (2–10 wt. %), quartz (1–5 wt. %), calcite (1–6 wt. %), feldspars (0–1 wt. %), and volcanic glass [39].

It is required now to clarify that the adsorption data reported in Figure 3.3 was determined volumetrically in a homemade Pyrex glass vacuum system, consisting of a sample holder, a dead volume, a dose volume, a U-tube manometer, and a thermostat [3,26]. Finally, the osmotic plot allowed the calculation of the maximum adsorption capacity of this zeolite, which is $m = N_a = 5.07$ mmol/g, and $b = 1/K = -0.92$ [(Torr)$^{0.5}$].

3.4   LANGMUIR AND FOWLER–GUGGENHEIM-TYPE ADSORPTION ISOTHERM EQUATIONS

3.4.1   APPLICATION OF THE GRAND CANONICAL ENSEMBLE METHODOLOGY TO DESCRIBE ADSORPTION IN ZEOLITES

The grand canonical ensemble method (GCEM) (see Chapter 1) is now applied to handle the adsorption process in zeolites along with other microporous materials. Hence, in order to get, with the help of the GCEM, isotherm equations, we must consider the zeolite as a grand canonical ensemble (GCE), that is, the zeolite cavities or channels are viewed in this model description of adsorption in a zeolite as independent open subsystems belonging to the GCE (see Chapter 1). Thereafter, the zeolite is considered as a GCE by itself, where the cavities or channels are the systems conforming ensemble. In addition, it is accepted that the adsorption space is energetically homogenous, that is, the adsorption field is the same in any place within the adsorption space.

In the frame of this model, developed in different steps by different authors [23,24,58–60], the zeolite is considered as a system composed of M cavities or channels. Consequently, we have M-independent open subsystems belonging to the GCE. If each cavity or channel can accommodate a maximum of m molecules, where $m = w/b$, w is the channel or cavity volume, and b is the sorbate molecular volume. Then the cavity or channel could be considered as an independent subsystem, and the grand canonical partition function for the zeolite will be [3,12]

$$\Theta = [1 + \lambda Z(1) + \lambda^2 Z(2) + \ldots \ldots \lambda^m Z(m)]^M = \bar{Z}^M \qquad (3.11)$$

in which

$$\bar{Z} = \sum_{N=0}^{m} \lambda^N Z(N) \tag{3.12}$$

is the grand canonical partition function of the cavity or channel, whereas $Z(N)$ is the canonical partition function for N molecules in the cavity or channel $(0 < N < m)$, $\lambda = \exp(\mu/RT)$ is the absolute activity, and μ is the chemical potential. Now, giving that the expression for the average number of molecules adsorbed $\bar{N}$ in the GCE is [24]

$$\bar{N} = \frac{\partial \ln \Theta}{\partial \ln \lambda} = RT \left(\frac{\partial \ln \Theta}{\partial \mu} \right) \tag{3.13}$$

In the present case, we express $\bar{N}$ in molar terms, provided $R = N_A k$, where $N_A = 6.02214 \times 10^{23}$ [mol^{-1}] is the Avogadro number, $k = 1.38066$ [JK^{-1}] the Boltzmann constant, and finally, $R = 8.31451$ [JK^{-1}mol^{-1}] is the ideal gas constant. At this point, if the condition is satisfied, $m \gg 1$, it is possible to make the calculations applying the grand canonical partition function of the cavity or channel (GCPFCC). Hence, the average number of molecules $(\bar{N})$ in the zeolite cavity or channel is [21]

$$\bar{N} = \frac{\partial \ln \bar{Z}}{\partial \ln \lambda} = RT \left(\frac{\partial \ln \bar{Z}}{\partial \mu} \right) \tag{3.14}$$

Thereafter, the volume coverage can be calculated with the help of the subsequent expression:

$$\theta = \frac{\bar{N}}{Mm} = \frac{M\bar{N}}{Mm} = \frac{\bar{N}}{m} \tag{3.15}$$

where θ is the micropore volume recovery.

We have to case for the adsorption in these conditions, that is, immobile and mobile adsorption. Thereafter, in the case of immobile adsorption in a homogeneous field, considering lateral interactions between neighboring molecules, the canonical partition function for $N < m$, molecules in the cavity or channel is [23]

$$Z(N) = \frac{m!}{N!(m-N)!} \left(Z_a^I \right)^N \exp \left(-\frac{N(E_0^a + \eta E_i)}{RT} \right) \tag{3.16}$$

whereas it is very well known [61] that

$$Z(N) = \frac{m!}{N!(m-N)!} X^N \tag{3.17}$$

while:

$$X \approx Z_a^I \exp \left(-\frac{(E_0^a + \eta E_i)}{RT} \right) \tag{3.18}$$

in which, m is the number of adsorption sites in the cavity or channel, N is the number of adsorbed molecules, and Z_a^I is the canonical partition function for the internal degrees of freedom of the adsorptive in the adsorbed phase. Besides, E_0^a is the reference energy state for the adsorbed molecule in the homogeneous adsorption field inside the cavity or channel. In addition, $\eta E_i = cN/2mE_i$ is the interaction energy of an adsorbed molecule with the neighboring adsorbed molecules, assuming a random distribution of neighbors, where c is the number of nearest neighbors to an adsorption site in the cavity [24].

Now, in order to get the isotherm equation, it is necessary to carry out the following approximation: $(cN/2m)E_i \approx (c\overline{N}/2m)E_i$ in Equation 3.17 [12]:

$$Z(N) = \frac{m!}{N!(m-N)!}\left(Z_a^I\right)^N \exp\left(-\frac{\left[NE_0^a + \left(c\overline{N}/2m\right)E_i\right]}{RT}\right) \tag{3.19}$$

To calculate the adsorption isotherm, we use Equations 3.5, 3.8, and 3.10, and the Newton's binomial polynomial expansion [23]:

$$B = (1 + \lambda X)^m = \sum_{N=0}^{m} \frac{m!}{N!(m-N)}(\lambda X)^N$$

to get the following equation [24]:

$$\overline{N} = \frac{\partial \ln \overline{Z}}{\partial \ln \lambda} = \lambda \frac{\partial \ln \overline{Z}}{\partial \lambda} = \frac{A}{B}$$

where $A = m\lambda X(1+\lambda X)^{m-1}$ and

$$X = Z_a^I \exp\left(-\frac{N(E_0^a + \left(c\overline{N}/2m\right)E_i)}{RT}\right)$$

Consequently, the adsorption isotherm is [12]

$$\theta = \frac{\overline{N}}{m} = \frac{K_I P}{1 + K_I P} \tag{3.20}$$

$$K_I = \left\{\frac{Z_a^I}{Z_g^I}\right\}\left[\frac{1}{RT\Lambda}\right]\exp\left(\frac{[(E_0^g - E_0^a) + \Omega\theta)]}{RT}\right) \tag{3.21}$$

or

$$K_I = K_0^I \exp\left(\frac{\Omega\theta}{RT}\right) \tag{3.22}$$

where Z_g^I is the canonical partition function for the internal degrees of freedom of the adsorptive in the gas phase. Besides, E_0^g is the reference energy state for the gas molecule and $\Omega = cE_i/2$. In addition, $\Lambda = (2\pi MRT/h^2)^{3/2}$, where $M = N_A m$ is the molar mass of the adsorptive molecule, m is the mass of the adsorptive molecule, N_A is the Avogadro number, and h is the Planck constant.

Finally, it is necessary to affirm that the average adsorption field in the volume filled by the adsorbed molecules in the present case of immobile adsorption is $\xi(\theta) \approx -[(E_0^g - E_0^a) + \Omega\theta]$ [2,35].

Now, in the mobile case, we consider again the zeolite as a GCE, but accept the hypothesis of mobile adsorption in a homogeneous adsorption field, where lateral interactions between neighboring molecules are present. Hence, the canonical partition function for $N < m$ molecules in the cavity or channel could be expressed as follows [24]:

$$Z(N) = \frac{w^N}{N!} [\Lambda]^N (Z_a^I)^N \exp\left(-\frac{N\left(E_0^a - \alpha\left(N/w\right)\right)}{RT}\right)$$

(3.23)

in which $\Lambda = (2\pi MRT/h^2)^{3/2}$, where M is the molar mass of the adsorptive molecule, h is the Planck constant, $\alpha \approx B_2 RT$ and B_2 is the second virial coefficient [8].

To calculate the adsorption isotherm, we use Equations 3.12, 3.14, and 3.23, and the exponential series expansion [12,23]:

$$B = \sum_{N=0}^{m} \frac{1}{N!} (\lambda X)^N \approx \exp(\lambda X) \approx \left(1 + \frac{\lambda X}{m}\right)^m$$

To get

$$\bar{N} = \frac{\partial \ln \bar{Z}}{\partial \ln \lambda} = \lambda \frac{\partial \ln \bar{Z}}{\partial \lambda} = \frac{A}{B}$$

where $A = \lambda X\left(1 + \frac{\lambda X}{m}\right)^{m-1}$

and, after that, get the following isotherm equation:

$$\theta = \frac{K_M P}{1 + K_M P}$$

(3.24)

where:

$$K_M = \left\{\frac{Z_a^I}{Z_g^I}\right\}\left[\frac{b}{RT}\right]\exp\left(\frac{[(E_0^g - E_0^a) + \Phi\theta)]}{RT}\right)$$

(3.25)

$$K_M = K_0^M \exp\left(\frac{\Phi\theta}{RT}\right) \tag{3.26}$$

in which $Z_a{}^I$ and $Z_g{}^I$ are the canonical partition functions for the internal degrees of freedom of the adsorptive in the adsorbed phase and the gas phase, respectively. Moreover, E_0^g is the reference energy state for the gas molecule, E_0^a is the reference energy state for the adsorbed molecule in the homogeneous adsorption field inside the cavity or channel. Meanwhile, $\Phi = \alpha/b$ is a parameter characterizing lateral interactions, where, b, the sorbate volume, is expressed in molar units.

Finally, we conclude that for mobile adsorption, the average adsorption field in the volume filled by the adsorbed molecules is $\xi(\theta) \approx -[(E_0^g - E_0^a) + \Phi\theta]$ [2,35].

3.4.2 SOME REMARKS IN RELATION WITH THE LANGMUIR TYPE AND FOWLER–GUGGENHEIM-TYPE ADSORPTION ISOTHERM EQUATIONS

At this point, it is necessary to assert that, in the event where $c \approx 0$ or $\alpha \approx 0$, Equations 3.21 and 3.25 reduces to a LT adsorption isotherm equation:

$$\theta = \frac{K_L P}{1 + K_L P} \tag{3.27}$$

In which, depending on the case, that is, the immobile $K_L = K_0^I$ or the mobile $K_L = K_0^M$.

Equations 3.20 and 3.24 are of the Fowler–Guggenheim Type (FGT) adsorption isotherm equations *describing a volume filling rather than a surface coverage.* Similarly, Equation 3.25 is of the LT adsorption isotherm equation, also *describing a volume filling rather than a surface coverage.* This is also the case for the adsorption isotherm Equation 3.9, which also reduces to a LT isotherm equation but *describing a volume filling rather than a surface coverage.*

In addition, an adsorption isotherm equation for the description of the adsorption process in zeolites and related materials, which was obtained with the help of the modified lattice gas model and quantum statistical methods [61]:

$$
\frac{\theta}{1-2\theta} = \left(1 - \frac{1}{m}\right)\left[\frac{1}{(P_1/P)\left(\exp[\beta\varepsilon_2(1-2\theta)]-1\right)}\right]
$$
$$
+ \left(\frac{1}{m}\right)\left[\frac{1}{(P_1/P)\left(\exp[\beta\varepsilon_1(1-2\theta)]-1\right)}\right] \tag{3.28}
$$

In which $P_1 = P_0 \exp(\beta\varepsilon_0)$, $\varepsilon_0 = \varepsilon + U/2$, $\varepsilon_1 = t + U/2$, $\varepsilon_0 = (U/2) - (t/m - 1)$, $\beta = 1/RT$, m is the maximum number of molecules adsorbed in a cavity or channel, ε is the interaction energy between the zeolite framework and the molecule located in one adsorption site, U is the interaction energy between adsorbed molecules, and t is the probability of jumping between the adsorption sites. It is evident that Equation 3.28 reduces to a LT isotherm equation if the condition, $\varepsilon_1 = \varepsilon_2 = 0$, is observed, that is, in the case of immobile adsorption without interactions, because $U = 0$ and $t = 0$ [61]:

$$\theta = \frac{1}{1 + (P_1/P)} \tag{3.29}$$

Consequently, the discussion carried out so far shows that diverse approaches for describing the adsorption process in zeolites, all lead to LT and FGT isotherms, describing adsorption as a volume-filling effect. Consequently, it is possible to conclude that the LT and FGT adsorption isotherm equations describing volume filling must be useful in the characterization of the adsorption properties of zeolites and related materials [12]. For systems where the lateral interactions are not considered between the adsorbed molecules isotherm of the following LT type are obtained: $\theta = K_L P / 1 + K_L P$, and for systems where they are considered the lateral interactions: $\theta = K_{FG} P / 1 + K_{FG} P$, in which K_L, and K_{FG} are functions of temperature and temperature and magnitude of adsorption, respectively, whose explicit form depends on the model.

As a result, the precedent isotherm equations, that is, Equations 3.9, 3.20, 3.24, 3.27 along with Equation 3.29 obtained with completely different model descriptions of the adsorption process in microporous materials all have a similar mathematical form. Thus, it is possible to conclude that the previously discussed adsorption isotherm equations describing the volume filling in micropore materials are suitable for the assessment of the micropore volume of adsorbents.

To demonstrate, the effectiveness of the deduced isotherm equations was carried out on an experimental test, using Ar adsorption at 87 K in the ensuing commercial zeolites, that is, NaY (CBV100, $SiO_2/Al_2O_3 = 5.2$) provided by the PQ Corporation, NaX (13X, $SiO_2/Al_2O_3 = 2.2$) provided by Micromeritics, and NaY (SK-40, SiO/AlO $= 4.8$) provided by the Linde Division of Union Carbide [12], if the experiments are made with a Micromeritics, ASAP 2000 equipment for adsorption isotherm determination [18]. In this regard, in Figure 3.4, it is reported the fitting of the experimental Ar at 87 K data with the linear form of the FGT isotherm equation type, that is [12],

$$\ln\left(\frac{\theta}{1 - \theta}\right) = \ln K + \frac{k\theta}{RT} \tag{3.30}$$

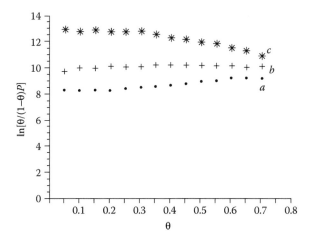

FIGURE 3.4 Plot of ln $[\Theta/(1-\Theta)P]$ versus $\Theta = n_a/N_m$ for the zeolites (*a*) Na–Y (CBV-100), (*b*) Na–Y (SK-40), and (*c*) Na–X (13 X).

which reduce to the LT isotherm equation type for $k = 0$. In this sense, the LT isotherm equation fits the Ar adsorption data in the range $0.01 < \Theta < 0.4$ for the tested Na–Y (SK-40), Na–Y (CBV-100), and Na–X (13 X) zeolites.

Thereafter, giving that Equation 3.30 for $k = 0$ describe the adsorption of Ar at 87 K in the above-referred zeolites, then the linear form of the LT isotherm is

$$P = N_a \left(\frac{P}{n_a}\right) + \frac{1}{K} \tag{3.31}$$

Permit a reliable measurement of the micropore volume (W^{Ar}) for the studied zeolites (Table 3.1 [12]), where the calculation of W^{Ar} was made with the help of the following equation: $W^{Ar} = N_a \times b$, where N_a is determined with Equation 3.31, provided the value for $b = 32.19$ cm³/mol, that is, the volume occupied by 1 mole of adsorbed *Ar* molecules.

TABLE 3.1

Micropore Volume (W^{Ar}) Measured with the Adsorption Isotherm of *Ar* at 87.3 K

Zeolite	W^{Ar} (cm³/g)
Na–Y (CBV-100)	0.319
Na–Y (SK-40)	0.311
Na–X (13 X)	0.192

3.5 THE *t*-PLOT METHOD

Halsey, De Boer, and coworkers developed the *t*-plot method, a procedure based on the concept introduced by Frenkel–Halsey–Hill (FHH), which states that it is possible to calculate *t*, that is, the width in Angstrom of the adsorbed layer [2,17,19,62]. The method considers the adsorbed phase like a liquid adhered film over the solid surface similar to what is shown in Figure 3.1, provided the model is valid for a multilayer adsorption, that is, specifically when $n_a/N_m > 2$, where N_m is the monolayer capacity.

To make the calculations, it is considered that the surface liquid film has a density equal to the bulk liquid adsorbate, ρ_L. Moreover, since it is in contact to a surface that create an attraction adsorption field over the solid surface. Subsequently, the film is influenced by this field. At this point, accordingly with the preceding assumptions the adsorption magnitude can be calculated by $n_a = \rho_L t$, whereas the adsorption field can be described by a model similar to those proposed in the Polanyi theory [20], where it is considered that the entropy contribution to the free energy is small in comparison with the large change of enthalpy. Hence [2], $\mu - \mu_L = RT \ln(P/P_0)$; supposing now that [1,8]

$$\mu - \mu_L = V(z) \tag{3.32}$$

because the adsorption process contemplated in the model is a multilayer one. Thus, $V(z) \approx -B/z^m$; accordingly, $RT \ln(P/P_0) = -B/z^m = -C/t^m$, where this equation shows a correlation between *t* and the relative pressure $x = P/P_0$.

The *t*-plot methodology suggests that for multilayer adsorption, the shape of the adsorption isotherm is less dependent on the adsorbent structure than in the case of monolayer adsorption [4]. This means that *t* depends fundamentally on $x = P/P_0$ and hardly on the nature of the adsorbent surface. Therefore, this thickness could be evaluated after normalizing an adsorption isotherm for an adsorbent that do not possess micropores or mesopores. The multiplayer thickness *t* can be calculated by the following relation: $t = n_a/N_m d_0$, where d_0 is the effective thickness of monolayer. Now, assuming, as was previously done that the surface liquid film is considered of uniform width *t*, and its density is equal to the bulk liquid adsorbate ρ_L, then [2,17,19] $d_0 = M/\sigma N_A \rho_L$, in which N_A is the Avogado number, whereas σ is the cross-sectional area, to be precise, the average area occupied by each molecule in a completed monolayer. As, for example, that $\sigma(N_2) = 0.162$ nm² for N_2 at 77 K, $M(N_2) = 28.1$ g/mol, and $\rho_L(N_2) = 0.809$ g/cm³ we will obtain $d_0 = 0.354$ nm.

Lippens and de Boer experimentally showed by measuring the following: *t* versus $x = P/P_0$ plots, that is, *t*-plots of nitrogen at 77 K, in different nonporous oxides, the existence of a *universal multilayer thickness curve*, as a consequence of the similarity of the obtained *t*-plots [62]. In practice, at the present moment, the *universal multilayer thickness curve* is not accepted [2]. Conversely, for example, the following relations between *t* and $(1/x)$ are used to carry out the *t*-plot [17]:

$$t = 3.54 \left[\frac{5}{2.303 \log(P_0/P)} \right]^{-\frac{1}{3}} \tag{3.33}$$

which is the Halsey equation, valid for N_2 at 77 K or the equation used by De Boer:

$$t = \left(\frac{13.99}{\log(P_0/P) + 0.034} \right)^{\frac{1}{2}} \tag{3.34}$$

Or, in more general terms the equation [27] is

$$t = a \left(\frac{1}{\ln(P_0/P)} \right)^{\frac{1}{b}} \tag{3.35}$$

valid for other adsorbates and/or temperatures, in which $a = 6.053$ and $b = 3$ for N_2 at 77 K, provided that in Equations 3.33 through 3.35, P is the equilibrium adsorption pressure and P_0 is the vapor pressure of the adsorptive at temperature T of the adsorption experiment.

To carry out the calculation of the microporous volume, and the area of the secondary porosity, or outer area with the help of the t-plot method, it is necessary to plot n_a versus t in Angstrom. In this regard, in Figure 3.5 a typical t-plot ($0.01 < P/P_0 < 0.3$) for the adsorption of N_2 at 77 K in a silica material is presented, specifically in the sample 70bs2 [16], where the t-plot is carried out with the help of Equations 3.33 or 3.34 to calculate t for a concrete value of $x = P/P_0$, and the amount adsorbed is taken from the experimental isotherm for the same value of $x = P/P_0$, and then the plot of n_a versus t is carried out.

Hence, to calculate microporous volume W in cm^3/g and the outer surface S in m^2/g using the t-plot method, we proceed as follows: at first, the points which do not fit a linear plot are eliminated (Figure 3.5), thereafter applying the linear equation: $y = n_a = Rt + N_a = mx + b$ is carried out the linear regression, then the intercept is calculated, $b = N_a$, and the slope $m = R$. Finally, applying the Gurvich rule through the following relation: $W^{MP} = N_a V_L$, in which V_L is the molar volume of the adsorptive at the temperature of the adsorption experiment T is calculated the volume of the micropores, whereas the outer surface will be [17,18] $S = RV_L$. As $n_a V_L = RV_L t + N_a V_L$, the term $RV_L t = St$ is equal to the contribution of the adsorption in the outer surface

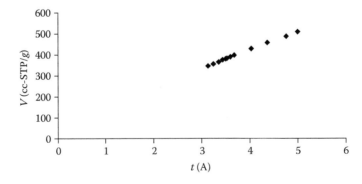

FIGURE 3.5 t-plot 70bs2-25C.

TABLE 3.2
Micropore Volume Calculated with the
Help of the *t*-Plot Method

Zeolite	W^N_2 (cm³/g)
Na–Y (CBV-100)	0.305
Na–Y (SK-40)	0.303

---

to the volume of the whole adsorbed phase. Finally, in Table 3.2 [14] values of the micropore volume, W^{MP} of two commercial zeolites are reported, specifically Na–Y (CBV100, SiO_2/Al_2O_3 = 5.2), provided by the PQ Corporation and Na–Y (SK-40, SiO/AlO = 4.8) provided by the Linde Division of Union Carbide.

The reported micropore volumes were calculated with the help of the *t*-plot method using N_2 adsorption isotherms at 77 K. The experiments were carried out with a Micromeritics ASAP 2000 equipment for adsorption isotherm determination [18].

To conclude, it is possible to state that comparing the results reported in Tables 3.1 and 3.2 a similarity between the results obtained by both methodologies is evident. Moreover, these results are consistent with the crystallographic void volume corresponding to the framework FAU of zeolite Y [63].

3.6 COMMENTS ABOUT DUBININ, OSMOTIC, LANGMUIR TYPE, FOWLER–GUGGENHEIM TYPE ISOTHERMS, AND THE *t*-PLOT METHOD

Normally, in porous materials the porosity such as micropores, mesopores, and macropores is integrated in the material as a whole. Afterward, the correct value of the micropore volume is extremely difficult to establish correctly, given that in the majority of cases it is problematical to determine the point where adsorption has finished in the micropores, and the adsorption in the mesopores has started [9].

For zeolites, which are primarily composed of micropores to calculate the micropore volume, the author proposed a method involving the fitting of the experimental adsorption data to the Dubinin isotherm equation, along with the LT, osmotic or FGT isotherm equations using a least square fitting computer program [3]. During the fitting process, different parameters are calculated, for instance, n, E, K, and B, in the case where the Dubinin and osmotic isotherm equations were used. The previously described methodology for the determination of the micropore volume in zeolites can be justified using NH_3 as adsorptive at 300 K of temperature, because this gas is perfect for this measurement, since their boiling point is 240 K and their molecular kinetic diameter, σ = 3.08 A facts ensuring that it is not adsorbed in the outer surface, and easily penetrates through the microporosity [63].

Some examples of the use of the Dubinin (n = 2) adsorption isotherm equations for the calculation of the micropore volume, and the characteristic energy of adsorption for the adsorption of NH_3 at 300 K on different natural and synthetic zeolites are reported in Table 3.3 [25].

TABLE 3.3

Parameters for the Adsorption of NH$_3$ on Natural and Synthetic Zeolites

Sample	N_m (mmol/g)	W^{MP} (cm^3/g)	E (kJ/mol)
HC	6.2	0.130	28
MP	6.8	0.143	25
AP	6.9	0.145	31
CMT	6.1	0.128	22
Na-A	9.8	0.204	23
Na-X	10.1	0.210	24

The maximum adsorption magnitude, N_m, and the microporous volume, W^{MP}, were measured in a Pyrex glass vacuum system consisting of a sample holder, a dead volume, a dose volume, a U-tube manometer, and a thermostat and are reported in mmol of NH$_3$ adsorbed per gram of dehydrated zeolitic rock, and cm^3/g of dehydrated zeolitic rock, respectively. Meanwhile, the errors in N_a, W, and E are ± 0.2 (mmol/g), ± 0.005 (cm^3/g), and ± 0.4 (kJ/mol), respectively [26]. Moreover, in Tables 3.4 and 3.5 are reported the chemical composition (in oxide wt. %.), and the mineralogical composition in wt. % of the natural zeolite rocks employed to illustrate the micropore volume measurement method [25]. The sample identification (label: deposit name, location) is HC: Castillas, Havana, Cuba; CMT and CMTC–C: Tasajeras, Villa Clara, Cuba; C1-C6: Camaguey, Cuba; MP: Palmarito, Santiago de Cuba, Cuba; SA, San Andres, Holguin, Cuba; AD: Aguas Prietas, Sonora, Mexico. CZ: Nizni Harabovec, Slovakia; GR: Dzegvi, Georgia. The synthetic zeolite sample NaA was provided by Degussa, and the sample NaX was provided by Laporte [37,47]. In Table 3.5, others are Montmorillonite (2–10 wt. %), quartz (1–5 wt. %), calcite (1–6 wt. %), feldspars (0–1 wt. %), magnetite (0–1 wt. %), and volcanic glass (3–6 wt. %).

In the case of natural zeolites this information could be very helpful to find out the quantity of zeolite present in the natural zeolite rock [3]. As the micropore volume for pure clinoptilolite and mordenite is $W_{HEU} \approx W_{MOR} \approx 0.16$ cm^3/g [28], whereas the microporous volume of erionite could be estimated to be $W_{ERI} \approx 0.18$–0.19 cm^3/g [63].

TABLE 3.4

Chemical Composition (in Oxide wt. %) of Some Natural Zeolite Rocks Used to Illustrate Distinct Properties and Applications of These Materials

Sample	SiO$_2$	Al$_2$O$_3$	Fe$_2$O$_3$	CaO	MgO	Na$_2$O	K$_2$O	H$_2$O
HC	66.8	13.1	1.3	3.2	1.2	0.6	1.9	12.1
MP	66.9	11.6	2.7	4.4	0.8	1.8	0.8	12.1
AP	59.6	14.2	2.3	2.2	1.5	2.4	3.3	13.8
CMT	66.6	12.5	2.0	2.7	0.7	1.7	0.8	12.9

TABLE 3.5

Composition (in wt. %) of the Natural Zeolite Used in the Adsorption Tests

Sample	Clinoptilolite	Mordenite	Erionite	Others
HC	85	0	0	15
MP	5	80	0	15
AP	0	0	85	15
CMT	42	39		19

Consequently, with the reported values for W and the micropore volume of the zeolitic phases present in a rock (W_{XXX}), it is feasible to calculate a fairly accurate value for the fraction of zeolitic phases present in the rock with the help of the equation: $f = W^{MP}/W_X$. The relation is valid if all the zeolite phases are present in the rock exhibit, approximately, the same value for W_X and also if the quantity of molecules adsorbed by the impurity phases could be neglected. On the other hand, adsorption isotherms of N_2 at 77 K are also used to measure the micropore volume.

Some examples of the use of this methodology for the calculation of the micropore volume of amorphous silica materials are reported in Table 3.6.

During this test, the silica samples tested were degassed at 200°C, for 3 h, in high vacuum (10^{-6} Torr), previous to the analysis and the micropore volume accessible to the N_2 molecule at 77 K (W^{MP} [cm^3/g]) was measured using the t-plot method [2]. The Gurvich rule was also used for the calculation of the micropore volume. However, the adsorbate in the micropore does not necessarily have the same density as the adsorptive in the liquid state, as is needed for the fulfillment of this rule [19]. This is one of the facts, which make difficult to arrive to an unequivocal measurement of the micropore volume. Other factor that makes it difficult to get an unequivocal evaluation of the micropore volume, as was previously stated, is that in the present case, that is, adsorption of N_2 at 77 K in an amorphous silica, it is

TABLE 3.6

Micropore Volume of Some Silica Measured with N_2 Adsorption at 77 K

Sample	W^{MP} (cm^3/g)
70bs2	0.18
68bs1E	0.27
75bs1	0.16
79BS2	0.21
74bs5	0.14
68C	0.00
MCM-41	0.00

very difficult to determine the point where adsorption in the micropore finishes and begins the external surface coverage [16], since pore filling is observed at pressures very close to the pressure range where monolayer–multilayer formation on the pore walls occurs [9].

3.7 THE BRUNAUER–EMMETT–TELLER METHOD

The BET theory of multilayer adsorption for the evaluation of specific surface area S, was developed by Brunauer, Emmett, and Teller [64]. However, here for the deduction of the isotherm equation the GCE approach applying a methodology developed by Hill will be used [21].

The adsorption process, within the frame of the BET theory, is contemplated as a layer-by-layer process. Moreover, the surface is supposed to be energetically homogeneous. Specifically, the adsorption field is the same in any place within the surface. In addition, the adsorption process is considered immobile, that is, each molecule is adsorbed in a concrete adsorption site in the surface. Hence, the first layer of adsorbed molecules has an energy of interaction with the adsorption field, E_0^a, and the vertical interaction between molecules after the first layer is E_0^L, that is, similar to the liquefaction heat of the adsorbate (Figure 3.6), as well as adsorbed molecules do not interact laterally.

The construction of the grand canonical partition function is a complicated process where it will be applied the concepts of independent distinguishable systems for each of the s molecules, which conform a conglomerate (Figure 3.6), that is, each conglomerate has a variable number of molecules and do not interact laterally with the neighboring conglomerates. Hence, the adsorbed phase forms a GCE of conglomerates. So, giving that the conglomerates form the adsorbed phase, the adsorbed phase could be considered as a GCE, where the canonical partition function for a molecule adsorbed in an adsorption site the first layer is

$$Z_1 = K_1 \exp\left(-\frac{E_0^a}{RT}\right) = q_1 \tag{3.36}$$

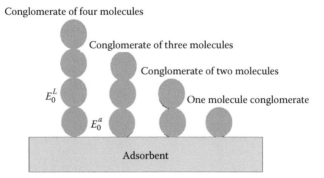

Conglomerate of four molecules

Conglomerate of three molecules

Conglomerate of two molecules

E_0^L One molecule conglomerate

E_0^a

Adsorbent

FIGURE 3.6   BET adsorption model.

Likewise, for the subsequent layers the molecular canonical partition function will be

$$Z_2 = Z_3 = \ldots\ldots = Z_s = K_L \exp\left(-\frac{E_0^L}{RT}\right) = q \tag{3.37}$$

Provided that a_s is the number of conglomerates with s molecules, then N, the number of adsorption sites in the surface is

$$N = \sum_{s=0}^{m} a_s \tag{3.38}$$

Therefore, the number of adsorbed molecules will be

$$N_a = \sum_{s=0}^{m} s a_s \tag{3.39}$$

and the grand canonical partition function of an arbitrary conglomerate having m molecules will be

$$\xi = q_0 + \lambda q(1) + \lambda^2 q(2) + \ldots\ldots + \lambda^m q(m) \tag{3.40}$$

where:
q_0 is the molecular canonical partition function of the empty site
$q(s)$ is the molecular canonical partition function of the site with a conglomerate
of s molecules adsorbed is

$$q(s) = \prod_{i=1}^{s} q_i \tag{3.41}$$

Finally, the grand canonical partition function of the adsorbed phase is

$$\Theta = \xi^N \tag{3.42}$$

Defining now,

$$C = \frac{K_1}{K_L} \exp\left(\frac{E_0^a - E_0^L}{RT}\right) \tag{3.43}$$

Subsequently, we will have $q_1 = Cq$ and $q(s) = q^s C$ for $s > 1$, then in this way

$$\Theta = (1 + \lambda Cq + \lambda^2 Cq^2 + \ldots\ldots + \lambda^m Cq^m)^N = \left(\frac{1 + (C-1)\lambda q}{1 - \lambda q}\right)^N \tag{3.44}$$

Thereafter, since $\bar{n}_a = \partial \ln \Theta / \partial \ln \lambda$ along with $\lambda q = P/P_0 = x$. Hence [21],

$$\frac{\bar{n}_a}{N_m} = \frac{Cx}{(1-x+Cx)(1-x)} \tag{3.45}$$

where:

$\bar{n}_a = n_a$ is the amount adsorbed

N_m is the monolayer capacity, and the other terms have the previously explained
     meaning

At this point, to apply to real adsorption data the BET isotherm equation, it is habitual to use Equation 3.45 in the linear form:

$$y = \frac{x}{n_a(1-x)} = \left(\frac{1}{N_mC}\right) + \left(\frac{C-1}{CN_m}\right)x = b + mx \tag{3.46}$$

where:

$b = (1/N_mC)$

$m = (C-1/CN_m)$

$y = x/n_a(1-x)$

$x = P/P_0$ in the region: $0.05 < x < 0.4$ [4]

If the term $(C-1/C) \approx 1$, then the slope m of the linear regression is $m \approx 1/N_m$. Accordingly, the monolayer capacity N_m is determined, and the specific surface area can be calculated as $S = N_m N_A \sigma$, where N_A is the Avogado number, and σ is the cross-sectional area, that is, the average area occupied by each molecule in a completed monolayer, where $\sigma(N_2) = 0.162$ nm^2 for N_2 at 77 K and $\sigma(Ar) = 0.138$ nm^2 for argon at 87 K [4]. In the general case where the condition, $C - 1/C \approx 1$, is not fulfilled, $b = (1/N_mC)$, $m = (C-1/CN_m)$ must be calculated. Thereafter, we will then have two equations with two unknowns that could be solved to get N_m and C. After that, S is calculated by following the procedure previously explained.

In Figure 3.7 the BET plot ($0.04 < P/P_0 < 0.3$) for the adsorption of N_2 at 77 K in a silica material is shown, explicitly in the sample 70bs2 (Table 3.3) [16].

The BET methodology must be carefully used to get proper results [1]. The method is very useful in cases where the sorbates do not penetrate in the primary porosity, that is, when the adsorption process occurs only in the outer surface. Therefore, the BET equation is truly valid for surface area analysis of nonporous and mesoporous materials consisting of pores of wide pore diameter. But it is not precisely applicable for microporous adsorbents as long as the BET theory describes a surface recovery, and adsorption in the primary porosity of zeolites is a volume-filling process [11]. It also appears that the BET method is inexact for the calculation of the surface area of mesoporous molecular sieves of pore widths less than about 4 nm. As pore filling is observed at pressures very close to the pressure range where monolayer–multilayer formation on the pore walls occurs, it may lead to a significant overestimation of the monolayer capacity in case of a BET analysis [3]. However, the BET surface area is widely taken as a reproducible parameter

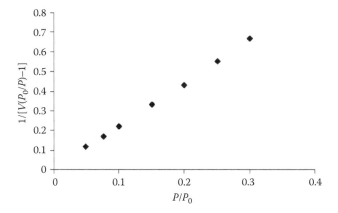

FIGURE 3.7 BET plot of sample 70bs2-25C.

for the characterization of the surface of porous materials, even though in some cases lacks a precise physical meaning. In this regard, in Table 3.7 the specific surface area of some silica and the mesoporous molecular sieve MCM-41 are reported [16].

Another source of error in the application of the BET equation is related with the surface chemistry of the sample under test. For example, in the case of silica samples, it is required to affirm that the cross-sectional area of nitrogen on hydroxylated surfaces [65], such as silica is not always $\sigma(N_2) = 0.162$ nm^2, as is normally considered for the calculation of the BET surface area [5].

One more cause of error in the application of the BET equation during adsorption experiments is the determination of the adsorbent mass. Consequently, the adsorbent mass must be very carefully measured with an analytical balance.

To end this section, it is required to state that there exist different causes for the scattering of the measured specific surface area data in an adsorption experiment. As a result, it has been estimated by measuring repeatedly the tested samples that the relative error in the BET surface area measurements of the adsorption parameters is normally around 20%–30% [1].

TABLE 3.7
BET-Specific Surface Area (S) of Some Silica and MCM-41

Sample	S (m^2/g)
70bs2	1600
68bs1E	1500
75bs1	1400
79BS2	1300
74bs5	1200
68C	320
MCM-41	800

3.8 HORVATH–KAWAZOE METHOD

The Horvath–Kawazoe (HK) method for determining the micropore size distribution (MPSD) was introduced by Horvath and Kawazoe in 1983 [35]. The procedure is founded on the idea that the relative pressure, $x = P/P_0$, required for the filling of micropores of a concrete size and shape is directly related to the adsorbate–adsorbent interaction energy [4,5]. This means that the micropores are progressively filled with an increase in adsorbate pressure. More concretely, in the HK method it is understood that only pores with dimensions lower than a particular unique value will be filled for a given relative pressure of the adsorbate. Then, the HK method allows the calculation of the PSD in the micropore range at low pressures [17].

As in the instance, previously analyzed by the FHH model, the entropy contribution to the free energy is small in comparison with the large change of enthalpy. Consequently, as well in the present case the following equation [36] is observed:

$$RT \ln\left(\frac{P}{P_0}\right) = U_0 + P_a \tag{3.47}$$

Thereafter, the HK method is based on Equation 3.47, including only van der Waals interactions calculated with the help of the Lennard–Jones (L–J) potential [36–38].

Halsey and coworkers applied the 6–12 L–J potential to the case of the interaction of one adsorbate molecule with an infinite layer plane of adsorbent molecules, obtaining [66]

$$\varepsilon(z) = \frac{N_{AS} A_{AS}}{2\sigma^4}\left[-\left(\frac{\sigma}{z}\right)^4 + \left(\frac{\sigma}{z}\right)^{10}\right] \tag{3.48}$$

Later Everett and Paul extended the previous result to two infinite lattice planes separated by a distance L, valid for the specific case of a slit pore (Figure 3.8) [67]:

$$E(z) = \frac{N_{AS} A_{AS}}{2\sigma^4}\left[-\left(\frac{\sigma}{z}\right)^4 + \left(\frac{\sigma}{z}\right)^{10} - \left(\frac{\sigma}{L-z}\right)^4 + \left(\frac{\sigma}{L-z}\right)^{10}\right] \tag{3.49}$$

where:
N_{AS} is the number of solid molecules/surface unit
L is the distance between the layers (Figure 3.8), [52]
$\sigma = 0.858d$, where $d = d_s + d_a/2$, and d_s is the diameter of the adsorbent molecule, and d_a is the diameter of the adsorbate molecule
z is the internuclear distance between the adsorbate and adsorbent molecules, $(L - d_s)$ is the effective pore width
A_{AS} is the dispersion constant, which takes into account the adsorbate–adsorbent interaction

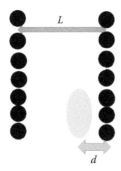

FIGURE 3.8 Adsorption in a slit pore.

Now, the term A_{AS} is calculated with the help of the Kirkwood–Muller formula [35,36]:

$$A_{AS} = \frac{6mc^2\alpha_S\alpha_A}{\left(\dfrac{\alpha_S}{\chi_S} + \dfrac{\alpha_A}{\chi_A}\right)}$$

where:

m is the mass of an electron
c is the speed of light
α_A and α_S are the polarizabilities of the adsorbate and the adsorbent molecules
χ_A and χ_S are the magnetic susceptibilities of the adsorbate and the adsorbent

Later, Horwath and Kawazoe proposed that the potential is increased by the adsorbate–adsorbate interaction, suggesting the following potential:

$$\Phi(z) = \frac{N_{AS}A_{AS} + N_{AA}A_{AA}}{2\sigma^4}\left[\left(-\left(\frac{\sigma}{z}\right)^4 + \left(\frac{\sigma}{z}\right)^{10}\right) - \left(-\left(\frac{\sigma}{L-z}\right)^4 + \left(\frac{\sigma}{L-z}\right)^{10}\right)\right] \quad (3.50)$$

where:

N_{AA} is the number of adsorbed molecules/surface unit
L is also the distance between the layers (Figure 3.8)
$\sigma = 0.858d$

As a final point, A_{AA} calculated with the help of the Kirkwood–Muller formula is the constant characterizing adsorbate–adsorbate interaction: $A_{AA} = 3mc^2\alpha_A\chi_A/2$. Finally, the average interaction energy is calculated, that is, the potential expressed by Equation 3.50 is volumetrically averaging as follows:

$$\xi(L) = \int_{d}^{L-d} \frac{\Phi(z)dz}{(L-2d)} = \left(\frac{N_{AS}A_{AS} + N_{AA}A_{AA}}{2\sigma^4(L-2d)} \right)$$

$$\times \left(\frac{\sigma^4}{3(L-d)^3} - \frac{\sigma^{10}}{9(L-d)^9} - \frac{\sigma^4}{3d^3} + \frac{\sigma^4}{9d^9} \right) \tag{3.51}$$

where:

$\zeta(L)$ is the average potential in a given pore obtained by the integration across the effective pore width

$\Phi(z)$ is the adsorption field inside the slit pore (Figure 3.8)

Finally, the average energy is related to the free energy change on adsorption, $\Delta G^{ads} = RT \ln(P/P_0)$, obtaining:

$$RT \ln\left(\frac{P}{P_0}\right) = N_A \left(\frac{N_{AS}A_{AS} + N_{AA}A_{AA}}{2\sigma^4(L-2d)} \right) \left(\frac{\sigma^4}{3(L-d)^3} - \frac{\sigma^{10}}{9(L-d)^9} - \frac{\sigma^4}{3d^3} + \frac{\sigma^4}{9d^9} \right) \tag{3.52}$$

In which, N_A is the Avogadro number provided in molar units: $RT \ln(P/P_0) = U_0 + P_a = N_A\xi(L)$.

The HK method is a tool for the characterization of microporous materials, which allow an estimation of the pore size of the studied materials. To calculate the MPSD, the relative pressure is first calculated as $x = P/P_0$, corresponding to a concrete pore width L, then Equation 3.52 with the help of an experimental adsorption isotherm it is estimated that the amount adsorbed, n_a, corresponds to this value of $x = P/P_0$. After that, by differentiating the amount adsorbed with respect to the pore width, dn_a/dL, the PSD in the micropore range is obtained [17].

In Table 3.8 a set of values for the parameters, α, χ, d, and N_s, for nitrogen and argon as adsorbates, and carbon and the oxide ion (e.g., a zeolite) as adsorbents are shown [17,18,35–37,68].

TABLE 3.8

Physical Properties of Nitrogen, as Adsorbate, and Carbon and the Oxide Ion as Adsorbents

Atomic Species	Polarizability α (10^{-24} cm³)	Magnetic Susceptibility χ (10^{-29} cm³)	Diameter d (nm)	Surface Density N_s (10^{18} atoms/m²)
Nitrogen	1.46	2.00	0.30	6.70
Argon	1.63	3.25	0.34	8.52
Carbon	1.02	13.5	0.34	38.4
Oxide ion	2.50	1.30	0.28	13.1

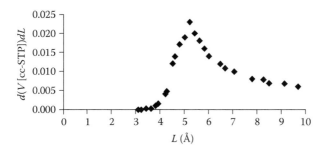

FIGURE 3.9 Horwath–Kawazoe micropore size distribution plot of a ZSM-5 zeolite.

In Figure 3.9 an H–K plot of MPSD for a zeolite ZSM-5-5020 is provided, provided by Zeolyst. The isotherm was obtained in a Quantachrome Autosorb-1 gas adsorption system, and the calculation is carried out by the equipment software [27].

It is evident from Figure 3.9 that the maximum of the PSD, fairly well, coincide with the crystallographic value of the pore diameter corresponding to the MFI framework of the analyzed ZSM-5 zeolite [41].

To finish, it is necessary to comment that the HK method was criticized by Lastoskie and collaborators, and a modified HK pore size analysis method was relatively recently presented, where the new HK model is generalized, so as to account for the effect of the temperature and the solid–fluid interaction potential strength on the adsorbed fluid density within a slit-shaped pore [69].

The pore-filling correlations forecasted by the two HK models were compared with density functional theory (DFT) results calculated using the same solid–fluid potential and potential parameters. It was found that the pore-filling correlation of the unweighted HK model was in accordance with the DFT correlation for argon and nitrogen adsorption at 77 K [68].

REFERENCES

1. K.S.W. Sing, D.H. Everett, R.A.W. Haul, L. Moscou, R.A. Pirotti, J. Rouquerol, and T. Siemieniewska, *Pure App. Chem.*, 57 (1985) 603.
2. F. Rouquerol, J. Rouquerol, K. Sing, P. Llewellyn, and G. Maurin, *Adsorption by Powders and Porous Solids* (2nd ed.), Academic Press, New York, 2013.
3. R. Roque-Malherbe, *Physical Chemistry of Materials: Energy and Environmental Applications*, CRC Press, Boca Raton, FL, 2009.
4. J.U. Keller and R. Staudt, *Gas Adsorption Equilibria: Experimental Methods and Adsorptive Isotherms*, Springer-Verlag, New York, 2004.
5. R.T. Yang, *Adsorbents: Fundamentals, and Applications*, John Wiley & Sons, New York, 2003.
6. R.M. Barrer, *Zeolites and Clay Minerals as Sorbents and Molecular Sieves*, Academic Press, London, UK, 1978.
7. S. Ross and J.P. Olivier, *On Physical Adsorption*, John Wiley & Sons, New York, 1964.
8. D.H. Everet and R.H. Ottewill, *Surface are Determination*, *International Union of Pure and Applied Chemistry*, Butterworth, London, UK, 2013.

9. S. Lowell, J.E. Shields, M.A. Thomas, and M. Thommes, *Characterization of Porous Solids and Powders: Surface Area, Pore Size and Density*, Kluwer Academic Press, Dordrecht, the Netherlands, 2004.

10. J.B. Condon, *Surface Area and Porosity Determination by Physisorption*, Elsevier, Amsterdam, the Netherlands, 2006.

11. M.M. Dubinin, *ACS Symposium Series*, 40 (1977) 1.

12. B.P. Bering and V.V. Serpinskii, *Ixv. Akad. Nauk, SSSR, Ser. Xim.*, 2427 (1974).

13. P.I. Ravikovitch and A.V. Neimark, *Colloids & Surf. A*, 187–188 (2001) 11.

14. R. Roque-Malherbe, *Mic. Mes. Mat.*, 41 (2000) 227.

15. M. Thommes, R. Kohn, and M. Froba, *J. Phys. Chem. B*, 104 (2000) 7932.

16. F. Marquez-Linares and R. Roque-Malherbe, *J. Nanosci. & Nanotech.*, 6 (2006) 1114.

17. Quantachrome, AUTOSORB-1, Manual, 2003.

18. Micromeritics, ASAP 2000, Description, 1992.

19. G.D. Halsey, *J. Chem. Phys.*, 16 (1948) 931.

20. D.M. Young and A.D. Crowell, *Physical Adsorption of Gases*, Butterworth, London, UK, 1962.

21. T.L. Hill, *An Introduction to Statistical Thermodynamics*, Dover Publications, New York, 1986.

22. V.A. Bakaev, *Dokl. Akad. Nauk SSSR*, 167 (1966) 369.

23. M. Dupont-Pavlovskii, J. Barriol, and J. Bastick, Colloques Internes du CNRS, No. 201 (Termochemie), 1972.

24. R. Roque-Malherbe, *KINAM*, 6 (1984) 35.

25. R. Roque-Malherbe, L. Lemes, L. López-Colado, and A. Montes, in *Zeolites'93 Full Papers Volume*, (D. Ming and F.A. Mumpton, Eds.), International Committee on Natural Zeolites Press, Brockport, New York, 1995, p. 299.

26. R. Roque-Malherbe, in *Handbook of Surfaces and Interfaces of Materials*, Vol. 5, (H.S. Nalwa, Ed.), Academic Press, New York, Chapter 12, 2001, p. 495.

27. M.M. Dubinin, *Prog. Surf. Memb. Sci.*, 9 (1975) 1.

28. M.M. Dubinin, E.F. Zhukovskaya, V.M. Lukianovich, K.O. Murrdmaia, E.F. Polstiakov, and E.E. Senderov, *Izv. Akad. Nauk SSSR*, (1965) 1500.

29. P.B. Balbuena and K.E. Gubbins, in *Characterization of Porous Solids* (I.J. Rouquerol, P. Rodriguez-Reynoso, K.S.W. Sing and K.K. Unger, Eds.), Elsevier, Amsterdam, the Netherlands, 1994, p. 41.

30. S.J. Gregg and K.S.W. Sing, *Adsorption Surface Area and Porosity*, Academic Press, London, UK, 1991.

31. D.W. Ruthven, *Principles of Adsorption and Adsorption Processes*, John Wiley & Sons, New York, 1984.

32. W. Rudzinskii, W.A. Steele, and G. Zgrablich, *Equilibria, and Dynamic of Gas Adsorption on Heterogeneus Solid Surfaces*, Elsevier, Amsterdam, the Netherlands, 1996.

33. J.B. Loos, *Modeling of Adsorption, and Diffusion of Vapors in Zeolites*, Coronet Books, Philadelphia, PA, 1997.

34. J.P. Fraissard (Ed.), *Physical Adsorption: Experiment, Theory and Applications*, Kluwer Academic Publishers, Dordrecht, the Netherlands, 1997.

35. G. Horvath and K. Kawazoe, *J. Chem. Eng. Japan*, 16 (1983) 470.

36. S.U. Rege and R.T. Yang, in *Adsorption: Theory, Modeling and Analysis* (J. Toth, Ed.), Marcel Dekker, New York, 2002, p. 175.

37. A. Saito and H.C. Foley, *A.I.Ch. E. J.*, 37 (1991) 429.

38. R. Roque-Malherbe, C. de las Pozas, and G. Rodriguez, *Rev. Cub. de Física*, 5 (1985) 107 (Chemical Abstracts 103 No. 221398h).

39. R. Roque-Malherbe, C. de las Pozas, and G. Rodriguez, *Rev. Cub. de Fisica*, 4 (1984) 143 (Chemical Abstracts 103 No. 221398).

40. R. Roque-Malherbe, A. Costa, C. Rivera, F. Lugo, and R. Polanco, *J. Mat. Sci. Eng. A*, 3 (2013) 263.
41. C.H. Baerlocher, W.M. Meier, and D.M. Olson, *Atlas of Zeolite Framework Types* (5th ed.), Elsevier, Amsterdam, the Netherlands, 2001.
42. A.A. Zagorodni, *Ion Exchange Materials: Properties and Applications*, Elzevier, Amsterdam, the Netherlands, 2007.
43. F. Marquez-Linares and R. Roque-Malherbe, *Facets-IUMRS J.*, 3 (2004) 8.
44. R. Roque-Malherbe, R. López-Cordero, J.A. González-Morales, J. Onate and M. Carreras, *Zeolites*, 13 (1993) 481.
45. E.M. Flanigen, R. Lyle-Patton, and S.T. Wilson, *Stud. Surf. Sci. & Catal.*, 37 (1988) 13.
46. H. Hattori and Y. Ono, *Solid Acid Catalysts: From Fundamentals to Applications*, CRC Press, Boca Raton, FL, 2015.
47. E.M. Flanigen, R.W. Broach, and S.T. Wilson, in *Zeolites in Industrial Separations and Catalysis* (S. Kulpathipanja Ed.), Wiley-WCH Verlag, Weinheim, Germany, 2010, p. 1.
48. T. Ishihara and H. Takinta, in *Catalysis*, Vol. 12, (J.J. Spivey Ed.), Royal Society of Chemistry, London, UK, 1996.
49. A. Suleiman, C. Cabrera, R. Polanco, and R. Roque-Malherbe, *RSC Advances*, 5 (2015) 7637.
50. R. Roque-Malherbe, F. Lugo, and R. Polanco, *App. Surf. Sci.*, 385 (2016) 360.
51. R. Roque-Malherbe, E. Carballo, R. Polanco, F. Lugo, and C. Lozano, *J. Phys. Chem. Solids*, 86 (2015) 65.
52. R. Roque-Malherbe, F. Lugo, C. Rivera, R. Polanco, P. Fierro, and O.N.C. Uwakweh, *Current App. Phys.*, 15 (2015) 571.
53. A. Rios, C. Rivera, G. Garcia, C. Lozano, P. Fierro, L. Fuentes-Cobas, and R. Roque-Malherbe, *J. Mat. Sci. Eng. A*, 2 (2012) 284.
54. R. Roque-Malherbe, O.N.C. Uwakweh, C. Lozano, R. Polanco, A. Hernandez-Maldonado, P. Fierro, F. Lugo, and J.N. Primera-Pedrozo, *J. Phys. Chem. C*, 115 (2011) 15555.
55. R. Roque-Malherbe, C. Lozano, R. Polanco, F. Marquez, F. Lugo, A. Hernandez-Maldonado, and J. Primera-Pedroso, *J. Solid State Chem.*, 184 (2011) 1236.
56. F. Marquez-Linares, O. Uwakweh, N. Lopez, E. Chavez, R. Polanco, C. Morant, J.M. Sanz, E. Elizalde, C. Neira, S. Nieto, and R. Roque-Malherbe, *J. Solid State Chem.*, 184 (2011) 655.
57. R. Roque-Malherbe, R. Polanco, and F. Marquez-Linares, *J. Phys. Chem. C*, 114 (2010) 17773 v.
58. V.A. Bakaev, *Dokl. Akad. Nauk SSSR*, 167 (1966) 369.
59. D.M. Ruthven, *Nature, Phys Sci.*, 232 (1971) 70.
60. W. Schirmer, K. Fiedler, and H. Stach, *ACS, Symposium Series*, 40 (1977) 305.
61. J. de la Cruz, C. Rodriguez, and R. Roque-Malherbe, *Surface Sci.*, 209 (1989) 215.
62. B.C. Lippens and J.H. de Boer, *J. Catalysis*, 4 (1965) 319.
63. D.W. Breck, *Zeolite Molecular Sieves*, John Wiley & Sons, New York, 1974.
64. S. Brunauer, P.H. Emmett, and E. Teller, *J. Amer. Chem. Soc.*, 60 (1938) 309.
65. A. Galarneau, D. Desplantier, R. Dutartre, and F. Di Renzo, *Mic. Mes. Mat.*, 27 (1999) 297.
66. J.R. Sams, G. Contabaris, and G.D. Halsey, *J. Phys. Chem.*, 64 (1960) 1689.
67. D.H. Everett and J.C. Powl, *J. Chem. Soc. Faraday Trans.*, 72 (1976) 619.
68. M.A. Parent and J.B. *Moffat Langmuir*, 11 (1996) 4474.
69. R.J. Dombrowski, C.H.M. Lastoskie, and D.R. Hyduke, *Colloids & Surf. A*, 187–188 (2001) 23.

4 Pore Size Distributions

4.1 INTRODUCTION

As is very well known, porous materials have applications, as: adsorbents, separation materials, catalysts along with others, because of their developed porosity and specific surface area [1–10]. Moreover, adsorption measurements are powerful tools for the characterization of adsorbents, porous separation materials, catalyst, and related materials. Concretely, in the previous chapters, we had explained the use of the adsorption methodology for the measurement of the micropore volume, that is, the surface area along with the micropore size distribution (MPSD) [11–22].

As was formerly affirmed, the International Union of Pure and Applied Chemistry (IUPAC) [6] classified pores by their inner pore width, in which mesoporous materials are those that show pores with an internal width between 2 and 50 nm, in our terminology named as the nanoporous region. Thereafter, in the present chapter the methodology for the characterization of nanoporous materials will be presented, applying basically the capillary condensation of vapors, which, as is very well known, is the primary method of assessment of the mesoporosity of nanoporous materials [1,23–32].

4.2 CAPILLARY CONDENSATION

Capillary condensation in nanoporous materials is normally related to a shift in the vapor–liquid coexistence in pores in comparison to the bulk fluid, that is, a fluid confined in a pore condenses at a pressure lower than the saturation pressure at a given temperature because the condensation pressure depends on the pore size and shape, and on the strength of the interaction between the fluid and pore walls [18,26,27] (Figure 4.1 [20]).

To be exact, pore condensation represents a confinement-induced shifted gas–liquid phase transition. Specially, condensation occurs at a pressure P less than the saturation pressure P_0 of the fluid [18]. Hence, the relative pressure ($x = P/P_0$) value in which the pore condensation takes place depends on the liquid–interfacial tension, the strength of the attractive interactions among the fluid, and pore walls, the pore geometry and the pore size [13,16,23,26]. Then it is assumed that for pores of a given shape and surface chemistry, there exists a one-to-one correspondence between the condensation pressure and the pore diameter. Thus, adsorption isotherms contain unequivocal information about the pore size distribution (PSD) of the sample under analysis [29–32].

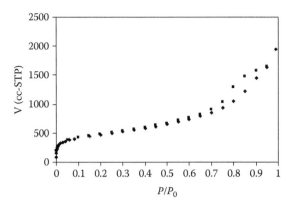

FIGURE 4.1 N_2 adsorption at 77 K in the silica 70bs2 (♦: adsorption and ■: desorption).

As a matter of fact, capillary condensation of vapors in the pores of solids is a prominent example of phase transitions in confined fluids [27]. In this regard, when a porous solid is contacted to the vapor of a wetting fluid, the latter condenses in pores at a vapor pressure lower than the saturation pressure at the given temperature, where P in P/P_0 is related to the liquid–vapor interfacial tension [28], the attraction interaction between the solid and the fluid, and the pore geometry, its size, and shape. Particularly, the capillary condensation is usually described by a step in the adsorption isotherm, particularly in materials with a uniform PSD, the capillary condensation step is notably sharp [13]; next, as an example in Figure 4.2.

In Figure 4.2 a graphic representation of a sorption isotherm is shown, as it is expected for adsorption/desorption of a pure fluid in a single mesopore of cylindrical shape. The isotherm shows a vertical pore condensation step. However, in a real system resulting in an adsorption process in a real porous material where there exist pores of different sizes, a less sharp pore condensation transition is observed (Figure 4.1).

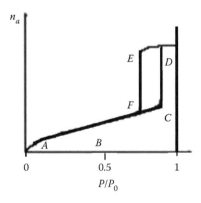

FIGURE 4.2   Adsorption isotherm for a single pore of cylindrical shape.

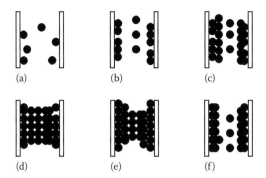

FIGURE 4.3   Representation of adsorption and pore condensation in a single mesopore (the processes described in the figure are explained in the text below).

In Figure 4.3 [18], it is shown that the adsorption mechanism in mesopores, which are at low increasing relative pressures, $(x_A = P/P_0)$ is similar to the adsorption process on planar surfaces, that is, after the completion of the monolayer formation (a), multilayer adsorption commences (b). Afterward, for a higher relative pressure, $x_C = P/P_0$ is observed after the attainment of a critical film thickness (c), when capillary condensation occurs basically in the core of the pore (d), that is, a transition from configuration (c) to (d) takes place.

Next, we have a desorption process in which the relative pressure decreases, that is, section (d) to (e), where the pore completely filled with liquid is separated from the bulk gas phase by a hemispherical meniscus that takes place in pore evaporation by thinning of a meniscus (e) at a pressure, which is less than the pore condensation pressure (Figures 4.2 and 4.3). Subsequently, the pressure where the hysteresis loop finishes $x_F = P/P_0$ is equivalent to a multilayer film in equilibrium with a vapor in the core of the pore, but in this case the evaporation takes place from a cylindrical meniscus. Therefore, in pores wider than approximately 5 nm, capillary condensation is connected with hysteresis [19], meaning that the vapor pressure decreases and desorption occurs at a pressure lower than the pressure of adsorption. It is needed to clarify now that the hysteresis loop is reproducible in adsorption experiments carried out with enough equilibration time. But as the pore size decreases, the experimental hysteresis loop gradually narrows and finally disappears for pores smaller than about 4 nm [6,21,22,24].

As a matter of fact, pore condensation implies a first-order phase transition between an inhomogeneous gas configuration, involving a vapor in the core region of the pore in equilibrium with a liquid-like adsorbed film (assembly [c] in Figure 4.3) and a liquid filling a pore (configuration [d] in Figure 4.3). Then at a critical point these two fluids (till now different fluids) become indistinguishable [18,27].

In practice, hysteresis is observed in materials consisting of slit-like pores, cylindrical-like pores, and a spherical pore, that is, ink-bottle pores [28–32]. In this regard, the IUPAC has made a classification of the different possible types of hysteresis loops (Figure 4.4) [6], where the H1 type is associated with porous materials formed of properly defined cylindrical-like pore channels or agglomerates of compacts of roughly homogeneous spheres [19,21–23].

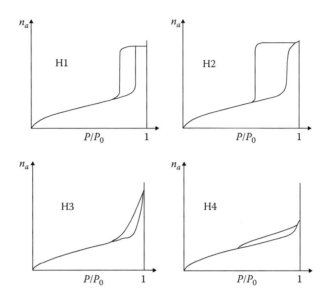

FIGURE 4.4   IUPAC classification of adsorption hysteresis.

It was as well established that materials that give rise to H2 hysteresis are frequently disordered, and their PSD is not well defined, whereas systems revealing H3 hysteresis type do not show any limiting adsorption at high relative pressure, in fact, observed in nonrigid aggregates of plate-like particles giving rise to slit-shaped pores [28–30]. Similarly, type H4 loops are also frequently associated with thin slit pores but also including pores in the micropore region [30–32].

4.3   MACROSCOPIC THEORIES TO DESCRIBE PORE CONDENSATION

4.3.1   THE KELVIN–COHAN EQUATION

During capillary condensation (Figure 4.3), the pore walls are first covered by a multiplayer adsorbed film at the beginning of pore condensation. Thereafter, at a certain critical thickness, t_c, pore condensation occurs in the core of the pore. For example, in pores of regular shape and width, such as slit-like or cylindrical pores (Figure 4.5), pore condensation can be described with the help of the Kelvin–Cohan equation [16].

This equation is based on the Kelvin equation, which is originated on the effect of the curvature of the surface on vapor pressure. This effect is best understood in terms of the pressure drop, ΔP, across an interface, which is described by the very well-known Young–Laplace equation [33], $P^{II} - P^{I} = \gamma\left(1/r_I + 1/r_{II}\right)$, in which γ is the surface tension, and r_I and r_{II} are the two curvature radii describing the surface (Figure 4.6).

A thermodynamic analysis allows the calculation of the effect of a change in mechanical pressure at constant pressure on the chemical potential of a substance, which is as follows [21,22]: $\Delta\mu = \int V_L dP$, where V_L is the molar volume of

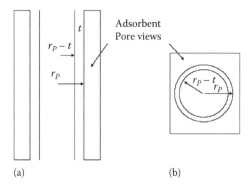

(a)   (b)

FIGURE 4.5   Slab-like (a) and cylindrical (b) pores during pore condensation.

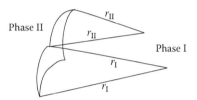

FIGURE 4.6   Sector of radius r_I and r_II separating phase I and phase II.

the fluid substance. If V_L is constant and applying the Young–Laplace equation as $\Delta\mu = \gamma V_L\left(1/r_\text{I} + 1/r_\text{II}\right)$. Hence, $\mu = \mu^0 + RT\ln(P)$; thereafter

$$RT\ln\left(\frac{P}{P_0}\right) = \gamma V_L\left(\frac{1}{r_\text{I}} + \frac{1}{r_\text{II}}\right) \tag{4.1}$$

is the Kelvin equation; now, it is possible to state that for a cylindrical pore the modified Kelvin equation or the Kelvin–Cohan equation is given by $RT\ln\left(P_A/P_0\right) = \gamma V_L/r_p - t_A$, whereas the evaporation/desorption process is associated with the formation of a hemispherical meniscus between the condensed fluid and vapor [17]:

$$RT\ln\left(\frac{P_D}{P_0}\right) = \frac{2\gamma V_L}{r_p - t_D} \tag{4.2}$$

where:

r_p is the pore radius

$x_A = P_A/P_0$ and $x_D = P_D/P_0$ are the relative pressures of adsorption and desorption, respectively

γ is the surface tension

V_L is the molar volume of the bulk liquid

t_A and t_D are the thickness of an adsorbed multilayer film, which is formed before the pore condensation at relative pressures x_A and x_D [13]

Hence, the Kelvin–Cohan equation gives a correlation between the pore diameter and the pore condensation pressure, predicting that pore condensation shifts to a higher relative pressure with an increasing pore diameter and temperature. Consequently, it serves as the basis for the traditional methods applied for mesopore analysis, such as the Barett–Joyner–Halenda (BJH) method [34]. In this regard, to account for the preadsorbed multilayer film, the Kelvin–Cohan equation is combined with a standard isotherm or *t*-curve, which typically refers to adsorption measurements on a nonporous solid, where the preadsorbed multilayer film is estimated using the statistical thickness of an adsorbed film on a nonporous solid with a surface chemistry similar to that of the sample under consideration [21,22].

Now, we will describe the details of the BJH method to determine the PSD, that is, a graphical representation of $\Delta V_p/\Delta D_p$ versus D_p, where V_p is the pore volume accumulated up to the pore of width D_p, measured in cc-STP/g. Å and specifically the amount adsorbed is measured in cubic centimeters at STP, that is, at standard temperature and pressure, that is, 273.15 K and 760 Torr, that is, 1.01325×10 Pa. Further, the pore volume denoted by W, is the sum of the micropore and mesopore volumes of the adsorbent measured in cm^3/g [6,9].

Now, it is necessary to remind as it was indicated in Figure 4.5 that the pore of radius, r_p, has a physically adsorbed layer of adsorbate molecules, normally nitrogen at 77 K of thickness t. Within this thickness, there is an inner capillary with radius $r_K = r_p - t$ from which evaporation occurs as $x = P/P_0$. In the frame of the BJH [34] method, in addition to the previous assumptions, related with the Kelvin–Cohan equation, it is supposed that at the initial relative pressure of the desorption process, which is close to unity, that is, in the range $0.9 < P/P_0 < 0.95$, all the pores are filled with the adsorbate fluid [26,27]. Therefore, the first step ($j = 1$) (Figure 4.7 [20]) in the desorption process only involves removal of the capillary condensate.

However, the next steps comprise both removal of condensate from the cores of a group of pores and the thinning of the multilayer in the larger pores, which is in the pores already emptied of condensate [18].

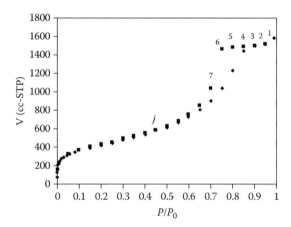

FIGURE 4.7 N_2 at 77 K adsorption on silica 68bs1E (♦: adsorption and ■: desorption).

Here, the core volume will be designated by V_{Kj} and the pore volume V_{pj}, whereas the radii are r_K and r_p, provided the calculations are normally related with the desorption of nitrogen at 77 K coming from that amount of nitrogen removed in each desorption step, j, is $\Delta n_a(j)$. Nevertheless, in the present case this amount is expressed as the volume $\Delta V(j)$ of liquid nitrogen. Now, in the first step ($j = 1$) the initial removal is the result of capillary evaporation alone. Accordingly, the volume of core space removed is equal to the volume of nitrogen removed, specifically $\Delta V_K(1) = \Delta V(1)$, where the relationship between the core volume V_{K1} and the pore volume V_p, if the pores of cylindrical shape, is given for the first group of mesopores by [34]

$$V_{p1} = \frac{V_{K1} \bar{r}_{p1}^{-2}}{\bar{r}_{K1}^{-2}} \tag{4.3}$$

where $\bar{r}_{p1}$ and $\bar{r}_{K1}$ are the mean pore and core radii, respectively. When the relative pressure, x, is lowered from $(P/P_0)_1$ to $(P/P_0)_2$, a volume, $\Delta V(1)$ will be desorbed from the pores, where this fluid volume, $\Delta V(1)$, represents not merely emptying the largest pore of its condensate, but a decrease in the thickness of the physically adsorbed layer by an amount, Δt_1. Subsequently, across this relative pressure reduction the average change in thickness is $\Delta t_1/2$. Thereafter, the pore volume of the largest pore may be expressed as

$$V_{p1} = \Delta V(1) \left(\frac{r_{p1}}{r_{K1} + (\Delta t_1/2)} \right)^2 \tag{4.4}$$

When the relative pressure, x, is again lowered to $(P/P_0)_3$, the volume of liquid desorbed comprises not only the condensate from the next larger size pores but also the volume from a second thinning of the physically adsorbed layer left behind in the pores of the largest size. Thus, the volume V_{p2} desorbed from pores of the smaller size is given by

$$V_{p2} = \left(\frac{r_{p2}}{r_{K2} + (\Delta t_1/2)} \right)^2 \left(\Delta V(2) - \Delta V_t(2) \right)$$

where $\Delta V(2) = \Delta V_K(2) + \Delta V_t(2)$, the volume desorbed from multilayer is $\Delta V_t(2) = \Delta t_2 Ac_1$, whereas Ac_1 is the area exposed by the previously emptied pores from which physically adsorbed gas is desorbed. Afterward, the previous equation could be generalized to symbolize any step of a stepwise desorption by writing

$$\Delta V_t(n) = \Delta t_n \sum_{j=1}^{n-1} Ac_j \tag{4.5}$$

This summation term in the previous equation is the sum of the average area in unfilled pores down to but not including the pore that was emptied in the concrete desorption step. Now, substituting the general value of $\Delta V_t(n)$ into the equation defining V_{p2} we will get

$$V_{pn} = \left(\frac{r_{pn}}{r_{Kn} + (\Delta t_n/2)}\right)^2 \left(\Delta V(n) - \Delta t_n \sum_{j=1}^{n-1} Ac_j\right) \qquad (4.6)$$

which is an exact expression for calculating the pore volumes at various relative pressures.

Recapitulating, in the BJH approach to the PSD calculation, is understood that all pores emptied of their condensate during a relative pressure decrement have an average radius $\bar{r}_p$ calculated from the Kelvin–Cohan equation radii at the upper and lower values of the relative pressure, x in the desorption step. Thenceforth the average core radius is expressed as [16]: $\bar{r}_c = \bar{r}_p - t_{\bar{r}}$, in which $t_{\bar{r}}$ is the thickness of the adsorbed layer at the average radius at intervals in the current decrement calculated from the equation [35]:

$$t(A) = \left(\frac{13.9}{\ln(P/P_0) + 0.034}\right)$$

which is an equation proposed by de Boer and collaborators for estimating t in the case of nitrogen at 77 K. Now, the term c could be calculated by

$$c = \frac{\bar{r}_p}{\bar{r}_p} = \frac{\bar{r}_p - t_{\bar{r}}}{\bar{r}_p}$$

Subsequently, with the previous equation and the expression for V_{pn}, it is possible to calculate the Barret–Joyner–Halenda-pore size distribution (BJH-PSD) [16,34]. In this regard, Figure 4.8 represents the BJH–PSD of an MCM-41 mesoporous molecular sieve [20], where a MCM-41 is a member of the M41S family of mesoporous ordered silicas, materials showing hexagonal together with cubic symmetry along with pore sizes ranging from 20 to 100 Å, that is, adsorbents characterized by an ordered, not crystalline, pore wall structure presenting sharp pore size dispersions, in the case of the MCM41 showing a hexagonal stacking of uniform diameter porous tubes, whose size can be varied from about 15 Å to more than 100 Å [36].

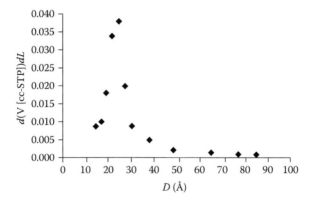

FIGURE 4.8 BJH–PSD of an MCM-41 MMS.

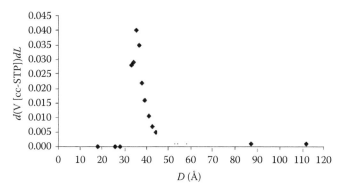

FIGURE 4.9 NLDFT–PSD corresponding to an MCM-41.

Notwithstanding the wide application of the BJH method for the PSD determination, it is required to acknowledge that it is founded on a simplified macroscopic description of the capillary condensation process with limitations at the microscopic level [23–32]. These limitations made the method inaccurate, specifically for materials with pores with 2–4 nm [13,14], that is, 20–40 Å. Specifically, the BJH method, as is evident from the comparison of Figures 4.8 and 4.9, underestimates the pore size by about 1 nm (10 Å) in relation with the nonlocal density functional theory (NLDFT) method [37,38]. This fact is made necessary to use more accurate methods as will be seen in the next sections, especially in Section 4.4 of this chapter [12–15].

4.3.2 THE DERJAGUIN–BROEKHOFF-DE BOER THEORY

The Derjaguin–Broekhoff-de Boer (DBdB) theory is equivalent to the Derjaguin approach. Within the frame of this theory, the equilibrium thickness of the adsorbed film t, in a cylindrical pore of radius r_P, is determined from the balance of capillary and disjoining pressures [39,40]:

$$RT \ln\left(\frac{P_0}{P}\right) = \Pi(t)V_L + \frac{\gamma V_L}{r_P - t} \qquad (4.7)$$

where:
P is the equilibrium adsorption pressure
P_0 is the vapor pressure of the adsorbate at the temperature of the adsorption experiment
γ and V_L are the surface tension and the molar volume of liquid
$\Pi(t)$ is the disjoining pressure of the adsorbed film, which is a parameter accounting for the sum of fluid–solid intermolecular interactions, that is, this pressure is equivalent to the action of the adsorption field [1]

Thereafter, Equation 4.3 could be justified as follows: since, the chemical potential difference, $\Delta\mu$, between the gas and the adsorbed phase could be given [18] as $\Delta\mu = \Delta\mu_a + \Delta\mu_c$, where the first term, $\Delta\mu_a$, related with multilayer adsorption rules the process during multilayer adsorption. In contrast, when the adsorbed film turns out to be thicker, the adsorption potential will become less important. Consequently, $\Delta\mu$ will be controlled virtually completely by the curvature contribution, $\Delta\mu_c$. In particular, the Laplace term given for a cylindrical pore [33] is $\Delta\mu_c = \gamma V_L/r_p - t$, whereas $\Delta\mu_a = \Pi(t)V_L$ is a relation calculated using the concept of disjoining pressure, Π introduced by Derjaguin in 1936, with the help of the relation $\Delta\mu = \int V_L dP$ [41].

Concretely, within the DBdB approach, in the case of dominant attractive potential, as was previously studied for the Frenkel–Halsey–Hill (FHH) model [7]: $\Pi(t) \propto 1/t^m$. Hence, to test the DBdB approach, Dubinin and coworkers suggested the standard N_2 isotherm at 77.4 K on nonporous oxides in the FHH form as [26].

$$\ln\left(\frac{P_0}{P}\right) = \frac{\Pi(t)V_L}{RT} = \frac{C}{t^m} \tag{4.8}$$

with the parameters $C = 44.54$ and $m = 2.241$ to get t in Å [13].

Besides, the stability condition for the wetting film was formulated by Derjaguin and coworkers as follows [31,32]:

$$\left(\frac{d\Pi(t)}{dt}\right) < 0 \tag{4.9}$$

where it is assumed that during adsorption, the isotherm draws a sequence of metastable state of the adsorption film, then capillary condensation takes place spontaneously when the film thickness approaches the limit of stability. Thereafter, the critical thickness of metastable films is determined from the condition [9]:

$$-\left(\frac{d\Pi(t)}{dt}\right)_{t=t_\alpha} = \frac{\gamma}{(r_P - t_{critical})^2} \tag{4.10}$$

in which $t_{critical}$ is the critical adsorbed-film thickness. Thus, the conditions of capillary condensation in a cylindrical pore are determined by Equations 4.7 and 4.10 [13]. Consequently, desorption from cylindrical capillary is determined by the condition of formation of the equilibrium meniscus given by the enhanced Kelvin–Cohan equation recognized as the Derjaguin equation [18]:

$$RT\ln\left(\frac{P}{P_0}\right) = \frac{2\gamma V_L + \dfrac{2V_L}{(r_P - t_e)^2}\displaystyle\int_{t_e}^{r_P}(r_P - t)^2\Pi(t)dt}{r_P - t} \tag{4.11}$$

where t_e is the thickness of the adsorbed film in equilibrium with the meniscus given by Equation 4.3.

Finally, as was previously commented, the chemical potential difference $\Delta\mu$ is given by $\Delta\mu = \Delta\mu_a + \Delta\mu_c$. Hence, when the adsorbed film turns out to be thicker, the adsorption potential becomes less important, and $\Delta\mu$ will be controlled virtually completely by the curvature contribution $\Delta\mu_c$. Then at a certain critical thickness, $t_{critical}$, pore condensation occurs in the core of the pore controlled by intermolecular forces in the core fluid [18].

4.3.3 Some Concluding Remarks about the Macroscopic Theories to Describe Pore Condensation

It is evident from our previous discussion that macroscopic theories do not describe pore condensation well. For instance, the Kelvin–Cohan methodology that contemplate pore condensation as a gas–liquid phase transition in the core of the pore between two phases, that is, bulk-like gas and liquid phases no-accurately predicts the pore condensation and hysteresis. Similarly, the DBdB theory [39,40,42,43] is described in essence with the mechanism of pore condensation and hysteresis as it is described in Figure 4.3. Anticipating a grow in the strength of the adsorbate–adsorbent interaction, a descent of the experimental temperature and diminishing the pore size will move the pore condensation transition to lower relative pressures. Notwithstanding a better description of pore condensation this theory does not make possible the totally accurate PSD calculations [23–27]. Meanwhile, microscopic approaches, such as the density functional theory (DFT) of inhomogeneous fluids, which propose that a fluid confined to a single pore can exist with two possible density profiles corresponding to inhomogeneous gas and liquid configurations in the pore, overpasses the gap between the molecular level and macroscopic approaches providing a very accurate description of the process along with good experimental PSD measurements [28–32].

4.4 DENSITY FUNCTIONAL THEORY

4.4.1 The Density Functional Theory Methodology in General

Molecular models, as for instance, the DFT provides a more comprehensive representation of phase transitions in pores than the classical thermodynamic methods [21,25–38,44–49]. As was previously explained, in Chapter 1, Section 1.10, the basic variable in DFT is the single particle density, $\rho(\bar{r})$. For a many particle classical system immersed in an external potential U_{ext} arising, for instance, from the adsorption field happening when a fluid is confined in a pore, a function, $\Omega[\rho(\bar{r})]$, which reduces to the grand potential $\Omega = \Omega[\bar{\rho}(\bar{r})]$ when $\rho(\bar{r}) = \bar{\rho}(\bar{r})$ is the equilibrium density, can be defined with the help of the following unique functional of the density [23]:

$$\Omega[\rho(\bar{r})] = F[\rho(\bar{r})] - \int d\bar{r}\rho(\bar{r})[\mu - U_{ext}(\bar{r})] \tag{4.12}$$

which is obtained by means of a Legendre transformation (see Appendix 1.1) of $F[\rho(\bar{r})]$, which as was previously defined, in Chapter 1, Section 1.10, is the intrinsic Helmholtz free energy [26,44], and μ is the chemical potential of the studied system. To simplify, Equation 4.12 is defined as $u(\bar{r}) = \mu - U_{ext}(\bar{r})$. Hence,

$$\Omega[\rho(\bar{r})] = F[\rho(\bar{r})] - \int d\bar{r}\rho(\bar{r})u(r) \tag{4.13}$$

As was formerly affirmed, the ground energy and all other properties of a ground state molecule are uniquely determined by the ground-state electron probability density of the system. Therefore, for a thermodynamic system of N classical particles, the true equilibrium density, $\bar{\rho}(\bar{r})$, is determined with the help of the Euler–Lagrange equation, that is, we must determine the minimum value of the functional $\Omega[\rho(\bar{r})]$:

$$\frac{\delta\Omega[\rho(\bar{r})]}{\delta\rho(\bar{r})} = 0 \qquad (4.14)$$

That is, $\bar{\rho}(\bar{r})$ is the solution of Equation 4.13, where $\delta/\delta\rho(\bar{r})$ indicates the functional derivative (see Appendix 1.4). Now, combining Equations 4.12 and 4.14 we will obtain [26]

$$u(\bar{r}) = \mu - U_{ext}(\bar{r}) = \frac{\delta F[\rho(\bar{r})]}{\delta\rho(\bar{r})} \qquad (4.15)$$

which is also the minimum condition. Thereafter, to get, $\bar{\rho}(\bar{r})$, as is normally done in the solution of statistical mechanical problems, the studied system is reduced to simpler components. Then, for a classical fluid with inhomogeneous density distribution, the functional, $F[\rho(\bar{r})]$, representing the intrinsic Helmholtz free energy can be expressed as [27] $F[\rho(\bar{r})] = F_{id}[\rho(\bar{r})] + F_{ex}[\rho(\bar{r})]$, where

$$F_{id}[\rho(\bar{r})] = kT\int d\bar{r}\rho(\bar{r})\{\ln\left(\rho(\bar{r})\Lambda^3\right) - 1\} \qquad (4.16)$$

is the ideal gas free-energy functional, resulting from a system of noninteracting particles, where $\Lambda = \left(h^2/2\pi mkT\right)^{1/2}$ is the thermal wavelength. Besides, $F_{ex}[\rho]$ represents the excess free energy for the classical system. Now defining [44]

$$c^1(\bar{r}) = \frac{\delta[\beta F_{ex}(\rho(\bar{r}))]}{\delta\rho(\bar{r})} \qquad (4.17)$$

Next, from the variational principle, equation for the classical system of N particles, it follows that the density profile satisfies [28]

$$\rho(\bar{r}) = \Lambda^{-3}\left[\exp(\beta\mu)\right]\left[\exp-\left(\beta U_{ext} + c^1\left(\bar{r},\rho(\bar{r})\right)\right)\right] \qquad (4.18)$$

where $c^1(\bar{r})$ is also a functional of $\rho(\bar{r})$ and $\beta = 1/kT$. This equation, when $c^1(\bar{r}) = 0$, that is, in the case of an ideal gas reduces to the barometric law for the density distribution in the presence of an external field. Moreover, to go on further, the studied system is additionally reduced to simpler components. Therefore, the excess intrinsic Helmholtz energy is split into contributions from the short-ranged repulsion and long-term attraction [29]: $F_{ex}[\rho(\bar{r})] = F_{rep}[\rho(\bar{r})] + F_{att}[\rho(\bar{r})]$.

4.4.2 CALCULATION OF THE PORE SIZE DISTRIBUTION

In the frame of this book we are interested in modeling capillary condensation and desorption of nitrogen and argon in slit and cylindrical pores [13,14] along with spherical cavities. Consequently, to compute the PSD of slit pores, cylindrical and spherical cavities, the experimental isotherm is described as a combination of theoretical isotherms in individual pores, where the theoretical isotherm is obtained, for example, with the help of a DFT approach, that is, the experimental isotherm is the integral of the single pore isotherm multiplied by the PSD explicitly, the Fredholm-type integral equation [44]:

$$N_{\exp}\left(\frac{P}{P_0}\right) = \int_{D_{\min}}^{D_{\max}} N_V^{ex}\left(D, \frac{P}{P_0}\right)\varphi_V(D)dD \tag{4.19}$$

Here, $N_V^{ex}(D, P/P_0)$ is a kernel of the theoretical isotherm in pores of different diameters, $\varphi_V(D)$ is the PSD function [30–32], and D is the pore diameter. Different kernels are used to carry out the calculations of the PSD from the experimental isotherms [10,12–15]. To find a PSD, from the experimental isotherm function $N_{\exp}(P/P_0)$, initially, $N_V^{ex}(D, P/P_0)$ is calculated with the help of $\overline{\rho}(\overline{r})$, as will be explained in the below-mentioned equation. Then, the integral equation is represented as a matrix equation with pores spanning logarithmically in different ranges, where the procedure to expand $\varphi_V(D_{in})$ with the help of, for example, the γ distribution with multiple modes is as follows [45]:

$$\varphi_V(D) = \sum_{i=1}^{m} \frac{\alpha_i(\gamma_i, D)^{\beta_i}}{\Gamma(\beta_i)D}\left(\exp\left[-\gamma_i D\right]\right)$$

where:
  m is the number of modes of the distribution
  $\Gamma(\beta_i)$ is the gamma function [50]
  α_i, β_i, and γ_i are adjustable parameters that define the amplitude, mean, and variance of the mode, i, of the distribution

Subsequently, a solution, that is, the PSD, $\varphi_V(D)$, is found using multilinear least square-fitting procedures of the parameters defining the assumed PSD, so as to match the experimental isotherm [51].

4.4.3 THE NONLOCAL DENSITY FUNCTIONAL THEORY FOR THE DESCRIPTION OF ADSORPTION IN SLIT PORES, CYLINDRICAL PORES, AND SPHERICAL CAVITIES

Evans and Tarazona NLDFT will be applied to find $\overline{\rho}(\overline{r})$, a reliable procedure, which produces quantitative agreement with molecular simulations, and experimental data on regular different model systems [25–32].

In the NLDFT approach, the adsorption and desorption isotherms in pores are calculated based on the intermolecular potentials of fluid–fluid and solid–fluid interactions, considering that each individual pore has a fixed geometry and is open, and in contact with the bulk adsorbate fluid. Besides, the local density of the adsorbate confined in a pore at a given chemical potential μ, volume V, and temperature T is determined by the minimization of the grand thermodynamic potential Ω [26–28]. Specifically, the NLDFT proceeds by considering that the adsorbed fluid in a pore is in equilibrium with a bulk gas phase, where the local fluid density of the adsorbate confined in the pore, in the presence of a spatially varying external potential, U_{ext}, at a given chemical potential, volume, and temperature, is determined by the minimization of the grand potential [29–32,37,38,47]:

$$\Omega[\rho(\bar{r})] = F_{\text{id}}[\rho(\bar{r})] + F_{\text{rep}} + F_{\text{att}}[\rho(\bar{r})] - \int d\bar{r}\rho(\bar{r})[\mu - U_{\text{ext}}(\bar{r})] \qquad (4.20)$$

Now, in Equation 4.20, the intrinsic free energy, is divided in a perturbative fashion, in the ideal gas free energy, F_{id} given by an exact expression and the excess free energy, F_{ex}, which takes into account the interparticle interactions, where the term, F_{ex}, consists of two components, F_{rep}, representing the repulsive forces between molecules and is commonly described as a reference system of hard spheres. Thenceforward, $F_{\text{rep}} = F_{\text{HS}}$, and the free energy arises from the attractive interactions, that is, F_{att}, where F_{att} is the mean-field free energy due to the Lennard–Jones (LJ) attractive interactions [23–32].

To take into consideration the repulsive forces several functionals for hard sphere fluids have been developed between them are the so-called weighted density approximation [52], the so-called fundamental measure theory [53], and the smoothed density approximations [54]. The smoothed density approximation, developed by Evans, Tarazona, and collaborators, is applied in nearly all NLDFT versions that are presently used for pore size characterization. Then the free energy of the hard sphere system is calculated with the help of [23]

$$F_{\text{rep}}[\rho(\bar{r})] = F_{\text{HS}}[\rho(\bar{r})] = kT \int \rho(\bar{r}) d\bar{r} f_{\text{ex}}[\bar{\rho}(\bar{r})] \qquad (4.21)$$

where the hard sphere diameter, d_{HS}, is [55]

$$d_{\text{HS}} = \int_0^\sigma \left\{1 - \exp\left[-\beta u_{\text{rep}}(r)\right]\right\} dr$$

and $f_{\text{ex}}[\bar{\rho}(\bar{r})]$, the excess free energy per molecule is [56]

$$f_{\text{ex}}[\bar{\rho}(\bar{r})] = \mu[\bar{\rho}(\bar{r})] - \frac{P[\bar{\rho}(\bar{r})]}{\bar{\rho}(\bar{r})} - kT\left\{\ln[\Lambda^3 \bar{\rho}(\bar{r})] - 1\right\}$$

In the previous equation, $\mu[\bar{\rho}(\bar{r})]$ and $P[\bar{\rho}(\bar{r})]$ are the chemical potential and the pressure, in that order, of a uniform hard sphere fluid, and $\bar{\rho}(\bar{r})$ is a smoothed density profile. The smoothed density profile, which is introduced into $f_{\text{ex}}[\bar{\rho}(\bar{r})]$, the functional is defined as follows [57]:

$$\bar{\rho}(\bar{r}) = \int d\bar{R}\rho(\bar{R})W\left(\left|\bar{r} - \bar{R}\right|, \bar{\rho}(\bar{r})\right) \qquad (4.22)$$

where $W\left(\left|\bar{r} - \bar{R}\right|, \bar{\rho}(\bar{r})\right)$ is a weighting function, which takes into account nonlocal effects. The idea behind the smoothed or weighted density approximations is to construct a smoothed density $\bar{\rho}(\bar{r})$, which is an average of the true density profile $\rho(\bar{r})$, over a local volume that is determined by the range of the interatomic forces. Several types of weighting functions have been proposed to study confined fluids, provided the Tarazona prescription for the weighting functions, use a power expansion in the smoothed density, truncating the expansion in the second order, which yields [23]

$$W_{\text{nonlocal}}\left(\left|\bar{r}-\bar{R}\right|\right)=W_0\left(\left|\bar{r}-\bar{R}\right|\right)+W_1\left(\left|\bar{r}-\bar{R}\right|\right)\bar{\rho}(\bar{r})+W_2\left(\left|\bar{r}-\bar{R}\right|\right)\left(\bar{\rho}(\bar{r})\right)^2 \quad (4.23)$$

The expansion coefficients, $W_0, W_1,$ and W_3 are [54,57] as follows:

$$W_0(r)=\frac{3}{4\pi\sigma^3} \qquad \text{for } r<\sigma$$

$$W_0(r)=0 \qquad \text{for } r>\sigma$$

$$W_1(r)=0.475+0.648\left(\frac{r}{\sigma}\right)+0.113\left(\frac{r}{\sigma}\right)^2 \qquad \text{for } r<\sigma$$

$$W_1(r)=0.288\left(\frac{\sigma}{r}\right)-0.924+0.764\left(\frac{r}{\sigma}\right)-0.187\left(\frac{r}{\sigma}\right)^2 \qquad \text{for } \sigma<r<2\sigma$$

$$W_1(r)=0 \qquad \text{for } r>\sigma$$

and

$$W_2(r)=\left(\frac{5\pi\sigma^3}{144}\right)\left(6-12\left(\frac{r}{\sigma}\right)+5\left(\frac{r}{\sigma}\right)^2\right) \qquad \text{for } r<\sigma$$

$$W_2(r)=0 \qquad \text{for } r>\sigma$$

where $r=\left|\bar{r}-\bar{R}\right|$.

To calculate the hard sphere, excess free energy is normally used as the Carnahan–Starling equation of state [58]:

$$P_{\text{HS}}[\bar{\rho}]=\bar{\rho}kT\left(\frac{1+\bar{\xi}+\bar{\xi}^2-\bar{\xi}^3}{\left(1-\bar{\xi}\right)^3}\right)$$

$$\mu_{\text{HS}}[\bar{\rho}]=kT\left[\ln(\Lambda^3\bar{\rho}+\left(\frac{8\bar{\xi}-9\bar{\xi}^2+3\bar{\xi}^3}{\left(1-\bar{\xi}\right)^3}\right)\right]$$

where $\bar{\xi}=(\pi/6)\bar{\rho}d_{\text{HS}}^3$ and the hard sphere diameter can be calculated with the help of the Barker–Handerson diameter [59] and $\left(d_{\text{HS}}/\sigma_{\text{ff}}\right)=\left(\left(\eta_1 kT/\varepsilon_{\text{ff}}\right)+\eta_2/\left(\eta_3 kT/\varepsilon_{\text{ff}}\right)+\eta_4\right)$, where $\eta_1=0.3837$, $\eta_2=1.305$, $\eta_3=0.4249$, and $\eta_4=1$ [54] are the fitting parameters. Besides

$$F_{\text{att}}[\rho(\bar{r})]=(1/2)\iint d\bar{r}d\bar{r}'\rho(\bar{r})\rho(\bar{r}')\Phi_{\text{atr}}\left(\left|\bar{r}-\bar{r}'\right|\right) \qquad (4.24)$$

where the attractive fluid–fluid potential, Φ_{attr}, is calculated according to the Weeks–Chandler–Andersen scheme (WCA) [60].

$$\Phi_{attr}\left(\left|\bar{r}-\bar{r}'\right|\right) = \phi_{ff}\left(\left|\bar{r}-\bar{r}'\right|\right) \quad \text{for } \left|\bar{r}-\bar{r}'\right| > r_m \tag{4.25}$$

$$\Phi_{attr}\left(\left|\bar{r}-\bar{r}'\right|\right) = -\varepsilon_{ff} \quad \text{for } \left|\bar{r}-\bar{r}'\right| < r_m \tag{4.26}$$

where $r_m = 2^{1/6}\sigma_{ff}$ and ϕ_{ff} is the Lennard–Jones potential of a pair of adsorbate molecules whose centers of mass are separated by a distance r and [24,30]

$$\phi_{ff} = 4\varepsilon_{ff}\left[\left(\frac{\sigma_{ff}}{r}\right)^{12} - \left(\frac{\sigma_{ff}}{r}\right)^{6}\right] \tag{4.27}$$

where:

r is the separation distance between a pair of molecules and the parameters

σ_{ff} and ε_{ff} are the Lennard–Jones molecular diameter, and the well depth for the adsorbate–adsorbate pair potential

r_m reflects the minimum of the potential [54,56] of the fluid–fluid Lennard–Jones 12–6 pair potential

In addition, the solid–fluid potential $U_{ext}(r)$, in the case of spherical silica pores, is the result of Lennar–Jones interactions with the outer layer of oxygen atoms in the wall of a spherical cavity of radius R measured to the centers of the first layer of atoms in the pore wall. Being granted, the potential at a distance x from the wall is obtained by integrating the Lennard–Jones 12–6 potential over the spherical surface [61]:

$$U_{ext}(x,R) = 2\pi\rho_s^O\varepsilon_{sf}\sigma_{sf}^2\left\{\frac{2}{5}\sum_{j=0}^{9}\left(\frac{\sigma_{sf}^{10}}{R^j x^{10-j}} + (-1)^j\frac{\sigma_{sf}^{10}}{R^j(x-2R)^{10-j}}\right)\right.$$
$$\left. -\sum_{j=0}^{3}\left(\frac{\sigma_{sf}^{4}}{R^j x^{4-j}} + (-1)^j\frac{\sigma_{sf}^{4}}{R^j(x-2R)^{4-j}}\right)\right\} \tag{4.28}$$

Here, ε_{sf} and σ_{sf} are the energetic and scale parameters of the potential, respectively. Moreover, ρ_s^O is the number of oxygen atoms per unit area of the pore wall or the oxygen surface density. The parameters of the adsorptive–adsorptive and adsorptive–adsorbent intermolecular potentials necessary to evaluate Equations 4.25 through 4.28 are reported in Tables 4.1 and 4.2 [29,31,62–64].

TABLE 4.1
Adsorptive–Adsorptive Intermolecular Potential Parameters

Gas	ε_{ff}/k (K)	σ_{ff} (nm)	d_{HS} (nm)
Nitrogen	94.45	0.3575	0.3575
Argon	118.05	0.3305	0.3380

TABLE 4.2
Adsorptive–Adsorbent Intermolecular Potentials Parameters

Gas/Solid	ε_{sf}/k (K)	σ_{sf} (nm)	Surface Density
Nitrogen/carbon	53.22	0.3494	Carbon: $\rho_s^C = 3.819 \times 10^{19}$ m^{-2}
Nitrogen/silica	147.3	0.3170	Silica: $\rho_s^O = 1.53 \times 10^{19}$ m^{-2}
Argon/silica	171.24	0.3000	Silica: $\rho_s^O = 1.53 \times 10^{19}$ m^{-2}

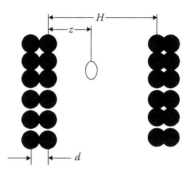

FIGURE 4.10 Schematic representation of a slit pore.

In a slit pore of physical pore width H (Figure 4.10) located between two homogeneous carbon slabs, the external potential U_{ext} depends only on one spatial coordinate z is given by [30] $U_{ext}(z) = \phi_{sf}(z) + \phi_{sf}(H - z)$, provided the potential of the adsorbate interaction with one of the bounding slabs, ϕ_{sf}, in the case of graphite layers are well described by the Steele potential [65]:

$$\phi_{sf} = 2\pi\rho_s\varepsilon_{sf}\sigma_{sf}^2\Delta\left[\frac{2}{5}\left(\frac{\sigma_{sf}}{z}\right)^{10} - \left(\frac{\sigma_{sf}}{z}\right)^4 - \left(\frac{\sigma_{sf}^4}{3\Delta([z + 0.61\Delta]^3)}\right)\right] \tag{4.29}$$

where:

ε_{sf} and σ_{sf} are the Lennard–Jones parameters for the adsorbate molecule interaction with a carbon atom in the porous solid

$\rho_s = 0.114$ Å^{-3} is the atomic bulk density of graphite

$\Delta = 3.34$ Å is the distance between graphite layers [24]

In Tables 4.1 and 4.2 [13,50–52] the parameters of the adsorptive–adsortive together with adsorptive–adsorbent intermolecular potentials along with the number of carbon atoms per unit area of the pore wall or the carbon surface density are reported. $\rho_s \times \Delta = \rho_s^C = 0.114$ Å$^{-3} \times 3.34$ Å $= 0.381$ Å$^{-2} = 3.81 \times 10^{19}$ m^{-2} required to evaluate Equations 4.27 through 4.29.

Theoretical adsorption isotherms for a pore of a geometry are constructed by solving the equilibrium adsorbate density profile. In this regard, the density profile that minimizes the grand potential function of the adsorbed fluid for a range of chemical potentials μ, that is, the equilibrium density profile that fulfills

$$u(\bar{r}) = \mu - U_{ext}(\bar{r}) = \frac{\delta F[\rho(\bar{r})]}{\delta \rho(\bar{r})} = \frac{\delta(F_{id} + F_{HS})}{\delta \rho}$$

$$+ \frac{\delta\left\{(1/2)\iint d\bar{r}.d\bar{r}'\rho(\bar{r})\rho(\bar{r}')\Phi_{atr}\left(\left|\bar{r}-\bar{r}'\right|\right)\right\}}{\delta \rho}$$

(4.30)

is obtained by introducing Equations 4.16 and 4.21:

$$\mu - U_{ext}(\bar{r}) = kT \ln(\Lambda^3 \rho(\bar{r})) + \frac{\delta\left\{kT \int \rho(\bar{r}) f_{ex}[\bar{\rho}(\bar{r})]d\bar{r}\right\}}{\delta \rho}$$

$$+ \frac{\delta\left\{(1/2)\iint d\bar{r}.d\bar{r}'\rho(\bar{r})\rho(\bar{r}')\Phi_{atr}\left(\left|\bar{r}-\bar{r}'\right|\right)\right\}}{\delta \rho}$$

(4.31)

where Equation 4.31 is an implicit expression, whose functional inversion yields the density profile in terms of the chemical potential, the attractive and external potentials, and the geometry of the system and particles.

To carry out the inversion of the Euler–Lagrange Equation 4.31, is required for the calculation of a series of convolution integrals, such as $\int f(\bar{r})\phi(|\bar{r}-\bar{r}'|)d\bar{r}$, where $\phi(|\bar{r}-\bar{r}'|)$ is an arbitrary isotropic kernel and $f(\bar{r})$ is an arbitrary function. Thereafter, the inversion of these convolution integrals depends on the symmetry of the concrete pore [30], which normally reduces the problem to a one-dimensional one, as is the case in the slit, cylindrical, and spherical pore geometries. Hence, these convolutions are calculated by repeated one-dimensional integration taking advantage of the Gaussian quadrature [50] to increase the velocity of the numerical evaluation, for example, $g(x) = \int \phi(z-x)f(z)dz$. The integral, in principle, can be evaluated by the Gaussian quadrature method. Then the numerical quadrature replaces the integral by a summation $g(x_i) = \sum_{k=1}^{n} B_k \phi(z_k - x_i)f(z_k)$; or in matrix notation: $g_i = \sum_{k=1}^{n} B_{ik} f_k$. Then inverting the matrix (B_{ik}) we will have $f(x_k) = f_k = \sum_{i=1}^{n} B_{ki}^{-1} g_i$, that allows the numerical evaluation of the unknown function, $f(x)$ [66].

For the case of a slit pore [30] the specific excess adsorption $N_V^{ex}(H, P/P_0)$ is a kernel of the theoretical isotherm in slit-like pores of different diameters:

$$N_V^{ex}\left(H, \frac{P}{P_0}\right) = \frac{1}{H}\int_0^H (\rho(z) - \rho_{bulk})dz$$

(4.32)

where:

$\rho(z)$ is the density profile that minimizes the grand potential functional of the adsorbed fluid over a range of chemical potential values

ρ_{bulk} is the bulk gas density at a given relative pressure P/P_0 at chemical potential μ

H and z have their meaning explained in Figure 4.10

We consider, for the case of cylindrical pores, a cylindrically symmetric density profile $\rho(r)$ that minimizes the grand potential functional of the adsorbed fluid over a range of chemical potential values. For this cylindrically symmetric density distribution, $\rho(r)$, where r is a cylindrical coordinate from the pore center, the excess adsorption per unit area of a cylindrical pore is calculated from [37,38,44,45]

$$N_V^{ex}\left(D, \frac{P}{P_0}\right) = \frac{2}{D}\int_0^{D/2} \rho(r)r^2 dr - \frac{D_{in}}{4}\rho_{bulk} \tag{4.33}$$

where ρ_{bulk} is the bulk gas density at a given relative pressure P/P_0, $D_{in} = D - \sigma_{ss}$ is the internal pore diameter, which is the diameter of the cylindrical layer formed by the centers of the atoms present in the pore wall D, less the effective diameter of the atom.

For the situation of spherical pores (specifically, ink-bottle pores, considered as spherical cavities connected by narrow cylindrical neck windows) spherically symmetric density distributions $\rho(r)$, where $r = R - x$ is a radial coordinate from the pore center, in this case, the specific excess adsorption per unit of internal pore volume is calculated as [26]

$$N_V^{ex} = \frac{3\int_0^R \rho(r)r^2 dr}{\left(R - \frac{\sigma_{OO}}{2}\right)^3} - \rho_{bulk} \tag{4.34}$$

where:
  $\sigma_{OO}/2$ is the effective radius of an atom in the pore wall
  ρ_{bulk} is the bulk gas density at a given relative pressure P/P_0

The pore diameters reported below are internal, $D_{in} = 2R - \sigma_{OO}$, when expressed in dimensional units and crystallographic when expressed in units of the molecular diameter of the adsorbate $2R/\sigma_{ff}$.

As was formerly expressed, the whole adsorption isotherm $N_{exp}(P/P_0)$, that is, the experimental isotherm, is the integral of the single-pore isotherm multiplied by the PSD $\varphi_V(D)$ [41], $\int_{D_{min}}^{D_{max}} N_V^{ex}(D, P/P_0)\varphi_V(D)dD$. In order to calculate the PSD, a theoretical single-pore isotherm, $N_V^{ex}(D, P/P_0)$, which correctly describes the relationship between the isotherm and the PSD is needed [26–32].

In Table 4.3, the mode of the DFT–PSD distribution (d) and the NLDFT–pore volume (W) of silica samples are reported [23], whose adsorption isotherms were measured in a Quantachrome Instruments Autosorb-1 [1], provided these parameters with the help of the nitrogen–silica at 77 K adsorption branch kernel included in the software of the Autosorb-1 are calculated [16].

Moreover, in Figure 4.11 the NLFDT–PSD corresponding to the sample labeled 70bs2 is shown [23].

TABLE 4.3
W and d Corresponding to Silica and the MCM-41 Mesoporous Material

Sample	W (cm³/g)	d (Å)
70bs2	3.0	65
68bs1E	2.4	81
75bs1	2.7	125
79BS2	1.6	31
74bs5	1.4	61
68C	0.46	21
MCM-41	1.7	35

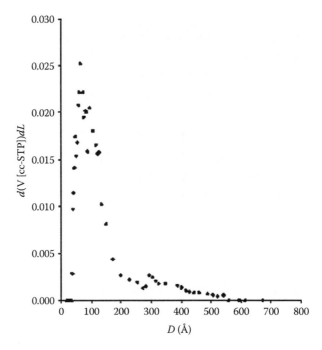

FIGURE 4.11   NLDFT–PSD (dV/dD) corresponding to the sample 70bs2.

4.4.4 SOME CONCLUDING REMARKS ABOUT THE MOLECULAR MODELS TO DESCRIBE ADSORPTION

The molecular models, specifically the DFT approaches, as was described in the previous section provide a comprehensive representation of physical adsorption in porous solids [21,23–32]. However, during the past 100 years the theory of adsorption was developed thermodynamically, kinetically, and statistically, and also using empirical isotherm equations. All these methods describe real adsorption systems for certain materials in a certain range of pressure and temperature. However, all the reported methods fail when they are applied out of their range of soundness. Nevertheless, the DFT approaches allow us to calculate the equilibrium density profile, and integrating the density profile over the pore geometry we can obtain a theoretical kernel, which allows us to generate the experimental isotherm if we know the PSD or vice versa to get the PSD if we know the experimental isotherm. On the other hand, in contrast to the classical methods the DFT methodology allows us to describe the complete range of pressures and temperatures. But the classical methods generally bring a quantifiable functional dependences between the magnitude of adsorption, and the equilibrium pressure, and temperature, whereas the molecular methods, such as the DFT approach only give a numerical kernel, which could only be calculable with the help of computers. Besides the kernel is only useful for one adsorbate–adsorbent system and for the adsorption or desorption branch of the isotherm [26].

Regarding capillary condensation the DFT approach gives the option to make a precise analysis of the process. In this regard, the NLDFT revealed that in principle, both pore condensation and pore evaporation can be associated with metastable states of the pore fluid, fact coherent with the classical van der Waals representation, which predicts that the metastable adsorption branch end at a vapor-like spinodal, where the threshold of stability for the metastable states is attained, and the fluid suddenly condenses into a liquid-like state [18]. Therefore, the desorption branch would end at a liquid-like spinodal, which is compatible to spontaneous evaporation; nevertheless, metastabilities takes place merely on the adsorption branch. Thereafter, assuming that a pore of limited length vaporization can take place by means of a thinning meniscus and a nucleation problem. And therefore, metastability is not expected to occur during the desorption. It can be obviously observed in Figure 4.12 that the experimental desorption branch is associated with the equilibrium gas–liquid transition, whereas the experimental condensation step corresponds to the spinodal spontaneous transition, and the spinodal evaporation is not experimentally observed [13,18].

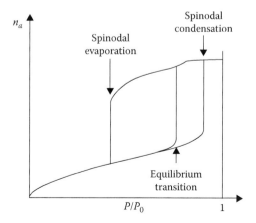

FIGURE 4.12 NLDFT description of a capillary condensation.

REFERENCES

1. R. Roque-Malherbe, *Physical Chemistry of Materials: Energy and Environmental Applications*, CRC Press, Boca Raton, FL, 2009.
2. S. Ross and J.P. Olivier, *On Physical Adsorption*, John Wiley & Sons, New York, 1964.
3. B.P. Bering, M.M. Dubinin, and V.V. Serpinskii, *J. Coll. Int. Sci.*, 38 (1972) 185.
4. M.M. Dubinin, *Prog. Surf. Memb. Sci.*, 9 (1975) 1.
5. R.M. Barrer, *Zeolites and Clay Minerals as Sorbents and Molecular Sieves*, Academic Press, London, UK, 1978.
6. K.S.W. Sing, D.H. Everett, R.A.W. Haul, L. Moscou, R.A. Pirotti, J. Rouquerol, and T. Siemieniewska, *Pure App. Chem.*, 57 (1985) 603.
7. T.L. Hill, *An Introduction to Statistical Thermodynamics*, Dover Publications, New York, 1986.
8. R. Roque-Malherbe, *J. Therm. Anal.*, 32 (1987) 1361.
9. Micromeritics, ASAP 2000, Description, 1992.
10. R. Roque-Malherbe, *KINAM*, 6 (1984) 35.
11. R. Roque-Malherbe, L. Lemes, L. López-Colado, and A. Montes, in *Zeolites'93 Full Papers Volume* (D. Ming and F.A. Mumpton, Eds.), International Committee on Natural Zeolites Press, Brockport, NY, 1995, p. 299.
12. R. Roque-Malherbe, *Mic. Mes. Mat.*, 41 (2000) 227.
13. M. Thommes, R. Kohn, and M. Froba, *J. Phys. Chem. B*, 104 (2000) 7932.
14. S.U. Rege and R.T. Yang, in *Adsorption: Theory, Modeling and Analysis* (J. Toth, Ed.), Marcel Dekker, New York, 2002, p. 175.
15. R.T. Yang, *Adsorbents: Fundamentals, and Applications*, John Wiley & Sons, New York, 2003.
16. Quantachrome, AUTOSORB-1, Manual, 2003.
17. J.U. Keller and R. Staudt, *Gas Adsorption Equilibria: Experimental Methods and Adsorptive Isotherms*, Springer-Verlag, New York, 2004.
18. S. Lowell, J.E. Shields, M.A. Thomas, and M. Thommes, *Characterization of Porous Solids and Powders: Surface Area, Pore Size and Density*, Kluwer Academic Press, Dordrecht, the Netherlands, 2004.
19. J.B. Condon, *Surface Area and Porosity Determination by Physisorption*, Elsevier, Amsterdam, the Netherlands, 2006.
20. F. Marquez-Linares and R. Roque-Malherbe, *J. Nanosci. Nanotech.*, 6 (2006) 1114.
21. F. Rouquerol, J. Rouquerol, K. Sing, P. Llewellyn, and G. Maurin, *Adsorption by Powders and Porous Solids* (2nd ed.), Academic Press, New York, 2013.
22. D.H. Everet and R.H. Ottewill, *Surface Area Determination, International Union of Pure and Applied Chemistry*, Butterworth, London, UK, 2013.
23. R. Evans, in *Fundamentals of Inhomogeneous Fluids* (D. Henderson, Ed.), Marcel Dekker, New York, 1992, p. 85.
24. C. Lastoskie, K.E. Gubbins, and N. Quirke, *J. Phys. Chem.*, 97 (1993) 4786.
25. Y. Rosenfeld, M. Schmidt, H. Löwen, and P. Tarazona, *Phys. Rev. E*, 55 (1997) 4245.
26. A.V. Neimark, P.I. Ravikovitch, M. Grun, F. Schuth, and K. Unger, *J. Colloid Interface Sci.*, 207 (1998) 159.
27. P.I. Ravikovitch, G.L. Haller, and A.V. Neimark, *Adv. Colloid Interface Sci.*, 77 (1998) 203.
28. K. Schumacher, P.I. Ravikovitch, A. Du Chesne, A.V. Neimark, and K. Unger, *Langmuir*, 16 (2000) 4648.
29. A.V. Neimark, P.I. Ravikovitch, and A. Vishnyakov, *Phys. Rev. E*, 62 (2000) R1493.
30. R.J. Dombrokii, D.R. Hyduke, and C.M. Lastoskie, *Langmuir*, 16 (2000) 5041.
31. A.V. Neimark, P.I. Ravikovitch, and A. Vishnyakov, *Phys. Rev. E*, 64 (2001) 011602.
32. P.I. Ravikovitch and A.V. Neimark, *Colloids Surf. A*, 187–188 (2001) 11.

33. A.W. Adamson and A.P. Gast, *Physical Chemistry of Surfaces* (6th ed.), John Wiley & Sons, New York, 1997.
34. E.P. Barret, L.G. Joyner, and P.H. Halenda, *J. Amer. Chem. Soc.*, 73 (1951) 373.
35. J.H. de Boer, B.C. Lippens, J.C.P. Broekhoff, A van den Heuvel, and T.V. Osinga, *J. Colloid Interface Sci.*, 21 (1966) 405.
36. T.J. Barton, L.M. Bull, G. Klemperer, D. Loy, B. McEnaney, M. Misono, P. Monson et al. *Chem. Mater.*, 11 (1999) 2633.
37. A.V. Neimark, P.I. Ravikovitch, and A. Vishnyakov, *J. Phys. Condens. Matter.*, 15 (2003) 347.
38. P.I. Ravikovitch and A.V. Neimark, *J. Phys. Chem. B*, 105 (2001) 6817.
39. J.C.P. Broekhoff and J.H. de Boer, *J. Catal.*, 9 (1967) 8.
40. J.C.P. Broekhoff and J.H. de Boer, *J. Catal.*, 9 (1967) 15.
41. B.V. Derjiaguin and M.M. Kusakov, *Proc. Acad. Sci. USSR, Chem. Ser.*, 5 (1936) 741.
42. J.C.P. Broekhoff and J.H. de Boer, *J. Catal.*, 10 (1968) 153.
43. J.C.P. Broekhoff and J.H. de Boer, *J. Catal.*, 10 (1968) 377.
44. J. Landers, G. Yu, and A.V. Neimark, *Colloids Surf. A.*, 437 (2013) 3.
45. G. Yu, M. Thommes, and A.V. Neimark, *Carbon*, 50 (2012) 1583.
46. J. Wu, *AIChE J.*, 52 (2006) 1169.
47. M. Thommes, K, Kaneko, A.V. Neimark, J.P. Olivier, F. Rodroguez-Reinoso, J. Rouquerol, and K.S.W. Simg, *Pure Appl. Chem.*, 87 (2015) 1051.
48. P.A. Monson, *Mic. Mes. Mat.*, 160 (2012) 47.
49. Y.-X. Yu, *J. Chem. Phys.*, 131 (2016) 024704.
50. G.B. Arfken and H.J. Weber, *Mathematical Methods for Physicists* (5th ed.), Academic Press, New York, 2001.
51. N.R. Draper and H. Smith, *Applied Regression Analysis*, John Wiley & Sons, New York, 1966.
52. W.A. Curtin and N.W. Ashcroft, *Phys. Rev. A*, 32 (1985) 2909.
53. E. Kierlik and M. Rosinberg, *Phys. Rev. A*, 42 (1990) 3382.
54. P. Tarazona, U.M.B. Marconi, and R. Evans, *Mol. Phys.*, 60 (1987) 573.
55. Y. Tang and J. Wu, *J. Chem. Phys.*, 119 (2003) 7388.
56. H. Pan, J.A. Ritter, and P.B. Balbuena, *Ind. Eng. Chem. Res.*, 37 (1998) 1159.
57. P. Tarazona, *Phys. Rev. A*, 31 (1985) 2672.
58. N.F. Carnahan and K.E. Starling, *J. Chem. Phys.*, 51 (1969) 635.
59. J.A. Barker and D.J. Henderson, *J. Chem. Phys.*, 47 (1967) 4714.
60. J.D. Week, D. Chandler, and H.C. Andersen, *J. Chem. Phys.*, 54 (1971) 5237.
61. P.I. Ravikovitch and A.V. Neimark, *Langmuir*, 18 (2002) 1550.
62. A.V. Neimark and P.I. Ravikovitch, *Mic. Mes. Mat.*, 44–45 (2001) 697.
63. P.I. Ravikovitch, A. Vishnyakov, R. Russo, and A.V. Neimark, *Langmuir*, 16 (2000) 2311.
64. P. Ravikovitch, A. Vishnyakov, and A.V. Neimark, *Phys. Rev. E*, 64 (2001) 011602.
65. W.A. Steele, *The Interaction of Gases with Solid Surfaces*, Pergamon Press, Oxford, UK, 1974.
66. S. Figueroa-Gerstenmaier, Development and application of molecular modeling techniques for the characterization of porous materials, Ph.D. Dissertation, Departament d'Enginyeria Quimica, Universitat Rovira I Virgili, Tarragona, Spain, 2002.

5 Molecular Transport in Porous Media

5.1 INTRODUCTION

Diffusion can be defined as the random movement of atoms, molecules, or small particles owing to thermal energy [1–12]. This effect is a universal property of matter linked to the tendency of systems to occupy all accessible states [13]. In other terms, diffusion is a natural propensity of all systems to level the concentration, if there is no whichever external influence to hinder this process, that is, atoms, molecules, or any particle that chaotically moves in the direction where less elements of their own species are placed [14].

In the case of porous materials, diffusion is a very important topic [15], given that this effect is essential in catalysis [16], gas chromatography [17], and separation processes [18], as for example, for the description of mass transfer through the packed-bed reactors used in the chemical industries [8,11]. Thereafter, a better understanding diffusion will help in the optimization and development of the industrial applications of these materials in separation, catalytic processes, along with kinetic-based pressure swing adsorption. Moreover, membrane-based separations rely on the diffusion properties of the applied membrane [19]. Consequently, to improve practical applications, diffusion must be precisely depicted.

5.2 FICK'S LAWS

The quantitative study of diffusion commenced in 1850–1855 with the works of Adolf Fick and Thomas Graham. Fick understood that diffusion obeys a law that is isomorphic to the Fourier law of heat transfer [2,4]. He proposed his first equation in order to macroscopically describe the diffusion process, provided the Fick's first equation is a linear relation between the matter flux $\bar{J}$ and the concentration gradient $\bar{\nabla}C$:

$$\bar{J} = -D\bar{\nabla}C \tag{5.1}$$

in which D is the Fickean diffusion coefficient or the transport diffusion coefficient, which is the proportionality constant, provided the units of these parameters are D ([longitude]^2/time), C (moles/volume), and J (moles/area-time) whose units in the International System (SI) for D are m^2/s, for the concentration C (mol/m^3), whereas the flux is expressed in (mol/m^{2}s).

It is necessary to recognize now that the flux along with the diffusion coefficient has to be chosen relative to a frame of reference, since the diffusion flux $\bar{J}$ gives the number of species crossing a unit area in the medium per unit of time [20]. Moreover, under the influence of external forces the particles move with an average drift velocity, v_F, which gives rise to a flux Cv_F, where C is the concentration of diffusing species. Then the total flux in this case is given by the relation [4]:

$$J = -D\nabla C + Cv_F \qquad (5.2)$$

Since, diffusion is the macroscopic expression of the tendency of a system to move toward equilibrium. Thereafter, the real driving force should be the gradient of the chemical potential μ. Hence, using the irreversible thermodynamics Onsager equation as follows [21]:

$$\overline{J} = -L\overline{\nabla}\mu \qquad (5.3)$$

in which L is the phenomenological Onsager coefficient, provided it is well identified in this equation as the cause for diffusive flow.

Moreover, the Fick's second equation:

$$\frac{\partial C}{\partial t} = -D\nabla^2 C \qquad (5.4)$$

is an expression of the law of conservation of matter, that is, $\partial C/\partial t = -D\overline{\nabla} \cdot \overline{J}$, if D do not depend on C.

Now, we can make a simple deduction of the Fick's first law (Figure 5.1), based on the random walk problem in one dimension [5]. Since, in this instance, $J_x(x,t)$, that is, the number of particles that move across unit area in unit time could then be defined as follows: $J_x(x,t) = N/A\tau$.

Since, the number of particles at position x and $(x + \Delta)$, at time t is $N(x)$, and $N(x + \Delta)$ correspondingly. Consequently, half of the particles at x at time t move to the right (i.e., to $x + \Delta$), along with the other half moving to the left (i.e., to $x - \Delta$) during the time step τ. Similarly, half of the particles at $x + \Delta$, at time t

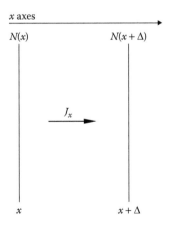

FIGURE 5.1   One-dimensional diffusion.

move to the right (i.e., to $x + 2\Delta$), whereas the other half moves to the left (i.e., to x). Afterward, the net number of particles that move from x to $x + \Delta$, when $t = t + \tau$ is $N = [N(x+\Delta,t) - N(x,t)]/2$, now

$$J_x(x,t) = -\frac{\Delta^2}{2\tau}\left(\frac{1}{\Delta}\right)\left(\frac{N(x+\Delta,t)}{A\Delta} - \frac{N(x,t)}{A\Delta}\right) = -D\frac{[C(x+\Delta)-C(x,t)]}{\Delta} \quad (5.5)$$

Then, for $\Delta \to 0$: $J_x(x,t) = \lim_{\Delta \to 0} -D[C(x+\Delta) - C(x,t)]/\Delta = -D\,\partial C(x,t)/\partial x$.

5.3 TRANSPORT, SELF, AND CORRECTED DIFFUSION COEFFICIENTS

5.3.1 TRANSPORT DIFFUSION AND SELF DIFFUSION

In Figure 5.2a the mechanism of transport diffusion is schematically described, which simply results from a concentration gradient. Whereas in Figure 5.2b the self-diffusion process is graphically represented that occurs in the absence of a chemical potential gradient describing the uncorrelated movement of a particle [12], provided the self-diffusion process is portrayed by the molecular trajectories of a large number of molecules and then determining their mean square displacement (MSD) [13].

The differences in the microphysical situations between transport and self-diffusion implies that D is the transport diffusion coefficient, and D^* is the self-diffusion coefficient, are generally different [8,12].

5.3.2 INTERDIFFUSION AND THE FRAME OF REFERENCE FOR POROUS MATERIALS

Given that the flux has to be chosen relative to a frame of reference. Hence, a porous solid media is a suitable and unambiguous frame of reference with respect to the measure of the diffusive flux. Hence, as, the framework atoms are not transported the coordinates to measure the molecular motion of the mobile species are the fixed coordinates of the solid [15]. In particular, interdiffusion occurs when two atomic or molecular species mix together. Hence, interdiffusion in one dimension of two components A and B could be described as follows [5]:

$$J_A = -D_A\frac{\partial C_A}{\partial x} \quad \text{and} \quad J_B = -D_B\frac{\partial C_B}{\partial x} \quad (5.6)$$

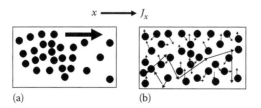

(a) (b)

FIGURE 5.2 Transport (a) and self (b) diffusion mechanisms.

If the partial molar volume of the species A and B are different ($V_A \neq V_B$), thereafter the interdiffusion of both species show the way to a net flow relative to a fixed coordinate frame of reference. Consequently, the total volumetric flux can be calculated with the following equation [22]:

$$J = V_A D_A \frac{\partial C_A}{\partial x} + V_B D_B \frac{\partial C_B}{\partial x} \tag{5.7}$$

where the plane across which there is no net transfer of volume is given by

$$J = 0 \tag{5.8}$$

Now, in the case where there is no volume change during the mixing process [4,8]:

$$V_A C_A + V_B C_B = 0 \quad \text{and} \quad V_A \frac{\partial C_B}{\partial x} + V_B \frac{\partial C_B}{\partial x} = 0 \tag{5.9}$$

After that, to fulfill Equation 5.7, together with Equation 5.8, under the condition Equation 5.9, provided that V_A and V_B are finite, results in $D_A = D_B$. Accordingly, interdiffusion can be depicted by a single diffusivity, provided that the fluxes are defined relative to the plane of no net volumetric flow, hence, considering diffusion into a porous solid, as a special case of binary diffusion, where the diffusivity of the framework atoms is zero. Thenceforth, the frame of reference is the fixed coordinate of the porous solid. For this reason, the interdiffusion coefficient is simply the diffusivity of the mobile species.

5.3.3 TRANSPORT D, CORRECTED D_0, DIFFUSION COEFFICIENTS ALONG WITH THEIR RELATIONS

As the driving force of diffusion is the gradient of the chemical potential [17], the chemical potential can be related to the concentration by considering the equilibrium vapor phase:

$$\mu_A = \mu_A^0 + RT \ln P_A \tag{5.10}$$

in which, P_A is the partial pressure of the component A, and μ_A^0 is the chemical potential of the standard state of component A. Since mass transport is described by an atomic or molecular mobility b_A, which is defined by [7]

$$\bar{v}_A = b_A \bar{F}_A \tag{5.11}$$

where $\bar{v}_A$ is the average drift velocity, and

$$\bar{F}_A = -\nabla \mu_A \tag{5.12}$$

is the force on particle A, in the case where the only driving force is a concentration gradient. In addition, it is very well known that

$$J_A = v_A C_A \tag{5.13}$$

With the help of the previous relations we could get for one-dimensional diffusion [20]:

$$J_A = RTb \left(\frac{d \ln P_A}{d \ln C_A} \right) \left(\frac{dC_A}{dx} \right) \tag{5.14}$$

Consequently

$$D_A = RTb \left(\frac{d \ln P_A}{d \ln C_A} \right) = D_0 \left(\frac{d \ln P_A}{d \ln C_A} \right) = D_0 \Psi \tag{5.15}$$

where $D_0 = RTb$.

Now, we discuss two extreme situations, that is, microporous and macroporous materials. In this regard, for a microporous adsorbent there is no clear distinction between molecules on the surface, and the molecules in the gas phase, since adsorption in microporous adsorbents is a volume-filling process [23]. Thereafter, D_0 is usually referred to as the corrected diffusion coefficient, and Ψ is called the thermodynamic correction factor, which rectifies for the nonlinearity between the pressure and the concentration of the microporous adsorbent, in which P_A is the sorptive gas pressure and C_A is the concentration of the sorbed phase in the microporous material [24].

Nevertheless, for macroporous materials, it is a distinction between molecules on the surface, and molecules in the gas phase. Subsequently, it is possible to consider that adsorption do not affect the diffusion process. Therefore, considering that the gas phase is an ideal gas, we will have

$$C_A = \frac{n_A}{V} = \frac{P_A}{RT}$$

in which n_A is the number of moles of the species A in the volume V at constant temperature T and partial pressure P_A. So, it is very easy to show that $\psi = 1$ and

$$D = RTb = D_0 \tag{5.16}$$

5.3.4   TRANSPORT D, CORRECTED D_0, SELF-COEFFICIENTS IN ZEOLITES ALONG WITH THEIR RELATIONS

Experimental studies of diffusion in zeolites have been carried out by different methods [23–36]. Concretely, the Fickean diffusion coefficients can be measured with the help of steady-state methods [24] and uptake methods [26–33]. On the other hand, the self-diffusion coefficient can be measured directly with the help of microscopic methods [36].

The corrected diffusivity is calculated in experimental studies where the transport diffusion is measured [31]. Thereafter, the calculated Fickean diffusion coefficients, D, are rectified to obtain the corrected diffusion coefficients, D_0. That is, the Fickean diffusion coefficient, D, is determined using a particular solution of the Fick's second equation. After that using this parameter is possible to applying Equation 5.18, the corrected diffusion coefficient, D_0. After that, since the corrected diffusion coefficient is approximately equal to the self-diffusion coefficient, that is, $D^* \approx D_0$ [34]. Then, the calculations are carried out with the help of the following equation [35]:

$$D^* = D(1-\theta) \tag{5.17}$$

where $\theta = (n_a/N_a)$ is the fractional saturation of the adsorbent, in which n_a is the amount adsorbed, and N_a is the maximum amount adsorbed, where the calculation of θ could be carried out, for example, with the help of the osmotic adsorption isotherm equation.

5.4 MEAN SQUARE DISPLACEMENT, THE RANDOM WALKER, AND GASEOUS DIFFUSION

5.4.1 THE MEAN SQUARE DISPLACEMENT

The general solution of the Fick's second equation in one dimension [6]:

$$\frac{\partial C}{\partial t} = D\frac{\partial^2 C(x,t)}{\partial x^2} \tag{5.18}$$

with the following initial and boundary conditions: $C(\infty,t)=0$, $C(-\infty,t)=0$, and $C(x,0) = M\delta(x)$ is [6] $C(x,t) = (1/(4\pi Dt))^{1/2} \exp(-x^2/4Dt)$. Now, it is possible to define $P(x,t) = C(x,t)/M$, where $M = \int_{-\infty}^{\infty} C(x,t)dx$ as the probability to find a diffusing particle at the position, x, during the time, t, if this particle was at $x = 0$ at $t = 0$. Thus, it is straightforward to demonstrate that the one-dimensional MSD is given by the Einstein equation:

$$\left\langle x^2 \right\rangle = \int x^2 P(x,t)dx = 2Dt \tag{5.19}$$

Moreover, the same result could be obtained for isotropic diffusion from a point source in three-dimensional space, that is, if $P(\bar{r},t)$ is the probability to find a diffusing particle at the position $\bar{r}$, during the time t, if this particle was at $\bar{r} = 0$ at $t = 0$ is given by

$$P(\bar{r},t) = \frac{C(\bar{r},t)}{M} = \left(\frac{1}{4\pi Dt}\right)^{3/2} \exp\left(-\frac{\bar{r}^2}{4Dt}\right) \tag{5.20}$$

that is, a Gaussian function known as the propagator, so, it is easy to corroborate that the three-dimensional MSD can be calculated as follows [3]:

$$\langle \bar{r}^2 \rangle = \int \bar{r}^2 P(\bar{r},t) dxdydz = 6Dt \tag{5.21}$$

5.4.2 GASEOUS DIFFUSION AND THE RANDOM WALKER

Now, we will make the calculations with the help of the one-dimensional random walker (ODRW); hence, within the framework of this model, when the steps of the ODRW are $\pm l$ in the x direction [5] together with the fact that the time between jumps is τ. For that reason, the jump frequency is $\Gamma = 1/\tau$. Subsequently, if $\Gamma = $ constant, together with the fact that the direction of the jumps is not correlated, that is, $\sum \langle l_i l_j \rangle = 0$, thenceforward for N steps, in which, $N = t/\tau$, the MSD is given by [12]

$$\langle x^2(N) \rangle = \left\langle \sum l_i \right\rangle^2 = \sum \langle l_i^2 \rangle + \sum \langle l_i l_j \rangle = \sum \langle l_i^2 \rangle \tag{5.22}$$

and so

$$\langle x^2(N) \rangle = Nl^2 = \frac{tl^2}{\tau} \tag{5.23}$$

Henceforth:

$$\langle x^2(N) \rangle = 2D^*t \tag{5.24}$$

In which, $D^* = l^2/2\tau$ is the self-diffusion coefficient, in general, different to the transport diffusion coefficient, specifically, $D \neq D^*$, since the portrayed processes are physically different, as was previously explained (Figure 5.2).

Now, it is possible to state that during gaseous diffusion, given that the consecutive displacement of the diffusing particle between collisions is statistically independent, we can denote the z component of the ith displacement of the diffusing particle by ξ_i [11]. Then, if the starting point of the diffusion process is $z = 0$, at that point the z component of the gaseous molecule during the diffusion process is [15]

$$z = \sum_{i=1}^{N} \xi_i \tag{5.25}$$

Therefore, as a consequence of the random direction of each displacement the mean of the distances between collisions is $\langle \xi_i \rangle = 0$. Hence, the mean displacement is $\langle z \rangle = 0$. In a quantitative fashion, the MSD in the z direction is given by

$$\langle z^2 \rangle = \left\langle \sum_1^N \xi_i \right\rangle^2 = \sum_1^N \langle \xi_i^2 \rangle + \sum_i \sum_j \langle \xi_i \xi_j \rangle = \sum_1^N \langle \xi_i^2 \rangle \tag{5.26}$$

Then owing to the statistical independence, we have $\langle \xi_i \xi_j \rangle = \langle \xi_j \xi_i \rangle = 0$. Consequently, $\langle \xi_i^2 \rangle = \langle \xi^2 \rangle$, where $\langle \xi^2 \rangle$ is the MSD per step, therefore:

$$\langle z^2 \rangle = N \langle \xi^2 \rangle \tag{5.27}$$

But provided that, $\xi = v_z t$, thenceforward

$$\langle \xi^2 \rangle = \langle v_z^2 \rangle \langle t^2 \rangle \tag{5.28}$$

Moreover, at this point, within the framework of the collision time approximation, we could state that the mean time between collisions $\langle t \rangle$, or the relaxation time of the molecule $\tau = \langle t \rangle$, can be calculated with the following expression [12]:

$$\langle t \rangle = \tau = \int_0^\infty \frac{t \exp(-t/\tau)}{\tau} dt \tag{5.29}$$

Accordingly [15],

$$\langle t^2 \rangle = \int_0^\infty \frac{t^2 \exp(-t/\tau)}{\tau} dt = 2\tau^2 \tag{5.30}$$

Further, by symmetry [3]: $\langle v_z^2 \rangle = (1/3) \langle v^2 \rangle$; for this reason,

$$\langle \xi^2 \rangle = \frac{2}{3} \langle v^2 \rangle \tau^2 \tag{5.31}$$

But given that $N = t/\tau$, applying Equation 5.27 we have

$$\langle z^2 \rangle = \left(\frac{2}{3} \langle v^2 \rangle \tau \right) t \tag{5.32}$$

Applying now the definition of MSD for the self-diffusion of a gaseous molecule, that is, $\langle z^2 \rangle = \int z^2 P(z,t) dz = 2D^* t$ the self-diffusion coefficient for a diffusing gas is given by [12]

$$D^* = \frac{1}{3} \langle v^2 \rangle \tau \tag{5.33}$$

5.5   TRANSPORT MECHANISMS IN POROUS MEDIA

In accordance with the classification scheme proposed by the International Union of Pure and Applied Chemistry (IUPAC), pores are distributed into three groups based on size, that is, (1) the pores that are bigger than 50 nm are labeled macropores, (2) whereas entities in the range between 2 and 50 nm are called mesopores, and (3) the pores that are less than 2 nm are named micropores [9]. This array reflects the dissimilarity in the

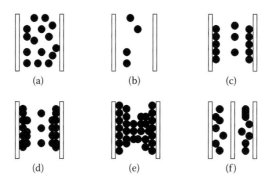

FIGURE 5.3 Gaseous flow (a), Knudsen flow (b), surface diffusion (c), multilayer diffusion (d), capillary condensation (e), and configurational diffusion (f).

kind of forces controlling the adsorption. Concretely, in macropores, the influence of pore walls is small, whereas in the mesopore range, surface forces along with capillary forces turn out to be significant, whereas for micropores, the overlapping surface forces of opposed pore walls produce an intense interaction [8,10].

To be definite it is necessary to state now that there exist various diffusion types, that is, gaseous [36], Knudsen [37], and surface [38], multilayer [39] along with capillary condensation and configurational diffusion [40] (Figure 5.3).

Gaseous flow, which takes place when the pore diameter is larger than the mean free path of the transported molecule, is schematically described in Figure 5.3a. Then collisions between molecules are more frequent than those between molecules and the pore surfaces [41], as the pore dimension decreases (Figure 5.3b), or the mean free path increases, due to pressure lowering. Then the total permeation is a linear combination of gas flow along with the diffusing species that are likely to strike more with the pore walls that lie between them where molecules exchange place during diffusion. Thereafter, molecules are transported virtually individually from one another according to the Knudsen flow [42]. Moreover, surface flow (Figure 5.3c) is achieved when the diffusing molecules can be adsorbed on the pore surfaces [25]. Besides, an annex of this mechanism is a multilayer diffusion (Figure 5.3d), a transition flow regime between the capillary and surface flows, that is, if capillary condensation is reached (Figure 5.3e), then the transported molecule is condensed within the pore, filling the pore, then evaporating at the other end of the pore [26]. The final mechanism (Figure 5.3f) is configurational diffusion taking place when pore diameters are small enough to let only small molecules to diffuse along the pores while preventing the larger one to get into the pores [15,25]. The most important diffusion mechanism in microporous materials is the configurational one, characterized by very small diffusivities ($10^{-12}-10^{-18}$ m^2/s) with a strong dependence on the size and the shape of the guest molecules, high activation energies (10–100 kJ/mol), along with strong concentration dependence [7,12].

To conclude, in this section, in practical terms, the relation between diffusion and pore size will be analyzed. In this regard, normally we will have fundamentally three regimes with different diffusivities according to the pore diameter (Figure 5.4) [43], that is, for macropores, in general the collisions between the molecules occur

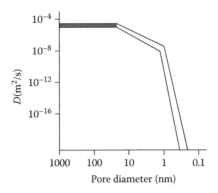

FIGURE 5.4   Diffusivity versus pore diameter.

much more frequently than collisions with the wall, then molecular diffusion is the dominant mechanism. For this case, the values of the diffusion constants are around $10^{-5}\,\text{m}^2/\text{s}$ [24]. At the same time as the size of the pores decreases, the number of collisions with the wall increases; then, Knudsen diffusion takes control. Consequently, the mobility turns to depend on the dimensions of the pore [36]. At even smaller pore sizes, in the range of 2 nm or less, that is, when the pore diameter turns out to be similar to the size of the molecules, the molecules will constantly experience the interaction with the pore surface. Consequently, diffusion in the micropores of a zeolite or related materials, as was previously stated, typically takes place in the configurational diffusion regime [11,12].

5.6   VISCOUS, KNUDSEN, AND TRANSITION FLOWS

The mean free path of the molecules λ can be related to the temperature T, and pressure P, by means of the following equation [44]:

$$\lambda = \frac{kT}{\sqrt{2}\pi\sigma_c^2 P} = \frac{1}{\sqrt{2}\pi\sigma_c^2 C} \tag{5.34}$$

where:
   σ_c is the collision diameter of the molecules
   k is the Boltzmann constant
   C is the gas concentration

In this regard, to categorize the diverse types of gas-phase flow, the ratio between the mean free path λ, and the characteristic length of the flow geometry L, commonly referred as the Knudsen number K_n, expressed as $K_n = \lambda/L$ is used. That is, according to the magnitude of K_n, three main flow regimes can be described, that is, viscous flow when $K_n \ll 1$, Knudsen flow for $K_n \gg 1$, and transition flow in the case when $K_n \approx 1$. More precisely, for $K_n < 10^{-2}$, the continuum hypothesis is correct. On the other hand, for $K_n > 10$, the continuum approach fails completely. Hence, the regime can be described as a free molecular flow [4]. Thereafter, since the mean free path of the molecules is far greater than the characteristic length scale, molecules are reflected from a solid surface travel, on average, many length scales previous to colliding with other molecules. In brief, in the large K_n region, continuum models, such as the compressible Navier–Stokes equation do not hold. In other words, when λ becomes comparable to L, the linear transport relationship for mass, diffusion, viscosity, and thermal conductivity is no longer valid. For this reason, discrete models are proposed to examine the behavior of the rarefied gas flow [14].

Now, we will analyze the physical reasons for the transition from the gaseous to the Knudsen flow. In this sense, to estimate the mean free path the kinetic theory of gases supposes that molecules collide basically with other molecules, not with the walls of the gas container. This hypothesis is valid if the gas concentration is sufficiently large. However, if the gas is made to dilute, this condition is no longer valid. Therefore, if the total collision probability per unit time is [2] $1/\tau_T = 1/\tau + 1/\tau_S$ where $1/\tau$ is the collision probability per unit time between molecules, and $1/\tau_S$ is the collision probability per unit time between molecules and the container surface, the kinetic theory of gases states that [4]

$$\frac{1}{\tau} = \frac{\langle v \rangle}{\lambda} \tag{5.35}$$

While:

$$\frac{1}{\tau_S} = \frac{\langle v \rangle}{L} \tag{5.36}$$

where L is the smallest dimension of the container, or as was previously affirmed, the characteristic length of the flow geometry. If we define now a resultant mean free path, λ_0, as

$$\lambda_0 = \langle v \rangle \tau_0 \tag{5.37}$$

Then, substituting Equations 5.35 through 5.37 in $1/\tau_T = 1/\tau + 1/\tau_S$, we will get

$$\frac{1}{\lambda_0} = \frac{1}{\lambda} + \frac{1}{L} = \sqrt{2}\pi C \sigma_c + \frac{1}{L}$$

In the expression for the viscosity of an ideal gas:

$$\eta = \frac{1}{3} C \langle v \rangle M \lambda = \frac{M \langle v \rangle}{3(2)^{1/2} \pi \sigma_c} \tag{5.38}$$

It is evident that, η is independent on C. However, if we use λ_0 instead of λ in Equation 5.38, an approximate description of the role of the collision on the walls will be obtained, in which, when $C \to 0$, $\lambda_0 \to L$, and subsequently, $\eta \propto C$, fact meaning that for Knudsen gas the concept of viscosity tends to lose its meaning [3].

5.7   VISCOUS AND KNUDSEN FLOWS IN MODEL POROUS SYSTEMS

5.7.1   VISCOUS FLOW IN A STRAIGHT CYLINDRICAL PORE

The gaseous self-diffusion coefficient is $D^* = 1/3\langle v^2 \rangle \tau = 1/3 \langle v \rangle \lambda$, provided that, $\langle v^2 \rangle = \langle v \rangle^2$ and $\langle v \rangle \tau = \lambda$. Moreover, since $\langle v \rangle = (8kT/\pi M)^{1/2}$ is the mean velocity, where M is the molecular weight of the gas molecule and $\lambda = kT/2^{1/2}\pi\sigma_c^2 P = 1/2^{1/2}\pi\sigma_c^2 C$ is the mean free path. Thereafter [36]:

$$D^* = \frac{2}{3\pi\sigma_c^2}\left(\frac{kT}{P}\right)\left(\frac{kT}{\pi M}\right)^{1/2} \tag{5.39}$$

Now if the previously discussed conditions for viscous diffusion are satisfied for a cylindrical macropore, that is, a pore with a diameter larger than 50 nm, in this situation, collisions between diffusing molecules will take place considerably more often than collisions between molecules and the pore surface. Then, the pore surface effect is negligible. Consequently, diffusion will take place by basically the same mechanism as in the bulk gas. Therefore, the pore diffusivity is subsequently equal to the molecular gaseous diffusivity (Equation 5.39).

The previous situation for gaseous flow is always valid for a liquid phase. Consequently, since the diffusion coefficient in liquid phase is [10]

$$D^* = \left(\frac{8kT}{\pi M}\right)^{1/2}\left(\alpha_T \sigma_m T\right) \tag{5.40}$$

where:

σ_m is the Lennard–Jones length constant for the diffusing molecule
α_T is the thermal expansion coefficient

As a result, the pore diffusivity is subsequently equal to the molecular liquid diffusivity (Equation 5.40).

5.7.2   KNUDSEN FLOW IN A STRAIGHT CYLINDRICAL PORE

In the Knudsen regime, the rate at which momentum is transferred to the pore walls surpasses the transfer of momentum between the diffusing molecules. The rate at which molecule collides with the unit area of the pore wall is $\omega = C\langle v \rangle/4$. Now if the average velocity in the flow direction is $\langle v_z \rangle$, subsequently the momentum flux per unit time to an element of area of the wall in the z direction is (Figure 5.5) [12] $F_z = c\langle v \rangle/4(m\langle v_z \rangle)(2\pi r dz)$, as this force must be equalized:

$$F_z = -\pi r^2 dP$$

where P is the gas pressure in the element of volume (Figure 5.5). At this time, it is possible to calculate the flux in the pore with the following expression:

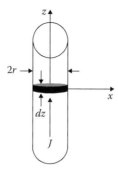

FIGURE 5.5   Knudsen flow in a straight cylindrical pore of diameter $d_P = 2r$.

$J = C \langle v_z \rangle = -D_K \, \partial C / \partial z$, where D_K is the Knudsen diffusivity. It is simple to calculate D_K knowing from the kinetic theory of gases the value of $\langle v \rangle = \left(8kT / \pi M \right)^{1/2}$, and that $P = kCT$. Accordingly [4],

$$D_K^* = \frac{d_P}{2} \left(\frac{\pi kT}{2M} \right)^{1/2}$$

(5.41)

in which d_P is the pore diameter. Thereafter, if the previously discussed conditions for Knudsen diffusion are satisfied for a mesopore, then the diffusion coefficient for the Knudsen flow in a straight cylindrical mesopore is described by Equation 5.41.

5.8   TRANSPORT IN REAL POROUS SYSTEMS

5.8.1   MEMBRANES

Membranes have been applied for the treatment of a variety of fluids ranging from gases, waste water, seawater, milk, yeast suspensions along with other fluids [26,38,39,45–49], where a membrane is a perm-selective barrier between two phases that are able of being permeated owing to a driving force, such as pressure, concentration, or electric field gradient [26]. Membranes are classified as organic or inorganic; porous and microporous; and symmetric and asymmetric. In particular, porous inorganic membranes are made of alumina, silica, carbon, zeolites, and other materials and are generally prepared by the slip-coating method, ceramic technique, and sol-gel method [46]. These elements are employed in catalytic reactors [16], gas separation [17–19], gasification of coal together with water decomposition, and other applications [45–48], where the materials applied for inorganic porous membrane synthesis experience phase transformations, structural changes, and sinterization at high temperature [38].

Concretely, microporous along with mesoporous membranes are used in microfiltration and ultrafiltration, and it supports for microporous material layer such as a zeolite thin film. In addition macroporous and mesoporous membranes are used as a support for the synthesis of asymmetric membranes with a dense thin film [8,39], where the mechanical strength of self-supported microporous inorganic membranes is normally inadequate. Consequently, mechanically robust porous substrates are applied as supports, that is, the macro- and/or mesopores of the support are covered with films of a micropore material, where the support gives mechanical strength, whereas the zeolite is intended to carry out selective separations, provided diverse procedures are used to deposit microporous thin films, that is, sol-gel, pyrolysis, and deposition techniques [41,46].

Nowadays, the synthesis along with the characterization of membranes is a growing activity in materials science. However, much work remains to be done for the commercial utilization of membranes; it est., is required to increase the membrane permeance, solve the brittleness problems together with the growth of the membrane area per unit volume, to be capable to scale-up the process, and to increase the membrane area per unit volume. Moreover, it is also essential to know and model the process of gas transport through macroporous, mesoporous, and microporous membranes [41–49].

5.8.2 PERMEATION MECHANISMS IN POROUS MEMBRANES

The permeation rate together with the selectivity of porous membranes is determined by their microstructure, that is, pore size, pore size distribution (PSD), tortuosity along with characteristics of the permeating molecules such as mass and size in conjunction with their interaction with the membrane pore walls [47,48]. In this regard, during transport of gases through porous membranes, given that the pressure is the driving force of the process, the gaseous molecules will be transported from the high-pressure to the low-pressure side of the membrane. Particularly, in Figure 5.6 [19], the scheme of a permeation cell is represented, where the membrane compartment is coupled with two-pressure transducers, $M1$ and $M2$, to measure the reject pressure (P_1) along with the permeate pressure (P_2), and a mass flow meter (F) to measure the flux (J) passing through the membrane.

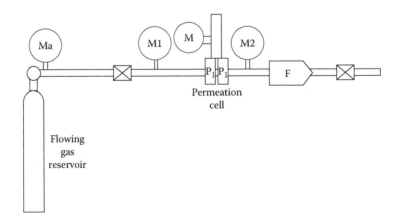

FIGURE 5.6 Permeation test facility diagram.

For one component gases accompanied by a linear pressure drop across the porous media, this transport process follows the Darcy Law [39]:

$$J = B\left(\frac{\Delta P}{l}\right) = \Pi\,\Delta P \tag{5.42}$$

$$J = \frac{Q}{V_m A} \tag{5.43}$$

$$\Pi = \frac{B}{l} \tag{5.44}$$

where:
A is the effective membrane area
B is the permeability [mol/m s Pa]
J is the molar gas flow [mol/m^2 s]
l is the membrane thickness
$\Delta P = P_1 - P_2$ is the transmembrane pressure [Pa]
Π is the gas permeance [mol/m^2 s Pa]
Q is the gas filtrate flux [m^3/s]
V_m is the molar volume of the flowing gas [m^3/mol], since for an ideal gas
$\quad V_m = V/n = RT/P$

At this point, it is necessary to state that during transport of gases through porous membranes, various mechanisms occur depending on the temperature, pressure, and membrane pore diameters. In this regard (Figure 5.3), gaseous laminar flow takes place in wide pores, whereas Knudsen flow is in narrower pores; further, surface diffusion, multilayer diffusion, and capillary condensation [41]. Finally, in microporous membranes, configurational diffusion is the transport mechanism [38].

To predict the flow regime, a dimensionless parameter labeled Reynold Number $Re = vL/\eta$, in which L is the characteristic length, v is the flow velocity, and η is the dynamic viscosity. Subsequently, if $Re < 2000$ the flow is laminar. Meanwhile, for a Knudsen flow in a pore, the number $K_n = \lambda/L$, where λ is the mean free path, and L is the characteristic length of the flow geometry, for $K_n \gg 1$ is established the Knudsen flow [44].

5.8.3 Viscous Flow in Membranes

In the simplest case of a flow through a straight cylindrical pore, the form of the Darcy law, based on the Hagen–Poiseuille equation describe the process by means of the following equation [14]:

$$J_v = \left(\frac{r^2}{8\eta V_m}\right)\left(\frac{\Delta P}{l}\right) \tag{5.45}$$

where r is the pore radius. Thereafter, the Darcy Law for laminar flow in a real macroporous membrane is described by the following equation [48]:

$$J_v = B_v \left(\frac{\Delta P}{l} \right) \tag{5.46}$$

where:

$$B_v = \frac{k}{\eta V_m} \tag{5.47}$$

in which, k is the permeation factor in m^2, and η is the dynamic viscosity of the gas in Pa.s. Now, for the description of this flow, the Carman–Kozeny equation can be employed as the Hagen–Poiseuille equation is not valid, given that normally inorganic macroporous and mesoporous membranes are obtained by the sinterization of packed quasi-spherical particles, then forming a random pore structure. In that case, the Carman–Kozeny equation for the permeability factor for a membrane formulated with pressed spherical particles is [50]

$$k = \frac{\varepsilon \, d_v^2}{16 \, C} \tag{5.48}$$

where:
  $\varepsilon = 1 - \rho_A / \rho_R$; therefore, $k \approx \varepsilon d_v^2 / 77$
  $C = 4.8 \pm 0.3$ is the Carman–Kozeny constant
  d_v is the membrane pore diameter
  ε is the membrane porosity
  ρ_A is the apparent membrane density (g/cm^3)
  ρ_R is the real membrane density (g/cm^3) [19]

Moreover, the *Dusty Gas Model* (DGM) also accounts for the viscous mechanisms in real porous systems. In the framework of this model, the permeability for the viscous flow is given by the following expression [51]: $B_v = \varepsilon d_p^2 / 8\tau\eta V_m$, in which ε is the porosity, τ is the tortuosity, and d_p is the average pore diameter of the porous medium, in view of the fact that the viscous diffusion mechanism is acceptable for the transport process in the liquid phase. Then if we have a liquid filtration process through a porous membrane, the following form of the Carman–Kozeny equation [50]:

$$J_v = \left(\frac{\varepsilon^2 \rho}{K \eta S^2 (1 - \varepsilon)^2} \right) \left(\frac{\Delta P}{l} \right) \tag{5.49}$$

in which ε is the porosity, S is the pore area, K is a constant, and ρ is the molar density.

5.8.4   KNUDSEN FLOW IN MEMBRANES

While the membrane pore dimensions diminish, or the mean free path of the molecules augment, the permeating particles are inclined to strike more with the pore walls than among themselves. Thenceforth, the Knudsen flow regime is established [3]. Therefore, the molar gas flow J for the Knudsen flow in a straight cylindrical mesopore of length l and $\Delta P = P_1 - P_2$ transpore pressure is given by [24]

$$J_K = D_K \left(\frac{\Delta P / kT}{l} \right) \tag{5.50}$$

Now, given that $D_K^* = d_P / 2 (\pi kT / 2M)^{1/2}$ expresses the diffusivity of a Knudsen gas, in the simple case of a straight cylindrical mesopore, then for a real mesoporous membrane, which is formed by a complicated pore network, the expression for the permeation flux across the membrane is given by [41]

$$J_K = \left(\frac{G}{(2MkT)^{1/2}} \right) \left(\frac{\Delta P}{l} \right) \tag{5.51}$$

where G is a geometrical factor. Thereafter, if M is expressed in molar units thenceforward:

$$J_K = \left(\frac{G}{(2MRT)^{1/2}} \right) \left(\frac{\Delta P}{l} \right) \tag{5.52}$$

In this case, the geometrical factor, G, could be calculated with the help of a simple model where it is assumed that the diffusivity in a porous material D, can be linked to the diffusivity (in analogous physical conditions) inside a straight cylindrical pore D_K with diameter equal to the mean pore diameter of the pore network by a simple factor (ε_P / τ), explicitly:

$$D = \left(\frac{\varepsilon}{\tau} \right) (D_K) \tag{5.53}$$

in which ε is the porosity that takes into consideration the fact that transport only takes place throughout the pore and not through the solid matrix. Meanwhile, the other effects are grouped together into a parameter called the tortuosity factor, τ; accordingly,

$$G = \frac{d_P \varepsilon (\pi)^{1/2}}{2\tau} \tag{5.54}$$

5.8.5 Transitional Flow

Once the mean free path of the gas molecules is similar with the pore diameter, transference of momentum between diffusing molecules and the pore wall are both significant. Thereafter, in this instance, a transitional flow occurs by the joint effects of both the viscous and Knudsen mechanisms, in which for one-component gases together with a linear pressure drop across the porous media, the total flux J_t can be written as follows [39]:

$$J_t = J_v + J_K = \left(\frac{kP}{\eta RT} \right)\left(\frac{\Delta P}{l} \right) + \left(\frac{G}{(2MRT)^{1/2}} \right)\left(\frac{\Delta P}{l} \right) \tag{5.55}$$

where $V_m = V/n = RT/P$. Afterward, the expression for the total permeability B_t can be obtained from

$$B_t = \frac{J_t}{(\Delta P/l)} = aP + b \tag{5.56}$$

Therefore, from Equation 5.56, the relation between B_t and P has a positive intercept, which is related with the Knudsen flow permeability, whereas the slope is regulated by the gaseous viscous flow. As a result, the increase in the permeability with pressure is a sign that gaseous viscous flow might be responsible for mass transfer.

5.8.6 Surface Flow in the Adsorbed Phase

As soon as the temperature of the gas drops, hopping is the main diffusion mechanism, since it takes into account that molecules jumps between sites on the surface with a specific velocity. Nevertheless, the mostly applied mechanism is the random walk model, which uses the two-dimensional form of Fick's law along with the hydrodynamic model, which supposes that the adsorbed gas can be considered as a liquid layer that slides along the surface under the effect of a pressure gradient, provided it is for a low surface concentration in which the surface flux can generally be [8,12,25]

$$J_s = -D_s \left(\frac{dC_s}{dx} \right) \tag{5.57}$$

where D_s is the surface diffusivity, which is a function of the surface concentration C_s, which also follows the Arrhenius equation, where the activation energy for the surface diffusion process is E_s.

Moreover, in general, it is recognized that the surface flux makes an additional contribution to the gas-phase transport, then the total permeation is a linear combination of gas along with surface permeation, where in the case of very low pressure, in which gaseous viscous flow is very small and the adsorption isotherm is linear, the total flux is the sum of the Knudsen flux together with the surface diffusion flux [52]:

$$J_{Ks} = J_K + J_s = \left(\frac{G}{(2MRT)^{1/2}}\right)\left(\frac{\Delta P}{l}\right) + D_s K \frac{\Delta P}{l} \tag{5.58}$$

where the permeability for this combined Knudsen and surface diffusions is

$$B_{Ks} = \frac{J_{Ks}}{(\Delta P/l)} = \left(\frac{G}{(2MRT)^{1/2}}\right) + D_{sK} \tag{5.59}$$

In Figure 5.7 [53] a diagram of the total permeation as a function of temperature for the combination of gas phase along with surface flows is shown. In the graphics it is observed that the reduction of the surface diffusivity with temperature is more rapid than that of Knudsen diffusion result occurring in view of the fact that the heat of adsorption is bigger than the activation energy for surface diffusion. Consequently, if the surface diffusion is to be removed, experiments are usually carried out at high temperatures.

5.8.7 EXPERIMENTAL STUDY OF ZEOLITE-BASED POROUS CERAMIC MEMBRANES

Porous zeolite-based membranes were prepared by applying the ceramic method by the thermal transformation of a natural clinoptilolite at 700°C–800°C.

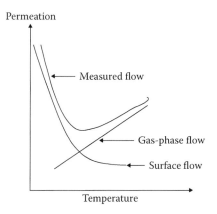

FIGURE 5.7   Total flow as function of temperature for gas and surface flow.

Thereafter, in the membranes produced, configurational diffusion is not feasible, because during the thermal treatment of the zeolite to obtain the ceramic membrane, the clinoptilolite framework collapsed [8]. Further, the particles used for the sintering process to make the tested membranes have a size of 220–500 mm. Subsequently, in these membranes only macropores are present [19]. Moreover, adsorption and capillary condensation on the surface of the membrane will be weak, owing to the relatively high temperature (300 K) and rather low pressures (0.2–1.4 MPa). Therefore, in this kind of permeation, only Knudsen or gaseous flow can take place.

To perform the permeation test of these membranes, permeability (B) and permeance (P) of H_2 were measured using Darcy's law along with the Carman–Kozeny equations [46–48] in the facility shown in Figure 5.6. A more detailed description of the permeation cell has been depicted in Figure 5.8.

Now, applying Equations 5.42 through 5.48; finally, in Figure 5.9 the results of two H_2 permeation experiments in membranes are reported, obtained using two-particle diameters, that is, 500 µm (Figure 5.9a) and 220 µm (Figure 5.9b) [19].

Meanwhile, in Tables 5.1 and 5.2 are reported B, P, and d_r.

While, in Table 5.2 the estimated membrane pore diameter (d_v) is presented.

Further, in Table 5.3 the calculated values for the mean free path (λ) of the hydrogen molecules are displayed, resulting from the data reported in Tables 5.2 and 5.3, facts that allowed us to conclude that Knudsen flow was not possible, since $d_v \gg \lambda$.

Consequently, the determining process is gaseous laminar flow through the membrane pores, a feature that made the application of the Darcy law for gaseous laminar flow feasible, along with the Carman–Kozeny relation equation because the Hagen–Poiseuille equation is not suitable [19].

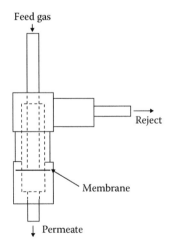

Feed gas

Reject

Membrane

Permeate

FIGURE 5.8   Permeation cell.

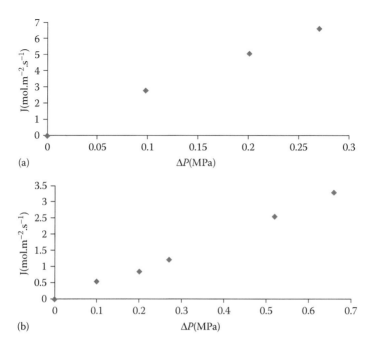

FIGURE 5.9 Permeation study of H_2 in membranes produced with powders of 500 μm (a) and 220 μm (b) of particle diameter.

TABLE 5.1
Hydrogen Permeability [B] and Permeance [P] in the Studied Membranes

Sample d_p (μm)	Treatment Temperature (°C)	Treatment Time (h)	$B \times 10^8$ (mol.m^{-1}s^{-1}.Pa)	$\Pi \times 10^6$ (mol.m^{-2}.s^{-1}.Pa)
220	700	2	1.1	4.1
500	700	2	4.9	18.1
220	800	1	1.4	5.2
500	800	1	6.8	25.1

TABLE 5.2
Estimated Membrane Pore Diameters, d_v

Sample d_p (μm)	Treatment Temperature (°C)	Treatment Time (h)	d_v (μm)
220	700	2	35
500	700	2	79
220	800	1	36
500	800	1	82

TABLE 5.3
Mean Free Path (λ) of Hydrogen at
$T = 300$ K and Different Pressures

Pressure (MPa)	λ (nm)
0.2	55.4
0.4	27.7
0.6	18.4
0.8	13.9
1.0	11.1

5.9 DIFFUSION IN ZEOLITES AND RELATED MATERIALS

Zeolites and related materials are inorganic, microporous, and crystalline solids that are widely used in the chemical and the petroleum industries as catalysts, sorbents, ion exchangers, and cation conductors [8,54,55]. Transport or diffusion of adsorbed molecules through the pores together with the cavities within the crystals of these materials plays a dominant role [12,15,24,25,27,56,57], given that the crystal lattice of the zeolites and the related materials are formed by four connected tridimensional frameworks of T atoms bridged by oxygen atoms [58–60].

Concretely, zeolites form frameworks with cavities and channels of molecular dimensions [58], which are normally negatively charged. Hence, to compensate the negative charge arising in the zeolite frameworks, cations are placed at different sites [59] within the cavities and channels of the zeolite [60], provided these properties are attractive for many new industrial processes [54]. However, our capability to select and adapt zeolites for particular processes is restricted by a deficient knowledge of the molecular-level interactions and their effect on macroscopic phenomena [55].

In many processes, where zeolites are applied, the rate of diffusion of adsorbed molecules inside the zeolite pore system plays a significant, even critical, role in determining the overall observed performance of the whole process [61]. Nevertheless, diffusion in zeolites is in general poorly understood owing to the sensitivity of zeolite diffusivities on the dimensions of the diffusing molecules, and zeolite pores and cavities, as well as on energetic interactions, such as those between adsorbates and the zeolitic framework, and charge compensating cations [31,32,34]. Particularly multicomponent diffusion has not received the necessary consideration in comparison to single-component diffusion [62], even though it is, certainly, such a multicomponent behavior that is of importance in the practical applications of zeolites and related materials [33,63].

5.9.1 MODEL TO DESCRIBE DIFFUSION IN ZEOLITES

In the case of microporous materials, specifically zeolites, configurational diffusion is the term coined to describe molecular transport, where this diffusion regime is characterized by very small diffusivities (10^{-12} to 10^{-18} m²/s), a strong dependence on the size and shape of guest molecules, high activation energies (10–100 kJ/mol), and a strong concentration dependence [24,43], where the strong interactions between the adsorbed species and the zeolite lattice give rise to configurational diffusion [64].

Mass transport in microporous media takes place in an adsorbed phase. Consequently, when a molecule diffuses inside a zeolite channel, it becomes attracted to and repelled by different interactions, such as the dispersion energy, repulsion energy, polarization energy, field dipole energy, field gradient quadrupole, sorbate–sorbate interactions, and the acid–base interaction with the active site if the zeolite contains hydroxyl bridge groups [54]. Accordingly, this transport can be pictured as an activated molecular hopping between fixed sites [24,65,66]. Therefore, during the transport of gases through zeolites, both diffusion between localized adsorption sites and the activated gas translation diffusion will contribute to the overall process. Hence, it is possible to consider that the adsorbate–adsorbent interaction field inside these structures is characterized by the presence of sites of minimum potential energy for the interaction of adsorbed molecules with the zeolite framework and the charge compensating the cations [8], a simple model of the zeolite–adsorbate system is that of the periodic array of interconnected adsorption sites, where molecular migration at adsorbed molecules through the array is assumed to proceed by thermally activated jumps from one site to an adjacent site and can be envisaged as a sort of lattice gas. In this sense, in Figure 5.10 it is schematically described that the adsorption together with the diffusion in zeolites within the modified lattice-gas model, taking into consideration the crystalline structure of the zeolite, the interaction between adsorbed molecules, and the

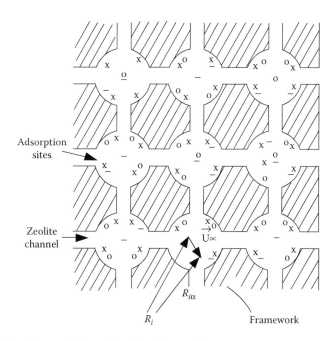

FIGURE 5.10 Representation of a zeolite framework.

transition of adsorbed molecules between different adsorption sites in the same unit cell and between different unit cells [67].

The model describes the zeolite as a three-dimensional array of N identical cells i, centered at R_i, each containing N_O identical sites localized at $R_{i\alpha} = R_i + U_\alpha$, where the potential energy is a minimum. If a molecule is localized at $R_{i\alpha}$, its energy would be $-\varepsilon$ and the interaction energy between molecules localized at different sites, that is, at $R_{i\alpha}$ and $R_{i\beta}$ is $-U_{\alpha\beta}$. Thereafter, the movement of molecules between sites, that is, jumping of molecules inside a cell, can be taken into account along with the transport between cells, through zeolite channels and cavities [67]. Thereafter, the solution of the motion equation for a molecule, in the analyzed system, points to the existence of energy bands consisting on the one hand of $N(N_0 - 1)$-fold degenerate state with energy E_2, in which molecules are adsorbed in a site, together with N delocalized states in a state with energy E_1, where molecule moves throughout the zeolite framework, jumping from site-to-site through the zeolite channels and cavities. Moreover, between the adsorption the diffusion states, there exists an energy gap, E_g. At low pressures, that is, in the Henry's law region of adsorption, $E_g = E_1 - E_2$, denoting that diffusion is an activated process within the present model [67,68]. In this regard, the delocalized state can be considered to be a transition state. Hence, applying the transition state theory [69–72], a well-known method to make chemical kinetics calculations [73–75].

Concretely, here the transition-state methodology for the calculation of the self-diffusion coefficient of molecules in zeolites with linear channels and different dimensionalities of the channel system is applied [32]. To be definite, the transition state is the delocalized state of movement as was previously established, when the zeolite was described as a three-dimensional array of N identical cells each containing N_O identical sites, hence, considering an arrangement of adsorption sites in a linear channel in which each site is an energy minimum with respect to the neighboring space. Further, the number of adsorbed molecules is supposed to be small relative to the number of sites so that each molecule could be considered to be an independent subsystem with enough free sites for jumping. Thereafter, the canonical partition function for a molecule considered to be an independent subsystem in one of the adsorption sites with energy E_2 is given by [74] $Z_a = Z_x^a Z_y^a Z_z^a Z_i^a \exp(-E_2/RT)$, in which Z_x^a, Z_y^a, and Z_z^a represent the canonical partition functions for the movement in the directions x, y, and z of the adsorbed molecule, whereas Z_i^a represents the partition function for the internal degrees of freedom of the molecule, and E_2 is the energy of the adsorption state.

Hence, at this point, we will deal with the translation of the molecule in the z direction through the transition state, that is, through the delocalized state of movement, for which the partition function for a molecule in the transition state becomes [32]

$$Z^* = Z_x^* Z_y^* \left(\frac{2\pi MRT}{h^2} \right)^{1/2} \Delta z Z_i^* \exp\left(-\frac{E_1}{RT} \right) \qquad (5.60)$$

where Z_x^* and Z_y^* are the partition functions for the movement in the x and y directions for the molecule in the transition state, whereas Δz is the distance of movement in the transition state, where $\Delta z = l$, l is the jump distance, M is the molecular weight, R is the gas constant, h is the Planck's constant, whereas T is the absolute temperature. Supposing at this point, as in the classical transition state theory, that molecules in the ground (adsorbed state) and transition states (delocalized movement state) are in equilibrium, it is easily shown that the equilibrium constant for this equilibrium process is [13]

$$K = \frac{N^*}{N_a} = \frac{L^*}{L} \exp\left(-\frac{E_g}{RT}\right)\left(\frac{2\pi MRT}{h^2}\right)^{1/2} \frac{\Delta z}{Z_z^a} \tag{5.61}$$

Provided that $(Z_x^* Z_y^* / Z_x^a Z_y^a)(Z_i^* / Z_i^a) \approx 1$, Equation 5.61 is a good approximation, N^* being the number of molecules in the transition state, N_a is the number of adsorbed molecules, L^* is the maximum number of transition states, L is the maximum number of adsorption sites, and Z_z^a is the partition function for the motion of the molecule in the z direction in the adsorption state. Now, given that the average velocity of translation of the molecule through the transition state is [69] $\langle v_z \rangle = (2RT/\pi M)^{1/2}$. Subsequently, the time of residence of the molecule in the transition state is $T = \langle v_z \rangle / \Delta z = \langle v_z \rangle / l$. Then, at this point it is possible to assert that the number of molecules passing through the transition state per unit time is $\vartheta = (dN^*/dt) = N^*/T$. Besides, the jump frequency of the molecules through the transition state is $\Gamma = \vartheta/N_a$. Bearing in mind now that the dynamic equilibrium between the adsorbed state and the diffusion state Γ, the jump frequency of molecules between sites is [32]

$$\Gamma = \frac{L^*}{L}\left(\frac{2RT}{h}\right)\exp\left(-\frac{E_g}{RT}\right)\frac{1}{Z_z^a} \tag{5.62}$$

where $L^*/L = 1, 2,$ or 3, depending on the dimensionality of the channel system [12]. Now, the self-diffusion coefficient for a molecule in a zeolite is [2–5] $D^* = (l^2/k\tau)$, where l is the jump distance; $k = 1, 2,$ or 3 is the dimensionality; and $\tau = 1/\Gamma$ is the time.

Taking into account now two extreme situations, the first one, very strong adsorption in the site; to be precise, localized adsorption, where the partition function for the movement in the z direction in the adsorption site is a vibrational partition function for a molecule in a potential energy well [76]:

$$Z_z^a = \left(\frac{kT}{h\upsilon}\right)^{1/2} \tag{5.63}$$

where υ is the vibration frequency and k is the Boltzmann constant, whereas the second case is that where the adsorption is delocalized, explicitly the molecule could move in the neighborhood of the site, and the partition function for the movement in the z direction in the adsorption site is the translational partition function [13]:

$$Z_z^a = \left(\frac{2\pi MRT}{h^2}\right)l \tag{5.64}$$

We can obtain a diffusion coefficient for the case of localized adsorption on the sites by introducing in $D^* = (l^2/2k\tau)$ the expression for τ, described in Equation 5.62 and using for Z_z^a, the vibrational partition function in Equation 5.63.

$$D_i^* = \nu l^2 \exp\left(-\frac{E_g}{RT}\right) \tag{5.65}$$

The diffusion coefficients for mobile adsorption can also be calculated by introducing in $D^* = (l^2/2k\tau)$ the expression for τ described in Equation 5.62 and using for Z_z^a, the translational partition function in Equation 5.65 [32]

$$D_i^* = \frac{1}{2}\left(\frac{RT}{\pi M}\right)^{1/2} l \exp\left(-\frac{E_g}{RT}\right) \tag{5.66}$$

Equations formally similar to Equations 5.65 and 5.66 for the self-diffusion coefficients were calculated using different approaches, and they result in [24]

$$D^* = gul \exp\left(-\frac{E}{RT}\right) \tag{5.67}$$

In which, $g = 1/z$, where z is the coordination number, u is the velocity at which the molecule travels, where $u = \nu l$ for localized adsorption, and $u = (8RT/\pi M)^{1/2}$ for mobile adsorption. Finally, l is the jump distance or the diffusional length.

5.9.2 ANOMALOUS DIFFUSION

When the geometry of the channels is enclosing the diffusing molecules, the individual molecules are incapable to pass each other. Hence, the single-file diffusion (SFD) effect takes place [77–80], a process leading to a high extent of mutual interaction of the diffusing molecules, producing significant divergence from normal diffusion, that is, the process described by the Fick's laws or Einstein relation [58,81–83]. The non-Fickean or anomalous behavior can be considered as a result of constraints imposed by the system, which forces the molecules to move in a highly correlated manner, that is, the case for the diffusion of molecules whose diameters are similar to the channel width of molecular sieves whose frameworks involve one-dimensional channel networks, provided examples of these structures: $AlPO_4$-5, $AlPO_4$-8, $AlPO_4$-11, SSZ-24, Omega, ZSM-12, ZSM-22, ZSM-23, ZSM-48, VPI-8, and MCM-41 [33].

We must analyze now that the proportionality between the observation time t, and the MSD, $\langle x^2 \rangle$, existing in the Einstein equation: $\langle x^2 \rangle = 2D^*t$, where D^* is the self-diffusivity, is a consequence of the essential assumption of normal diffusion, namely, that is possible to divide the total observation time into equal time intervals so that the probability distribution of molecular displacement is identical for each of these time intervals. In addition, the displacement probability is independent of previous displacements. Accordingly, the random walk of each individual particle may be taken into account as a Markovian process. Specifically, a process whose further progress is solely determined by the specified state, and not by the former or the past [2–5].

On the other hand, if the conditions for anomalous diffusion are working in the diffusion system, consecutive displacements are correlated. Subsequently, as an immediate consequence the MSD cannot be expected to increase in proportion to time t [84]. Thenceforth, in the case of SFD, this correlation conducts an increased probability for subsequent displacements to be directed reverse to each other [15]. Thereafter, a detailed study shows that under the conditions of SFD in the long-time limit, the MSD increases with the square root of the observation time, that is, $\langle x^2 \rangle = 2Ft^{1/2}$, where F denotes the single-file mobility [85] with a probability distribution given by a Gaussian function, in complete analogy to the case of normal diffusion [86,87]. Therefore, for SFD, the MSD, that is, $\langle x^2 \rangle$, for large time of observation is proportional to $t^{1/2}$, if the condition that molecules must be always in the same order during the transport process is satisfied. To be exact, if the mutual passage of random walkers is excluded [85], where this propensity may be understood by rationalizing that, as a consequence of SFD, molecular displacement in one direction will most probably then lead to an increased concentration of particles ahead it, somewhat than at the back of it [15].

Experimental diffusion studies with zeolitic systems where it was supposed that SFD took place have been carried out applying pulsed-field gradient nuclear magnetic resonance (PFG–NMR) [88], quasi-elastic neutron scattering (QENS) [89], the tracer zero-length-column (T-ZLC) [90] technique, and the frequency response [79] method. Nevertheless, there is a need for more experimental results to evidence this phenomenon. For example, for methane in $AlPO_4$-5 single-file behavior it was claimed to be experimentally observed by pulsed-field gradient-nuclear magnetic resonance PFG–NMR [88]. However, this was contradicted by other findings, which claim ordinary diffusion behavior for this system [89,91]. In addition, the self-diffusion transport of propane in $AlPO_4$-5 was studied with the help of the tracer zero-length-column (ZLC) method and was established a fast one-dimensional Fickian diffusion, with no evidence of single-file behavior in contrast with the PFG–NMR results [90].

In conclusion, the previously reported results are in part contradictory, and distant from complete. This fact is essentially originated by both the diverging temporal and spatial ranges of observation of these techniques and the deviations of the systems under study are ideal structures [15]. However, if the zeolite channel network embody communicated cages and/or interconnected channels, as is the case, for example, of the MFI, MEL, LTA and FAU framework types [60], where molecules could exchange place during the transport process [85] the combination of molecules during diffusion is produced [13]. Therefore, giving that this is the prerequisite, which leads to a MSD proportional to time, in this case, the transport process is not of the SFD type, because statistically the molecules have time enough to be exchanged during the transport process. Therefore, in this case, ordinary diffusion is the transport mechanism [33].

On the other hand, if the zeolite channel network embody noninterconnected channels, as is the case, for example, of the AFI, AET, AEL, MAZ, MTN, TON, and MTT framework types [60], where molecules, in general, could not exchange place during the transport process [85], the merge of molecules during diffusion and counter diffusion is not produced [23,33]. As a result, since this is the prerequisite, which leads to a SFD-type of diffusion, we should experimentally observe this regime. Nevertheless, as the crystals of these materials are finite, then statistically the molecules have time enough to be exchanged during the transport process. Subsequently, also in this case ordinary diffusion is the observed transport mechanism. In conclusion, single-file diffusion should be a rare effect in molecular transport in zeolites and related materials.

To recapitulate about normal and anomalous diffusion, it is possible to assert that in a variety of physical systems the simple scaling, that is, $\langle x^2 \rangle \propto t$, corresponding to Fickian diffusion is violated [92,93], then the MSD will grow as $\langle x^2 \rangle \propto t^\alpha$, where the coefficient $\alpha \neq 1$. Hence, a consistent generalization of the diffusion equation could still be second order in the spatial coordinates and have a fractional-order temporal derivative, for example [94]:

$$\frac{\partial^\alpha P(\bar{r}, t)}{\partial t^\alpha} = \kappa_\alpha \nabla^2 P(\bar{r}, t) \tag{5.68}$$

where $P(\bar{r}, t)$ is the probability of finding the diffusing particle in the point $(\bar{r}, t)$ where the particle concentration is $C(\bar{r}, t)$ and $P(r, t) \propto C(x, t)$. Thus, the diffusion equations holds when probabilities are substituted for concentrations. Then, anomalous transport phenomena could have subdiffusion if $\alpha < 1$, and superdiffusion if $\alpha > 1$.

The time derivative of Equation 5.68 is a fractional derivative, which is properly defined in a branch of mathematics called fractional calculus, where a fractional integral of order α in the variable x is defined as [95,96]

$$_aD_x^{-\alpha} = \frac{1}{\Gamma(\alpha)} \int_0^x (x - y)^{\alpha - 1} f(y) dy \quad \text{valid for } x > a$$

where $\Gamma(\alpha)$ is the gamma function, if the fractional derivative is defined then by Sokolov et al. [94] $_aD_x^\alpha = d^n(_aD_x^{\alpha - n})/dx^n$. In another notation $_0D_x^{-\alpha}(f(x)) = d^{-\alpha}f(x)/dx^{-\alpha}$ valid for $x > 0$ and $_0D_x^\alpha(f(x)) = d^\alpha f(x)/d^\alpha x$, where Equation 5.68 handles boundary problems in the same way as its normal counterpart does [94]. Consequently, it is a valuable tool for solving diffusion in some complex systems. In a previous work, the author following a Gaussian probability distribution treated analytically single-file diffusion on the basis of the Fick's relations of normal diffusion by merely considering its dependence on the square root of the time instead of time itself:

$$\frac{\partial C(x, t)}{\partial t^{1/2}} = F \frac{\partial^2 C(x, t)}{\partial t^2} \tag{5.69}$$

where F is the SFD mobility [84], and Equation 5.69 is within the spirit of Equation 5.68. However, it was affirmed [15] that Equation 5.69 is only valid for an infinitely extended single-file system, and as soon as this condition of infinite extension has to be abandoned, and boundary conditions are imposed and be attached to this analogy leads to erroneous conclusions. Nevertheless, if Equation 5.68 is a consistent generalization of the diffusion equation, as was previously firmly acknowledged [94], fact which is an expression of the existence of real physical systems, which exhibits anomalous diffusion. Subsequently, these systems should be finite systems, in order to be real systems, and consequently, as was previously stated, boundary conditions could be imposed to Equation 5.68, and then also to Equation 5.69.

As was previously discussed, SFD in zeolites should be a rare effect, because the zeolite crystals are very small in general, and the conditions for the establishment of the SFD regime are not fulfilled in actual systems. However, for a large enough crystal possibly the temporal and spatial conditions for SFD could be satisfied. If this is the case, an equation similar to Equations 5.68 and 5.69 with proper boundary conditions should describe the system.

5.9.3 Experimental Methods for the Study of Diffusion in Zeolites

The experimental studies of diffusion in zeolites have been carried out by different methods [8,25,79,90–92]. For example, the Fickean diffusion coefficients can be measured with the help of steady-state methods such as membrane permeation (MP) [37–40], and also by uptake methods [4,12], as the zero-length column (ZLC) method [27], the frequency response (FR) [28,29], and the Fourier transform infrared (FTIR) method of measuring the kinetics of sorption [30–35], whereas the self-diffusion coefficient can be measured directly with the help of PFG–NMR method, or with QENS [12,25,88,89,97]. The MP, ZLC, FR, PFG–NMR, and QENS methods are very well explained in literature [27–29,30–35,37–40,88,89,97] and will be not explained here. However, the FTIR method is not always broadly discussed in textbooks. Therefore, here we will discuss the measurement of the Fickean diffusion coefficient with the FTIR method developed by Karge and Niessen [30,31].

In Figure 5.11 the diagram of an experimental facility to make measurements of the Fickean diffusion coefficient by the FTIR method is shown. It comprises the IR cell, connected through stainless steel pipes to a manifold containing the thermostatted saturator, having also two gas inlets, for the carrier gas, that is, helium or

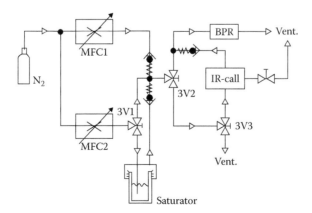

FIGURE 5.11 Experimental facility.

nitrogen, with grade of purity of 99.9999%, where, through inlet 1, the carrier gas bubbles through the saturator in which the adsorbate is located; then, the carrier gas coming from inlet 1 is saturated with the corresponding hydrocarbon is, thereafter, mixed with a measured flow of pure carrier gas coming (N_2 or He) from inlet 2, whereas sensitive mass flow controllers are used in both inlets to control the gas flux. Finally, the simultaneous variations of both flow controls enable the variation of the relative partial pressure of the hydrocarbon in the range $0.01 < P/P_0 < 0.9$.

Summarizing, the measurements were performed as follows: the compound to be tested was filled in the stainless steel saturator held thermostatically at 25°C. Then, a flow of the gas carrier was divided into two, been each of these streams passed through a flow controller [32], that is, one stream went through a tube with a sinter-plate (the saturator) in the base where the carrier gas bubbles in the liquid adsorbate, hence became saturated with the tested substance, which was then mixed with the bypass stream of pure carrier gas at the outlet of the saturator. Finally, the unified stream was passed through the IR cell [35]. The measurements are generally performed by monitoring the change in the absorbance of a band of the sorbate molecule in a FTIR spectrometer using spectra consisting of 1 scan per spectrum and 0.85 second per scan, without delay between scans, and analyzing the proper range for the tested adsorbate. For instance, from 1450 to 1550 cm^{-1}, and using the band around 1482 cm^{-1} for benzene; the range from 1477 to 1517 cm^{-1}, and the band around 1497 cm^{-1} for both toluene and ethylbenzene along with the range from 1550 to 1650 cm^{-1} and the band around 1613 cm^{-1} for m-xylene together with the double peak with bands at 1467 and 1497 cm^{-1}, and the range from 1420 to 1520 cm^{-1} for o-xylene were used [32].

The key equipment in the testing facility is the water-cooled IR high-temperature cell (Figure 5.11). In this demountable cell, the constituting parts are fitted with Viton O-rings to minimize leaking, whereas the temperature of the sample holder is controlled electronically with very low variation of temperature, normally $\Delta T < 1$°C.

Self-supported wafers obtained by pressing 7–9 mg/cm^2 of the zeolite sample powder at 400 MPa are located in the sample holder, then introduced into the cell. It is necessary to state that these wafers fulfill the condition of the absence of macroporous limitations for the transport of the diffusing molecules during the sorption process, enabling measurement of intracrystalline diffusion [30–35]. Hence, preceding the measurement of the diffusion coefficient, the samples were carefully degassed at 450°C during 2 h in a flow of the pure carrier gas. Afterward, the sample is cooled to the desired temperature and kept at this temperature with the help of the temperature control. Thenceforth, the flow rate is adjusted to get the desired relative partial pressure. Besides, a background spectrum of the pure degassed zeolite is obtained as a reference, then the flow coming from the saturator after mixing with the flow of pure carrier gas is admitted at a precisely defined pressure to the IR cell. Finally, the collection of the set of IR spectra is started at the same moment of the admittance of the diffusing molecule, and the spectra are stored as the difference spectra obtained by subtracting from each measurement [30–35].

The Fickean diffusion coefficient is evaluated using a solution of Fick's second law for the appropriate geometry. In this regard, for spherical crystals (Figure 5.12a), with variable surface concentration, and initial concentration inside the sphere equal zero.

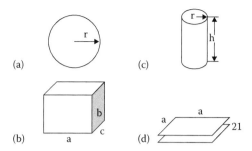

FIGURE 5.12   Sphere (a), coffin (b), cylinder (c), and plate (d) crystal geometries.

In this case, the solution of the Ficks second law is [6]

$$\frac{M_t}{M_\infty} = 1 - 3\frac{D}{\beta a^2}\exp(-\beta t)\left\{1 - \left(\frac{\beta a^2}{D}\right)^{1/2}\cot\left(\frac{\beta a^2}{D}\right)^{1/2}\right\}$$

$$+ \left(\frac{6\beta a^2}{D\pi^2}\right)\sum_1^\infty \left(\frac{\exp\left(-Dn^2\pi^2 t/a^2\right)}{n^2\left(n^2\pi^2 - \left(\beta a^2/D\right)\right)}\right)$$

(5.70)

M_t is proportional to the absorbance A (a.u.) is the amount of sorbate taken at time t, M_∞ is the equilibrium sorption, D is the Fickean diffusion coefficient, $r = a$ is the radius of the zeolite crystallite, β is a time constant describing the evolution of the adsorptive partial pressure in the dead space of the IR cell, that is, $P = P_0[1 - \exp(-\beta t)]$, where P_0 is the steady-state partial pressure, and P is the partial pressure at time [6]. An expression is obtained using only the first four terms, which is fitted to the experimental data (Figure 5.13a and b [33]).

Then, for each experiment the numerical values of the Fickean diffusion coefficient D, are calculated with the help of a nonlinear regression method, which allows the calculation of the best-fitting parameters of the approximation to Equation 5.70 (Table 5.4).

The use of Equation 5.69 is completely justified for a set of uniform spherical particles of radius a, as is the case of the diffusion in beta zeolite. However, this fact must be experimentally confirmed with the help of a scanning electron microscopy (SEM). For instance, a beta zeolite material has been studied showing round-shaped crystals with $a = 0.45$ μm [32], In the case of ZSM-5 and ZSM-11, it is possible to define an equivalent spherical radius by means of the equation [58]:

$$r = \frac{2}{3}\left(\frac{1}{a} + \frac{1}{b} + \frac{1}{c}\right)$$

(5.71)

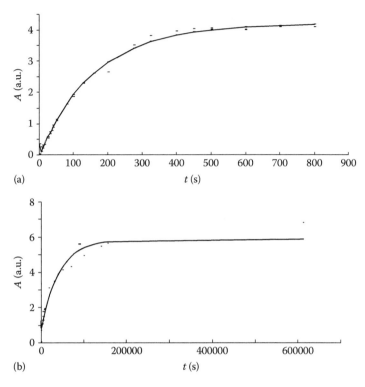

FIGURE 5.13   Diffusion kinetics (a) in *p*-xylene and (b) in *o*-xylene.

TABLE 5.4

Corrected Diffusion Coefficients ($D_0 \times 10^9$ [cm²/s]) of Benzene, Toluene, and Ethylbenzene in H-ZSM-5 and H-Beta Zeolites at Different Temperatures

Sortive	H-ZSM-5	H-Beta	T (°C)
Benzene	0.5	0.6	300
	2	3.2	350
	8	8.2	400
		13	450
Toluene	1	0.2	300
	4	0.9	350
	6	10	400
	16	18	450
Ethylbenzene	0.6	0.1	300
	1.5	0.2	350
	5	2.5	400
	17	26	450

Since the ZSM-5 and ZSM-11 zeolites, in general, shows very regular *coffin*-shaped crystals (Figure 5.12b) with the following average size: $a \times b \times c$ [32,33].

For a cylindrical geometry, which is the case if the zeolite crystals, which conform that the wafer are approximately cylindrical (Figure 5.12c), then for diffusion through the cylindrical surface of the cylindrical shape particle with variable surface concentration, and initial concentration inside the cylinder equal cero, the solution of the Ficks second law is [6]

$$\frac{M_t}{M_\infty} = 1 - \frac{2J_1\left(\left(\beta a^2/D\right)^{1/2}\right)\exp(-\beta t)}{\left(\beta a^2/D\right)^{1/2} J_0\left(\left(\beta a^2/D\right)^{1/2}\right)} + \left(\frac{4}{a^2}\right)\sum_1^\infty \left(\frac{\exp(-D\alpha_n t)}{\alpha_n^2\{(D\alpha_n^2/\beta)-1\}}\right) \tag{5.72}$$

where $J_0(x)$ is the Bessel function of the first kind of order zero, α_n are the roots of $J_0(a\alpha_n) = 0$, $J_1(x)$ is the Bessel function of the first order, M_t, which is proportional to the absorbance A (where A is reported in arbitraty units [a.u.]), that is, $M_t \sim A^T$, is the amount of sorbate taken at time t, and M_∞ is also proportional to the absorbance A, that is, $M_\infty \sim A^T_\infty$ is the equilibrium sorption value, D is the Fickean diffusion coefficient in the case of single component diffusion [30–35]. In addition, $r = a$ is the radius of the zeolite cylindrical crystallite and ß is a time constant, which describe the evolution of the sortive partial pressure in the dead space of the IR cell, α_n.

For a plate-like geometry, which is the case if the zeolite crystals, which conform that the wafer are approximately parallel slabs (Figure 5.12d), then for diffusion through the surface of the slab-shaped particles, with variable surface concentration and initial concentration inside the sphere equal cero, the solution of the Ficks second law is [5]

$$\frac{M_t}{M_\infty} = 1 - \exp\left(-\beta t\right)\left(\frac{D}{\beta l^2}\right)^{1/2}\left\{\tan\left(\frac{\beta l^2}{D}\right)^{1/2}\right\}$$

$$+ \left(\frac{8}{\pi^2}\right)\sum_1^\infty \left(\frac{\exp\left(-(2n+1)^2\pi^2Dt/4l^2\right)}{(2n+1)^2\left[1-(2n+1)^2(D\pi^2/4\beta l^2)\right]}\right) \tag{5.73}$$

where M_t, which is proportional to the absorbance, A (where A is reported in arbitraty units [a.u.]), that is, $M_t \sim A^T$, is the amount of sorbate taken at time t, and M_∞ also proportional to the absorbance A, that is, $M_\infty \sim A^T_\infty$ is the equilibrium sorption value, D is the Fickean diffusion coefficient in the case of single component diffusion [30–35]. In addition, $2l$ is the length of zeolite slab-like crystallite (Figure 5.12d), and β is a time constant, which describe the evolution of the sortive partial pressure in the dead space of the IR cell.

For the MCM-22 zeolite the shape of the crystallites is normally plate-like with dimensions $a \times a \times l = 0.5 \times 0.5 \times 0.05$ μm³ [32] (Figure 5.12d). For this zeolite, diffusion is impossible in the [001] crystallographic direction, because in this direction are stacked the cavities, corresponding to the MWW framework of the MCM-22 zeolite, which are joined by double 6-membered rings [60]. In this case, we should consider two possibilities for the disposition of the [001] direction, that is, parallel (parallel-sided slab model, Figure 5.12d) or perpendicular (cylindrical model, Figure 5.12c) to the surface of the plate-like crystal [32]. If we consider the [001] direction to be parallel to the crystallite face with area $a \times a - 0.5 \times 0.5$ μm (Figure 5.12d), the diffusion will be through the surface of the slab. Then, using the model of a parallel-sided slab with thickness $2l = 0.05$ μm for the solution of Fick's second law (Equation 5.73), the results obtained for the diffusion coefficients are very small. Nevertheless, if we consider the [001] direction to be perpendicular to the crystal face with area $a \times a$, and use a model of diffusion in a cylinder with radius $r = a/2 = 0.25$ μm, and use Equation 5.72 or Equation 5.70 with an equivalent spherical radius $a = (3/2)0.25 = 0.375$ μm, the results are in good agreement with the figures reported for the diffusion in H-ZSM-5 [32].

In all the situations, the obtained Fickean diffusion coefficients, D must be corrected in order to obtain the self-diffusion coefficients D^* with the help of Equation 5.14 [58], that is, $D^* = D(1-\theta)$, in which $\theta = n_a/N_a$ is the fractional saturation of the adsorbent, where n_a is the amount adsorbed and N_a is the maximum amount adsorbed. The calculation of θ could be carried out, for example, with the help of the osmotic adsorption isotherm equation: $n_a/N_a = K_0 P^B/1 + K_0 P^B$, where K and B are constants, fitting the experimental isotherms obtained with the equilibrium sorption values A, which are proportional to n_a [8].

In Table 5.4 [32], are reported the corrected diffusion coefficients D_0 of benzene, toluene, ethylbenzene, o-xylene, and m-xylene in H-ZSM-5, and H-Beta zeolites calculated with the help of Equation 5.70, and Equation 5.14 using the uptake data measured with the FTIR spectrometer, and taking into account the zeolite crystal geometry. At this moment, it is necessary to recognize that in the interpretation of the uptake data for all the considered systems, additional transport mechanisms superimposed on the intracrystalline diffusion were not considered [43].

On the other hand, the Eyring equation:

$$D^* = D_0^* \exp\left(-\frac{E_a}{RT}\right)$$

was used for the calculation of the activation energy (E_a) and the preexponential factor (D_0^*). The calculated values for the diffusional activation energy (E_a) and the preexponential factor are reported in Table 5.5 for all the studied systems.

The numerical evaluation of the preexponential factors in Equations 5.65 and 5.66 was possible because all the terms included in the equations are well defined, that is, $R = 8.3$ [kJ/(mol K], $T = 300$–400 (K), $M = 80$–100 (g/mol), $l = 10$ (Å) [28], and $v = 10^{12} - 10^{13}$ s⁻¹, where the calculations result in a preexponential term for localized

TABLE 5.5

Diffusional Activation Energies and Preexponential Factors of the Eyring Equation for the Diffusion of Benzene, Toluene, and Ethylbenzene in H-ZSM-5 and H-Beta Zeolites

Zeolite	Sortive	E_a (kJ/mol)	$D_0^* \times 10^4$ (cm²/g)
H-ZSM-5	Benzene	28	0.3
	Toluene	21	0.05
	Ethylbenzene	26	0.2
H-Beta	Benzene	32	2.2
	Toluene	36	3.3
	Ethylbenzene	41	9

adsorption in the range 10^{-1} [cm²/s] $< D_0^* < 10^{-2}$ [cm²/s], and a preexponential term for mobile adsorption in the range 4×10^{-4} [cm²/s] $< D_0^* < 6 \times 10^{-4}$ [cm²/s] [32].

If the approximation, $D^* = D_0$, [98], is made, then it is possible to compare the values reported in Table 5.5, and the calculated values, where the comparison indicates that the preexponential term for localized adsorption does not agree with the experimentally obtained preexponential terms. Consequently, we can conclude the diffusion of aromatic hydrocarbons in highly siliceous acid zeolites is not related to strong adsorption [32]. The conclusion was well reached by others for the diffusion of benzene and toluene in ZSM-5 zeolite [99,100].

REFERENCES

1. H. Mehrer, *Diffusion in Solids: Fundamentals, Methods, Materials, Diffusion-Controlled Processes*, Springer Series in Solid-State Sciences, Berlin, Germany, 2007.
2. P. Shewmon and M. Janssen (Eds.), *Diffusion in Solids* (2nd ed.), Springer, Berlin, Germany, 2016.
3. R. Reif, *Fundamentals of Statistical and Thermal Physics*, Waveland Press, New York, 2008.
4. W. Kauzmann, *Kinetic Theory of Gases, Dover Books on Chemistry*, Reprint Edition, London, UK, 2012.
5. J. R. Manning, *Diffusion Kinetics for Atoms in Crystals*, Van Nostrand, Princeton, NJ, 1968.
6. J. Crank, *The Mathematics of Diffusion* (2nd ed.), Oxford University Press, Oxford, UK, 1975.
7. B.S. Bokstein, *Diffusion in Metals*, Editorial Mir, Moscow, Russia, 1980.
8. R. Roque-Malherbe, *Physical Chemistry of Materials: Energy and Pollution Abatement Applications*, CRC Press, Boca Raton, FL, 2009.
9. K.S.W. Sing, D.H. Everett, R.A.W. Haul, L. Moscou, R.A. Pirotti, J. Rouquerol, and T. Siemieniewska, *Pure App. Chem.*, 57 (1985) 603.
10. F. Rouquerol, J. Rouquerol, K. Sing, P. Llewellyn, and G. Maurin, *Adsorption by Powders and Porous Solids* (2nd ed.), Academic Press, New York, 2013.

11. D.M. Ruthven, *Principles of Adsorption and Adsorption Processes*, John Wiley & Sons, New York, 1984.

12. J. Karger and D.M. Ruthven, *Diffusion in Zeolites and other Microporous Solids*, John Wiley & Sons, New York, 1992.

13. T.L. Hill, *An Introduction to Statistical Thermodynamics*, Dover Publications, New York, 1986.

14. R.B. Bird, W.E. Stewart, and E.N. Lightfoot, *Transport Phenomena* (2nd ed.), John Wiley & Sons, New York, 2002.

15. P. Brauer, S. Fritzsche, J. Karger, G. Schutz, and S. Vasenkov, *Lect. Notes Phys.*, 634 (2004) 89.

16. J.K.N. Nrskov, F. Studt, and F. Abild-Pedersen, *Fundamental Concepts in Heterogeneous Catalysis* (1st ed.), John Wiley & Sons, New York, 2014.

17. J.M. McNair and J.M. Miler, *Basic Gas Chromatography* (2nd ed.), John Wiley & Sons, New York, 2009.

18. J.D. Seader, E.J. Henley, and D.K. Roper, *Separation Process Principles with Applications using Process Simulators* (3rd ed.), John Wiley & Sons, New York, 2017.

19. R. Roque-Malherbe, W. del Valle, F. Marquez, J. Duconge, and M.F.A. Goosen, *Sep. Sci. Technol.*, 41 (2006) 73.

20. M. Kizilyalli, J. Corish, and R. Metselaar, *Pure Appl. Chem.*, 71 (1999) 1307.

21. D. Jou, G. Lebon, and J. Casas-Vazquez, *Extended Irreversible Thermodynamics* (4th ed.), Springer, Berlin, Germany, 2010.

22. P. Heitjans and J. Karger, *Diffusion in Condensed Matter*, Springer, Berlin, Germany, 2005.

23. R. Roque-Malherbe, *Mic. Mes. Mat.*, 41 (2000) 227.

24. J. Xiao and J. Wei, *Chem. Eng. Sci.*, 47 (1992) 1123.

25. J. Karger, S. Vasenkow, and S.M. Auerbach, in *Handbook of Zeolite Science and Technology* (S. Auerbach, K.A. Carrado, and P.K. Dutta, Eds.), Marcell Dekker, New York, 2003, p. 341.

26. A.J. Burggraaf, *J. Membr. Sci.*, 155 (1999) 45.

27. D.M. Ruthven and M. Eic, *ACS Symp. Ser.*, 388 (1988) 362.

28. Y. Yasuda, *J. Phys. Chem.*, 86 (1982) 1913.

29. N.G. Van den Begin and L.V.C. Rees, *Stud. Surf. Sci. Catal.*, 49B (1989) 915.

30. H.G. Karge and W. Niessen, *Catal. Today*, 8 (1991) 451.

31. W. Niessen and H.G. Karge, *Stud. Surf. Sci. Catal.*, 60 (1991) 213.

32. R. Roque-Malherbe, R. Wendelbo, A. Mifsud, and A. Corma, *J. Phys. Chem.*, 99 (1995) 14064.

33. R. Roque-Malherbe and V. Ivanov, *Mic. Mes. Mat.*, 47 (2001) 25.

34. G. Sastre, N. Raj, C. Richard, C. Catlow, R. Roque-Malherbe, and A. Corma, *J. Phys. Chem. B*, 102 (1998) 3198.

35. R. Wendelbo and R. Roque-Malherbe, *Mic. Mat.*, 10 (1997) 231.

36. J. Winkelman, *Diffusion in Gases: Liquids and Electrolytes*, Springer, Berlin, Germany, 2007.

37. E. Nagy, *Basic Equations of the Mass Transport through a Membrane Layer* (1st ed.), Elsevier, Amsterdam, the Netherlands, 2011.

38. R.W. Baker, *Membrane Technology and Applications*, John Wiley & Sons, New York, 2004.

39. M.F. Goosen, S.S. Sablani, and R. Roque-Malherbe, in *Handbook of Membrane Separations: Chemical, Pharmaceutical, and Biotechnological Applications* (A.K. Pabby, A.N. Sastre, and S.S. Rizvi, Eds.), Marcel Dekker, New York, 2007.

40. A.P. Sergiy, V. Divinski, T. Laurila, and V. Vuorinen, *Thermodynamics, Diffusion and the Kirkendall Effect in Solids*, Springer, Berlin, Germany, 2014.

41. G. Saracco and V. Specchia, *Catal. Rev. Sci. Eng.*, 36 (1994) 305.

42. M.R. Wang and Z.X. Li, *Phys. Rev. E*, 68 (2003) 046704.

43. M.F.M. Post, *Stud. Surf. Sci. Catal.*, 58 (1991) 391.
44. R.W. Barber and D.R. Emerson, *Advances in Fluid Mechanics IV* (M. Rahman, R. Verhoeven, and C.A. Brebbia, Eds.), WIT Press, Southampton, UK, 2002, p. 207.
45. H. Mizuseki, Y. Jin, Y. Kawazoe, and L.T. Wille, *J. App. Phys.*, 87 (2000) 6561.
46. H.P. Hsieh, *Inorganic Membranes for Separation, and Reaction, Membrane Science and Technology Series 3*, Elsevier, Amsterdam, the Netherlands, 1996.
47. S. Morooka and K. Kusakabe, *MRS Bulletin*, March 25, 1999.
48. M. Mulder, *Basic Principles of Membrane Technology*, Kluver Academic Publishers, Dordrecht, the Netherlands, 1996.
49. G. Saracco, H.W.J.P. Neomagus, G.F. Versteeg, and W.P.M. Swaaij, *Chem. Eng. Sci.*, 54 (1999) 1997.
50. F.J. Valdes-Parada, J.A. Ochoa-Tapia, and J. Alvarex-Ramirez, *Phys. A*, 388 (2009) 789.
51. E.A. Mason and A.P Malinauskas, *Gas Transport in Porous Media: The Dusty Gas Model*, Elsevier, Amsterdam, the Netherlands, 1983.
52. S.W. Rutherford and D.D. Do, *Ind. Eng. Chem. Res.*, 38 (1999) 565.
53. A.J. Burggraaf and L. Cot, *Fundamentals of Inorganic Membrane Science and Technology*, Elsevier, New York, 1996.
54. A. Corma, *Chem. Rev.*, 97 (1997) 2373.
55. F. Marquez-Linares and R. Roque-Malherbe, *Facets-IUMRS J.*, 2 (2003) 14, and 3 (2004) 8.
56. S.Y. Bhide and S. Yashonath, *J. Phys. Chem. B*, 104 (2000) 11977.
57. H. Ramanan, S.M. Auerbach, and M. Tsapatsis, *J. Phys. Chem. B*, 108 (2004) 17171.
58. R.M. Barrer, *Zeolite and Clay Minerals as Sorbents and Molecular Sieves*, Academic Press, London, UK, 1978.
59. W.J. Mortier, *Compilation of Extraframework Sites in Zeolites*, Butterworth, London, UK, 1982.
60. C. Baerlocher, W.M. Meier, and D.M. Olson, *Atlas of Zeolite Framework Types* (6th revised ed.), Elsevier, Amsterdam, the Netherlands, 2007.
61. N.Y. Chen, T.F. Degnan, Jr., and C.M. Smith, *Molecular Transport and Reaction in Zeolites*, VCH, New York, 1994.
62. R. Roque-Malherbe and V. Ivanov, *J. Mol. Catal. A*, 313 (2009) 7.
63. R. Roque-Malherbe and F. Diaz-Castro, *J. Mol. Catal. A*, 280 (2008) 194.
64. R. Snurr and J. Karger, *J. Phys. Chem. B*, 101 (1997) 6469.
65. P.H. Nelson, A.B. Kaiser, and D.M. Bibby, *J. Catal.*, 127 (1991) 101.
66. R.O. Snurr, A.T. Bell, and D.N. Theodorou, *J. Phys. Chem.*, 97 (1993) 13742.
67. J. de la Cruz, C. Rodriguez, and R. Roque-Malherbe, *Surf. Sci.*, 209 (1989) 215.
68. J. De la Cruz, C. Rodriguez, and R. Roque-Malherbe, *Surf. Sci. Lett.*, 209, 1989, A40.
69. S. Glasstone, K.J. Laidler, and H. Eyring, *The Theory of Rate Process*, McGraw-Hill, New York, 1964.
70. E.V. Anslyn and D.A. Doughtery, Transition state theory and related topics, in *Modern Physical Organic Chemistry*, University Science Books, 2006, p. 365.
71. K. Laidler and C. King, *J. Phys. Chem.*, 87 (1983) 2657.
72. K. Laidler, *The Chemical Intelligencer*, 4 (1998) 39.
73. J. Karger, H. Heifer, and R. Haberlandt, *J. Chem. Soc., Faraday Trans.*, 76 (1980) 1569.
74. D.M. Ruthven and R.I. Derrah, *J. Chem. Soc., Faraday Trans.*, 68 (1972) 2322.
75. R.L. Larry, A.T. Bell, and D.N. Theodorou, *J. Phys. Chem.*, 95 (1991) 8866.
76. W. Rudzinskii, and D.H. Everett, *Adsorption of Gases on Heterogeneous Surfaces*, Academic Press, New York, 1992.
77. C. Rodenbeck and J. Karger, *J. Chem. Phys.*, 110 (1999) 3970.
78. J.P. Hoogenboom, H.L. Tepper, N.F.A. Van der Vegt, and W.J. Briels, *J. Chem. Phys.*, 113 (2000) 6875.
79. L. Song and L.V.C. Rees, *Mic. Mes. Mat.*, 41 (2000) 193.

80. J. Karger, *Single-File Diffusion in Zeolites*, Springer, Berlin, Germany, 2008.
81. A.I. Skoulidas and D.S. Sholl, *J. Phys. Chem. B*, 105 (2001) 3151.
82. J. Valyon, G. Onyestyak, and L.V.C. Rees, *Langmuir*, 16 (2000) 1331.
83. D.W. Breck, *Zeolite Molecular Sieves*, John Wiley & Sons, New York, 1974.
84. K. Hahn and J. Karger, *J. Phys. Chem. B*, 100 (1996) 316.
85. K. Hahn and J. Karger, *J. Phys. A*, 28 (1995) 3061.
86. J. Karger, *Phys. Rev. A*, 45 (1992) 4173.
87. J. Karger, *Phys. Rev. E*, 47 (1993) 1427.
88. V. Kukla, J. Kornatowski, D. Demuth, I. Girnus, H. Pfeifer, L.V.C. Rees, S. Schunk, K. Unger, and J. Karger, *Science*, 272 (1996) 702.
89. H. Jobic, K. Hahn, J. Karger, M. Bee, A. Tuel, M. Noak, I. Girnus, and G. Kearly, *J. Phys. Chem.*, 110 (1997) 5834.
90. S. Brandani, D.M. Ruthven, and J. Karger, *Mic. Mat.*, 9 (1997) 193.
91. R. Roque-Malherbe, *Mic. Mes. Mat.*, 56 (2002) 321.
92. J.P. Bouchard and A. Georges, *Phys. Rep.*, 127 (1990) 127.
93. M.F. Schlesinger, G.M. Zaslavsky, and J. Klafter, *Nature*, 363 (1993) 31.
94. I.M. Sokolov, J. Klafter, and A. Blumen, *Phys. Today*, 55 (2002) 48.
95. K.B. Oldham and J. Spanier, *The Fractional Calculus*, Academic Press, San Diego, CA, 1974.
96. K.S. Miller and B. Ross, *An Introduction to Fractional Calculus and Fractional Differential Equations*, John Wiley & Sons, New York, 1993.
97. H. Pfeifer, in *NMR Basic Principles* (P. Diehl, E. Fluck, and R. Kosfeld, Eds.), Springer, Berlin, Germany, 1972, p. 53.
98. J. Karger, *Surf. Sci.*, 36 (1973) 797.
99. J. Xiao and J. Wei, *J. Chem. Eng. Sci.*, 47 (1992) 1143.
100. H. Karge and W. Niessen, *Mic. Mat.*, 1 (1993) 1.

6 Dynamic Adsorption

6.1 THE ADSORPTION REACTOR

To make use of adsorption unitary operations, usually a reactor (Figure 6.1) [27], where a dynamic adsorption process takes place, is packed with a concrete adsorbent. In this regard, the adsorbents mainly used in these applications are zeolites [1–3], silica [4–6], mesoporous molecular sieves [7–9], carbon adsorbents [10–15], alumina [16,17], clays [18,19], Prussian blue analogues [20–23], nitroprussides [24–26], porous polymers [27–29], metal organic frameworks (MOFs) [30–34], and akaganeites [35–41].

Concretely, dynamic adsorption is a mass-transfer process analyzed with complex calculations, where various parameters established by independent batch kinetic studies or estimated by suitable correlations are required [42–47].

At this point, how adsorbents are used to clean gas or liquid flows by the removal of a low concentrated impurity, using a plug-flow adsorption reactor (PFAR) (Figure 6.1) is explained, where the output of the reactor is the so-called breakthrough curve (Figure 6.2) [27] for which C_0 is the initial concentration, C_e is the breakthrough concentration, V_e is the fed volume of the aqueous solution of the solute A to breakthrough, and V_b is the fed volume to saturation.

To make quantitative inferences it is necessary to state that through the reactor a volumetric flow rate is passing: $F = \Delta V/\Delta t$ = Volume/Time of an aqueous solution with an initial concentration, C_0 [mass/volume] of a trace A, or a gas flow with a low concentration of the gas component, A, C_0 [mass/volume]. Moreover, the volume of the empty bed is $V_B = \varepsilon V$, where V is the bed volume and ε is the fraction of free volume in the bed, and the interstitial fluid velocity u is defined as [45] $u = F/S$, whereas the contact or residence time of the fluid passing though the reactor, τ, is calculated with the equation: $\tau = V_B/F$. Finally, some of the following conditions are needed for an appropriate performance of the adsorption reactor [27]:

1. *Residence time*: Provided that adsorption could be a slow process. Hence, the fluid contact or residence time must be long enough to occur molecular transport to the adsorption sites. Thus, as a rule of thumb it is conceivable to test residence times around the following values: 0.05 s $< \tau <$ 0.1 s for gaseous flows and 0.5 s $< \tau <$ 1 s for liquid flows.
2. *Particle size*: If the particle size is small enough, it will be a considerable pressure drop inside the reactor, so granular particles with big enough particle size must be packed in the reactor. Concretely, the particular particle size depends on the size of the reactor. Therefore, a rule of thumb is to construct the reactor following the subsequent approximate relation: $d_R/d_P \geq 10$, where d_R is the reactor diameter, and d_P is the particle size, where as for a laboratory test of a material, which is the principal aim of this book, $d_R/d_P \approx 10$ is a suitable selection.

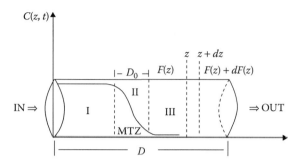

FIGURE 6.1   Adsorption reactor.

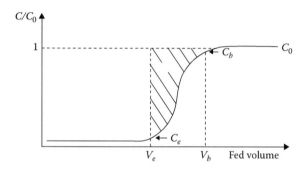

FIGURE 6.2   Breakthrough curve.

3. *Reactor longitude*: As residence time is relatively long, subsequently, large facilities are sometimes required to achieve the required treatment capacities. Afterward, in order to keep the proportions in the reactor dimensions, the following rule could be applied: $D/d_R \leq 10$, where D is the reactor length, and d_R is the reactor diameter. For a laboratory trial of a material, $D/d_R \approx 10$ is an outstanding option.

To conclude, it is required to affirm that the previously listed points are merely very rough designed criteria that are only justified for laboratory tests of materials, not, in general, for the design of industrial reactors, which is not the aim of this book.

6.2   THE PLUG-FLOW ADSORPTION REACTOR MODEL

The plug-flow model indicates that the fluid velocity profile is *plug shaped*, that is, it is uniform at all radial positions, events involving, in general, turbulent flow conditions, in such a way that the fluid constituents are well mixed [44]. Moreover, it is considered that the fixed-bed adsorption reactor is packed randomly using adsorbent particles, fresh or regenerated [48]. Further, in this adsorption separation process, a rate process and a thermodynamic equilibrium take place, in which individual parts of the system react so rapid that for practical purposes local equilibrium can be supposed. Obviously, the adsorption process is supposed to be very fast relative to the convection and diffusion effects, subsequently local equilibrium will exist close to the adsorbent beads [11]. Besides, no chemical reactions occur in the column. Hence, only mass transfer by convection is significant.

Ultimately, it is needed to accentuate that the adsorbent is contacted with a binary mixture in which one component is selectively adsorbed by the solid adsorbent, that is, in the flowing fluid a trace of a species is adsorbed from an inert carrier. In addition, the heat effects could be disregarded, so isothermal conditions will be taken, whereas the flow is fed to the top of the bed at a constant flow rate and under conditions, such that the mass transfer resistance is negligible [48].

The parameters characterizing the reactor are a cross-sectional area S, column length D, and adsorbent mass in the bed M (Figure 6.1). On the other hand, to describe the process, the adsorbent bed in the PFAR is normally divided in three zones: (1) the equilibrium zone, (2) mass transfer zone (MTZ) with a length D_0, and (3) the unused zone [4,10,11]. Besides, the length of the MTZ, D_0, can be calculated with the following expression (Figure 6.2) [49]: $D_0 = 2D(V_b - V_e / V_b + V_e)$, owing to the fact that the physical process of adsorption is so fast relative to other slow steps as diffusion. Thus in and near the solid adsorbent the general form for the equilibrium isotherm is [44] $q = KC^*$, where q is the equilibrium value of the adsorbate concentration expressed as moles solute adsorbed per unit volume of the solid particle, and C^* denotes the solute composition in moles of solute per unit volume of fluid, which could exist at equilibrium along with K, which is the linear partition coefficient. Now, in view of the previously recognized assumptions, the mass balance equation for the PFAR is IN $-$ OUT $=$ ACCUMULATION, which can be expressed as follows [11]:

$$FC(z) - FC(z + dz) = \varepsilon \frac{\partial C}{\partial t} S dz + (1 - \varepsilon) \frac{\partial q}{\partial t} S dz \qquad (6.1)$$

In which, the first term is linked to the fluid flow, and the two terms on the other side of the equation are associated with the accumulation in the fluid phase and the solid phase, respectively. Therefore, dividing by $S dz$, we will get

$$\frac{\partial C}{\partial t} + u \frac{\partial C}{\partial z} + \frac{1 - \varepsilon}{\varepsilon} \frac{\partial q}{\partial t} = 0 \qquad (6.2)$$

In which ε is the void fraction of the bed, u is the interstitial velocity of the carrier fluid, t is the operating time, z is the distance from the inlet of the mobile phase, $C(z,t)$ is the flowing solute composition, and q is the solute concentration in the stationary phase. At this point, to complete the required group of equations is needed to include the adsorption rate of the solute or contaminant, which can be depicted by the linear driving force model in terms of the overall liquid-phase mass-transfer coefficient [50]:

$$\frac{\partial q}{\partial t} = k'(C - C^*) \qquad (6.3)$$

In which, C^* is the mobile-phase concentration in equilibrium with the stationary-phase concentration q, beside, k' is the rate coefficient, which is a lumped mass-transfer coefficient, where $k' = k_c a/(1-\varepsilon)$, where k_c is the mass transfer coefficient per unit interfacial area, $k_c a$ is the mass transfer coefficient per unit volume, and a is the total interfacial area per unit volume of packed column [48]. Thereafter, the model description of the system encompasses three equations, that is $q = KC^*$ along with the relations, Equations 6.2 and 6.3, that is, three equations and three unknowns $(q, C,$ and $C^*)$. Accordingly, it is possible to eliminate q to get [44]

$$\frac{\partial C}{\partial t} + u\frac{\partial C}{\partial z} + \frac{1-\varepsilon}{\varepsilon}K\frac{\partial C}{\partial t} = 0 \tag{6.4}$$

$$\frac{\partial C^*}{\partial t} = \frac{k_c a}{(1-\varepsilon)K}(C - C^*) \tag{6.5}$$

The initial and boundary conditions associated with the simultaneously coupled partial differential equations (PDE) describing the operation of the PFAR are [48]

1. $C(z,0) = 0$ and $C^*(z,0) = 0$, initially clean interstitial fluid for $0 \le z \le D$.
2. $C(0,t) = C_0$: that is, constant composition at bed access.

To calculate the breakthrough curve, analytical solutions of the system of Equations 6.4 and 6.5 can be obtained using the Laplace transform method [3] (see Appendix 6.1). The Laplace method can be used directly to solve the coupled PDE describing the operation of the PFAR Equations 6.4 and 6.5. However, to simplify the solution it is better to change the form of these equations; in this sense, first, it is introduced with a new variable: $\theta = t - z/u$, which is the difference between the real time and the local fluid residence time. Now, it is necessary to introduce this new variable:

$$C(z,t) = C(z,\theta) \tag{6.6}$$

$$C(z,t) = C(z,\theta) \tag{6.7}$$

where:

$$\left.\frac{\partial C}{\partial z}\right|_t dz + \left.\frac{\partial C}{\partial t}\right|_z dt = \left.\frac{\partial C}{\partial z}\right|_\theta dz + \left.\frac{\partial C}{\partial \theta}\right|_z d\theta$$

Together with

$$d\theta = dt - \frac{dz}{u} \tag{6.8}$$

Then substituting Equation 6.8 in Equations 6.6 and 6.7, then equating the coefficients of dt and dz, we will get

$$\left.\frac{\partial C}{\partial z}\right|_t = \left.\frac{\partial C}{\partial z}\right|_\theta - \frac{1}{u}\left.\frac{\partial C}{\partial \theta}\right|_z \tag{6.9}$$

Together with

$$\left.\frac{\partial C}{\partial t}\right|_z = \left.\frac{\partial C}{\partial \theta}\right|_z \qquad (6.10)$$

Likewise, it is possible to demonstrate that

$$\left.\frac{\partial C^*}{\partial t}\right|_z = \left.\frac{\partial C^*}{\partial \theta}\right|_z \qquad (6.11)$$

Now, substituting in Equations 6.9 through 6.11 into Equations 6.4 and 6.5 we will get

$$\left.u\frac{\partial C}{\partial z}\right|_\theta = -\frac{k_c a}{\varepsilon}(C - C^*) \qquad (6.12)$$

$$(1-\varepsilon)K\frac{\partial C^*}{\partial \theta} = k_c a(C - C^*) \qquad (6.13)$$

To make the system of coupled Equations 6.12 and 6.13 more compact, the following two variables will be defined, that is, the dimensionless distance: $\xi = k_c a/\varepsilon(z/u)k_c a/\varepsilon(z/u)$ along with the dimensionless relative time: $\tau = [k_c a/K(1-\varepsilon)](\theta)$. Therefore, when these variables are substituted in Equations 6.12 and 6.13, the following system of coupled equations is obtained:

$$\frac{\partial C}{\partial \xi} = -(C - C^*) \qquad (6.14)$$

$$\frac{\partial C}{\partial \tau} = (C - C^*) \qquad (6.15)$$

where the initial and boundary conditions are $C(\xi,0) = 0, C^*(\xi,0) = 0$, and $C(0,\tau) = C_0$.
  Now, taking Laplace transforms with respect to τ, we get

$$\frac{dC(\xi,s)}{d\xi} = -\left[C(\xi,s) - C^*(\xi,s)\right] \qquad (6.16)$$

$$sC^*(\xi,s) = \left[C(\xi,s) - C^*(\xi,s)\right] \qquad (6.17)$$

To solve the system from Equation 6.16 C^* give up $C^*(\xi,s) = C(\xi,s)/s + 1$ along with

$$\frac{dC(\xi,s)}{d\xi} = -C(\xi,s) + \frac{C(\xi,s)}{s+1} = -C(\xi,s)\frac{s}{s+1} \qquad (6.18)$$

Now, integrating Equation 6.18 we will get

$$C(\xi,s) = A(s)\exp\left(-\frac{s}{s+1}\xi\right) = A\exp(-\xi)\exp\left(\frac{\xi}{s+1}\right) \qquad (6.19)$$

where $A(s)$ is the integration constant. The transform of the step change at the bed entrance is

$$L[C(0, \tau)] = L[C_0] = \frac{C_0}{s} \tag{6.20}$$

Consequently, it is possible to evaluate the integration constant $A(s)$:

$$A(s) = \frac{C_0}{s} \tag{6.21}$$

and the function that will be inverted is

$$C(\xi, s) = \left(\frac{C_0}{s}\right) \exp(-\xi) \exp\left(\frac{\xi}{s+1}\right) \tag{6.22}$$

This function could be inverted with the help of the shifting theorem except for the presence of the term, $1/s$. In a Laplace transform table [51–53] it is possible to find the following transforms:

$$L\left[J_0\left(2\sqrt{kt}\right)\right] = F(s) = \frac{e^{-\frac{k}{s}}}{s} = \int\limits_0^\infty e^{-st} J_0\left(2\sqrt{kt}\right) dt \tag{6.23}$$

where J_0 is the zero-order Bessel function of the first kind [54], which is applicable to the solution of the current problem, if we could replace s for $s + 1$ in Equation 6.22, to do this with the help of the following equation [51,52]:

$$\int\limits_0^\xi \exp(-\beta) \exp\left(\frac{\beta}{s+1}\right) d\beta = \left(\frac{s+1}{s}\right)\left[1 - \exp(-\xi)\exp\left(\frac{\xi}{s+1}\right)\right]$$

Then, it is possible to express the exponential in Equation 6.22 in the integral form:

$$C(\xi, s) = C_0\left[\frac{1}{s} - \int\limits_0^\xi \frac{\exp(-\beta)\exp\left(\frac{\beta}{s+1}\right)}{s+1} d\beta\right]$$

Hence, finally, using the shifting theorem the Laplace transform expressed in Equation 6.22, and noting that $J_0\left(2\sqrt{-kt}\right) = J_0\left(2i\sqrt{kt}\right) = I_0\left(2\sqrt{kt}\right)$ is possible to get $C(\xi, \tau) = C_0 u(\tau)\left[1 - \exp(-\tau)\int_0^\xi \exp(-\beta) I_0\left(2\sqrt{\beta\tau}\right) d\beta\right]$, where I_0 is the modified Bessel function, and $u(\tau)$ is the step function [55].

The above expressed analytical solution of the simultaneously coupled PDE describing the operation of the PFAR is an elegant example of the use of the Laplace transform methodology in the solution of PDEs. More significant, in the framework

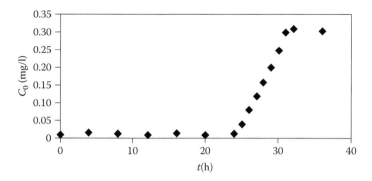

FIGURE 6.3   Dynamic adsorption of H_2O from CO_2–H_2O mix by Na–X synthetic zeolite.

of the use of adsorption, is a useful equation, if the Henry adsorption isotherm $q = KC^*$ is appropriate for the description of the adsorption process. Nevertheless, numerical solutions offer more rigorous results because of a smaller amount of simplifications implicated in the original model, and represent a more flexible line of attack, since their use is not limited by the type of adsorption isotherm, or initial and boundary conditions applied [56].

As an example of an application of a PFAR in a dynamic adsorption process of gas cleaning, in Figure 6.3, it is shown the breakthrough curve following from the dynamic adsorption of H_2O from a CO_2–H_2O mixture by a Na–X synthetic zeolite provided by Laporte [1,57].

To be definite, it is required to state that water concentration in the mixture prior to breakthrough was 0.32 mg/L after the passage of the gas flow through the adsorption reactor was reduced to 0.02 mg/L. Moreover, the reactor had a cross-sectional area $S = 10.2$ cm^2, column length $D = 6.9$ cm, adsorbent mass in the bed $M = 30$ g, the volume of the bed was 70 cm^3, the volume free of adsorbent in the bed was about 35 cm^3, and the volumetric flow rate was $F = 7$ [cm^3/s] [1].

APPENDIX 6.1   LAPLACE TRANSFORMS

The Laplace transform converts integral and differential equations into algebraic equations, simplifying, in some cases, the calculation of the solutions of a differential equation. There are two versions of Laplace transforms: the one-sided and the two-sided. Usually, the one-sided transform is used when we are dealing with causal systems and signals. The one-sided transform is the type of Laplace transform that will be used here. In this case, it is assumed that $f(t) = 0$, for $t < 0$ [44]. Then, Laplace transform of a function, $f(t)$, is a function, $L[f(t)] = F(s)$, of the complex variable $s = \sigma + i\omega$, and the transforms are defined as follows [51]:

$$L[f(t)] = F(s) = \int_0^\infty e^{-st} f(t)dt \tag{6A.1}$$

Of course, these transformations are only defined for those values of s for which the integrals converge. The condition for convergence is as follows [52]:

$$\int_0^\infty \left| e^{-st} f(t) \right| dt < \infty$$

The set of values, s, for which the above-mentioned condition is satisfied, is called the region of convergence (ROC) of the Laplace transform.

The operator L is linear, then [53]

$$L[ag(t) + bh(t)] = aL[g(t)] + bL[h(t)] \tag{6A.2}$$

The most elementary function where the operator L can be applied, that is, the Laplace transform is a constant, $f(t) = C$, consequently:

$$L[C] = \int_0^\infty Ce^{-st} dt = C\left(\frac{e^{-st}}{-s} \Big|_0^\infty \right) = \frac{C}{s} \tag{6A.3}$$

Another simple function is $f(t) = t$

$$L[Ct] = \int_0^\infty Cte^{-st} dt = C\left(\frac{e^{-st}}{s^2}(-st - 1) \Big|_0^\infty \right) = \frac{C}{s^2} \tag{6A.4}$$

A higher-level function we can consider the power function $f(t) = Ct^n$, where n is a natural number:

$$L[Ct^n] = \int_0^\infty Ct^n e^{-st} dt = \frac{Cn!}{s^{n+1}} \tag{6A.5}$$

Let now, $f(t) = Ce^{kt}$

$$L[Ce^{kt}] = \int_0^\infty Ce^{-st}e^{kt} dt = C\left(\frac{e^{-(s-k)t}}{-(s-k)} \Big|_0^\infty \right) = \frac{C}{s-k} \tag{6A.6}$$

Similarly [54]:

$$L[Ce^{-bt}] = \int_0^\infty Ce^{-st}e^{-bt} dt = C\left(\frac{e^{-(s+b)t}}{-(s+b)} \Big|_0^\infty \right) = \frac{C}{s+b} \tag{6A.7}$$

Now since:

$$f(t) = C\cosh kt = C\left(\frac{1}{2}(e^{kt} + e^{-kt}) \right)$$

and

$$f(t) = C senh kt = C\left(\frac{1}{2}(e^{kt} - e^{-kt})\right)$$

We then have [20,21]

$$L[C \cosh kt] = \int_0^\infty Ce^{-st}(\cosh kt) dt = \frac{Cs}{s^2 - k^2} \qquad (6A.8)$$

and

$$L[C senh kt] = \int_0^\infty Ce^{-st}(senh kt) dt = \frac{Ck}{s^2 - k^2} \qquad (6A.9)$$

Similarly [51]

$$L[C \cos \omega t] = \int_0^\infty Ce^{-st}(\cos \omega t) dt = \frac{Cs}{s^2 + \omega^2} \qquad (6A.10)$$

and

$$L[C \sin \omega t] = \int_0^\infty Ce^{-st}(\cos \omega t) dt = \frac{C\omega}{s^2 + \omega^2} \qquad (6A.11)$$

Besides [52]:

$$L\left[\frac{df}{dt}\right] = sF(s) - f(0) \qquad (6A.12)$$

and [53]

$$L\left[\frac{d^2 f}{dt^2}\right] = s^2 F(s) - sf(0) - f'(0) \qquad (6A.13)$$

In the case of a partial derivative of the variable t, we will have [54]

$$L\left[\frac{\partial f(x,t)}{\partial t}\right] = sF(x,s) - F(x,0) \qquad (6A.14)$$

In which,

$$F(x,s) = \int_0^\infty e^{-st} f(x,t) dt$$

In the case of a partial derivative of the variable x, the Laplace transform will be

$$L\left[\frac{\partial f(x,t)}{\partial x}\right] = \frac{dF(x,s)}{dx} \qquad (6A.15)$$

Similarly

$$L\left[\int_0^t f(\tau)d\tau\right] = \frac{F(s)}{s} \qquad (6A.16)$$

Other important result is the shifting theorem [54]:

$$L[e^{-at}f(t)] = F(s+a) \qquad (6A.17)$$

The Laplace transform of the unitary step function, which is a function that takes the value $u(t) = 1$ for $t > 0$, is

$$L[u(t)] = \frac{1}{s} \qquad (6A.18)$$

If the step function is delayed such that $u(t - T) = 1$, for $t > T$, then the Laplace transform will be

$$L[u(t-T)] = \int_0^T 0e^{-st}dt + \int_T^\infty 1e^{-st}dt = \frac{e^{-Ts}}{s} \qquad (6A.19)$$

Finally, it is necessary to define the inverse Laplace transform L^{-1}, as [51–54]

$$f(t) = L^{-1}\{F(s)\} = \frac{1}{2\pi i}\lim_{\omega\to\infty}\left[\int_{\sigma_0-i\omega}^{\sigma_0+i\omega} e^{st}F(s)ds\right] \qquad (6A.20)$$

REFERENCES

1. R. Roque-Malherbe, L. Lemes, L. López, and A. Montes, in *Zeolites'93 Full Papers Volume* (D. Ming and F.A. Mumpton, Eds.), International Committee on Natural Zeolites Press, Brockport, NY, 1995, p. 299.
2. R. Roque-Malherbe, Applications of natural zeolites in pollution abatement and industry, in *Handbook of Surfaces and Interfaces of Materials*, Vol. 5 (H.S. Nalwa, Ed.), Academic Press, New York, Chapter 12, 2001, p. 495.
3. R. Roque-Malherbe, A. Costa, C. Rivera, F. Lugo, and R. Polanco, *J. Mater. Sci. Eng. A*, 3 (2013) 263.
4. G.M.S. El Shafey, in *Adsorption on Silica Surfaces* (E. Papirer, Ed.), Marcel Dekker, New York, 2000, pp. 34–62.
5. E. Papirer, *Adsorption on Silica Surfaces*, CRC Press, Boca Raton, FL, 2000.
6. R. Roque-Malherbe, R. Polanco, and F. Marquez, *J. Phys. Chem. C*, 114 (2010) 17773.

7. L. Bonneviot, F. Béland, C. Danumah, S. Giasson, and S. Kaliaguine, Eds., *Stud. Surf. Sci. Catal.*, 117 (1998).

8. M. Kruk, M. Jaroniec, and A. Sayari, *Adsorption*, 6 (2000) 47.

9. P.S. Yaremov, N.D. Scherban, and V.G. Ilyin, *Theor. Exp. Chem.*, 48 (2013) 394.

10. V.N. Popov, *Mater. Sci. Eng. R-Rep.*, 43 (2004) 61.

11. D. Yi, Y. Xiao-bao, and N. Jun, *Front. Phys. China*, 1 (2006) 317.

12. F. Marquez, J. Duconge, C. Morant, J.M. Sanz, E. Elizalde, P. Fierro, and R. Roque-Malherbe, *Soft Nanosci. Lett.*, ID: 4600007 (2011).

13. F. Marquez, V. Lopez, C. Morant, R. Roque-Malherbe, C. Domingo, E. Elizalde, and F. Zamora, *J. Nanomater.*, ID: 189214 (2010) 7.

14. F. Marquez, O. Uwakweh, N. Lopez, E. Chavez, R. Polanco, C. Morant, J. M. Sanz, E. Elizalde, C. Neira, S. Nieto, and R. Roque-Malherbe, *J. Solid State Chem.*, 184 (2011) 655.

15. A. Suleiman, C. Cabrera, R. Polanco, and R. Roque-Malherbe, *RSC Adv.*, 5 (2015), 7637.

16. D.M. Ruthven, M. Hussain, and R. Desai, *Stud. Surf. Sci. Catal.*, 80 (1993) 545.

17. J.R. Chen, G.Y. Hsiung, Y.J. Hsu, S.H. Chang, C.H. Chen, W.S. Lee, J.Y. Ku, C.K. Chan, L.W. Joung, and W.T. Chou, *App. Surf. Sci.*, 169–170 (2001) 679.

18. H.A. Al-Abadleh and V.H. Grassian, *Langmuir*, 19 (2003) 341.

19. R.M. Barrer, *Zeolites and Clay Minerals as Sorbents and Molecular Sieves*, Academic Press, London, UK, 1978.

20. B. Zamora, J. Roque, J. Balmaseda, and E. Reguera, *Zeits. Anorgan. Allgem. Chem.*, 636 (2010) 2574.

21. F. Karadas, H. El-Faki, E. Deniz, C.T. Yavuz, S. Aparicio, and M. Atilhan, *Mic. Mes. Mat.*, 162 (2012) 91.

22. R. Roque-Malherbe, E. Carballo, R. Polanco, F. Lugo, and C. Lozano, *J. Phys. Chem. Solids*, 86 (2015) 65.

23. R. Roque-Malherbe, F. Lugo, and R. Polanco, *App. Surf. Sci.*, 385 (2016) 360.

24. L. Reguera, J. Balmaseda, C.P. Krap, and E. Reguera, *J. Phys. Chem. C*, 112 (2008) 10490.

25. R. Roque-Malherbe, C. Lozano, R. Polanco, F. Marquez, F. Lugo, A. Hernandez-Maldonado, and J. Primera-Pedroso, *J. Solid State Chem.*, 184 (2011) 1236.

26. R. Roque-Malherbe, O.N.C. Uwakweh, C. Lozano, R. Polanco, A. Hernandez-Maldonado, P. Fierro, F. Lugo, and J.N. Primera-Pedrozo, *J. Phys. Chem. C*, 115 (2011) 15555.

27. R. Roque-Malherbe, *Physical Chemistry of Materials: Energy and Pollution Abatement Applications*, CRC Press, Boca Raton, FL, 2009.

28. N.B. McKeown and P.M. Budd, *Polymers with Inherent Microporosity* (M.S. Silverstein, N. Cameron, and M. Hillmyer, Eds.), John Wiley & Sons, New York, 2011, p. 3.

29. R. Roque-Malherbe, Surface area and porosity characterization of porous polymers, in *Porous Polymers* (M.S. Silverstein, N. Cameron, and M. Hillmyer, Eds.), John Wiley & Sons, New York, 2011, p. 175.

30. T.P. Maji and S. Kitagawa, *Pure App. Chem.*, 79 (2007) 2155.

31. D.J. Tranchemontangne, J.L. Mendoza-Cortes, M. O'Keeffe, and O.M. Yaghi, *Chem. Soc. Rev.*, 38 (2009) 1257.

32. J-R. Li, C. Sculley, and H-C. Zhou, *Chem. Rev.*, 112 (2012) 869.

33. O.K. Farha and J.T. Hupp, *Acc. Chem. Res.*, 43 (2012) 1166.

34. A. Rios, C. Rivera, G. Garcia, C. Lozano, P. Fierro, L. Fuentes-Cobas, and R. Roque-Malherbe, *J. Mater. Sci. Eng. A*, 2 (2012) 284.

35. R.M. Cornell and U. Schwertmann, *The Iron Oxides: Structure, Properties, Reactions, Occurrence and Uses*, VCH, New York, 1996.

36. R. Zboril, L. Machala, and D. Petridis, *Chem. Mater.*, 14 (2002) 969.

37. A. Navrotsky, L. Mazeina, and J. Majzlan, *Science*, 319 (2008) 1635.

38. K. Stahl, K. Nielsen, J. Jiang, B. Lebech, J.C. Hanson, P. Norby, and J. Lanschot, *J. Corr. Sci.*, 45 (2003) 2563.

39. J.E. Post, P.J. Heaney, P.J.R.B. von Dreele, and R.B.J.C. Hanson, *Am. Mineral.*, 88 (2003) 782.

40. G.Z. Kyzas, E.N. Peleka, and E.A. Deliyanni, *Materials*, 6 (2013) 184.

41. R. Roque-Malherbe, F. Lugo, C. Rivera, R. Polanco, P. Fierro, and O.N.C. Uwakweh, *Curr. Appl. Phys.*, 15 (2015) 571.

42. A.J. Slaney and R. Bhamidimarri, *Water Sci. Technol.*, 38 (1998) 227.

43. A. Wolborska, *Chem. Eng. J.*, 37 (1999) 85.

44. R.G. Rice and D.D. Do, *Applied Mathematics and Modeling for Chemical Engineers*, John Wiley & Sons, New York, 1995.

45. R. Droste, *Theory and Practice of Water and Wastewater Treatment*, John Wiley & Sons, New York, 1997.

46. F. Rodriguez-Reinoso and A. Sepulveda, in *Handbook of Surfaces and Interfaces of Materials*, Vol. 5 (H.S. Nalwa, Ed.), Academic Press, New York, 2001, p. 309.

47. T.O. Salmi, J.-P. Mikkola, and J.P. Warna, *Chemical Reaction Engineering and Reactor Technology*, CRC Press, Boca Raton, FL, 2011.

48. J.-M. Chern and Y.-W. Chien, *Ind. Eng. Chem. Res.*, 40 (2001) 3775.

49. A.C. Michaels, *Ind. Eng. Chem.*, 44 (1952) 1922.

50. T.K. Sherwood, R.L. Pigford, and C.R. Wilke, *Mass Transfer*, McGraw-Hill, New York, 1975.

51. D.V. Widder, *The Laplace Transform*, Dover Books on Mathematics, New York, 2010.

52. N.B. McLachlan, *Laplace Transforms and Their Applications to Differential Equations*, Dover Books on Mathematics, New York, 2014.

53. F. Oberhettinger and L. Badii, *Tables of Laplace Transforms*, Springer-Verlag, New York, 1980.

54. G.B. Arfken and H.J. Weber, *Mathematical Methods for Physicists* (5th ed.), Academic Press, New York, 2001.

55. R.W. Churchill, *Modern Operational Mathematics in Engineering* (3rd ed.), MacGraw-Hill Companies, New York, 1971.

56. C. Tien, *Adsorption Calculation and Modeling*, Butterworth-Heineman, Boston, MA, 1994.

57. R. Roque-Malherbe, L. Lemes, M. Autie, and O. Herrera, Zeolite or the nineties, recent research reports, in *8th International Zeolite Conference* (J.C. Hansen, L. Moscou, and M.F.M. Post, Eds.), IZA, Amsterdam, the Netherlands, July 1989, p. 137.

7 Adsorption on Silica and Active Carbon

7.1 SILICA

The investigation of silica materials has been a field of vast interest for a long time, given that they are inert and very stable. In this regard, after the discovery of silica sols, and gels in the 1920s, and the invention of pyrogenic silica in the 1940s, finely dispersed, and porous silica became a subject of intensive research [1–9]. In this regard, porous silica is one of the different forms of amorphous silica. It can be synthesized by acidification of basic aqueous silicate solutions, and when reaction conditions are properly adjusted, porous silica gels are produced [1]. Then, if water is evaporated from the pores of silica hydrogels prepared porous xerogels are attained. Being produced with this methodology, high surface area catalysts and catalyst supports, chromatographic stationary phases, and adsorbents are industrially prepared [6,9].

A different route for the production of amorphous silica involves the reaction of alkoxides with water, where silicic acid is first produced by hydrolysis of a silicon alkoxide. Hence, it can either undergo self-condensation or condensation with the alkoxide [3]. Thereafter, the global reaction continues as a condensation polymerization to form high molecular weight polysilicates, which then connect together to form a network whose pores are filled with solvent molecules, that is, a gel is formed [3,9], henceforward the name sol-gel polymerization of alkoxides.

Other forms of silica, such as pyrogenic silica and mineral opals are nonporous [8]. In this regard, pyrogenic silica is composed of silica particles with a very narrow particle size distribution, provided this material is obtained by vaporizing SiO_2 in an arc or a plasma jet, or by the oxidation of silicon compounds [7].

Artificial opals are materials characterized by the presence of SiO_2 microspheres. They are obtained by self-assembly methods, that is, they are periodically arranged by forming close-packed structures, that is, they are colloidal *crystals*, which have been the subject of numerous investigations. Being such a big interest is because of the interesting behavior imposed by the order of the microspheres. Moreover, these materials are dielectrics with periodic structures where the refractive index varies in three dimensions [10–13].

7.2 AMORPHOUS SILICA MORPHOLOGY AND SURFACE CHEMISTRY

Silica is one of the most abundant chemical substances on earth found both in crystalline or amorphous form. The crystalline forms of silica are (1) quartz, (2) critobalite, and (3) tridymite, whereas the amorphous forms are (1) precipitated silica, (2) silica gel, (3) colloidal silica sols, and (4) pyrogenic silica [11], where amorphous silicas play an important role in different topics, in view of the fact that siliceous materials are used as adsorbents, ultrafiltration membrane synthesis, catalyst, nanomaterials support, and other large surface, and porosity-related applications [1,4,6,12–14].

Porous silica is one of the different forms of amorphous silica, these materials can be prepared by sol-gel methods, performed in a liquid and at low temperature, where the produced inorganic solids, mostly oxides or hydroxides, are formed by chemical transformation of chemical solutes termed as precursors, where the solid is formed as the result of a polymerization process, which involves the establishment of M–OH–M or M–O–M bridges between the metallic atoms M of the precursor molecules [1–3,6].

The drying process, after the gel formation, is carried out at a relatively low temperature to produce a xerogel [15] (Figure 7.1).

During this thermal drying or room-temperature evaporation, capillary forces produce stresses on the gel. An effect that increases the coordination numbers of the particles along with the collapse of the network, that is, particle agglomeration is fashioned. Meanwhile, aerogels (Figure 7.1), that is, dried gels showing a very high relative pore volume, are materials synthesized by traditional sol-gel chemistry, but are dried, basically, by supercritical drying. Therefore, the dry samples keep the porous texture, which they had in the wet stage [16,17]. In this regard, the

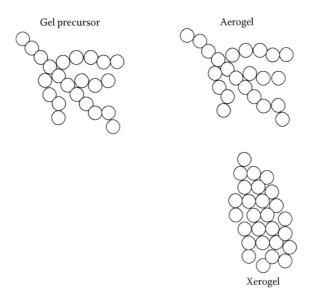

FIGURE 7.1   Aerogel and xerogel formation during the drying process of a precursor.

supercritical extraction of solvent from a gel does not induce capillary stresses due to the lack of solvent–vapor interfaces. Therefore, the compressive forces exerted on the gel network are appreciably reduced relative to those created during formation of a xerogel. Aerogels, therefore, retain a much stronger similarity to their original gel network structure than xerogels. Aerogels are materials with lower apparent densities and larger specific surface areas than xerogels [18]. Afterward, porous xerogels and aerogels are applied as catalysts, catalyst supports, chromatographic stationary phases, and adsorbents [4–9]. In the case of xerogels, there is no doubt that the amorphous framework is formed by very small globular units of size of 10–20 Å, provided that these particles are usually densely packed within secondary aggregates, where the structure is more open for aerogels, and the primary particles are not densely packed [6].

On the other hand, pyrogenic silica is a nonporous material produced by vaporizing SiO_2 in an arc or a plasma jet, or by the oxidation of silicon compounds, provided this form of silica is composed of particles with a very narrow particle size distribution, that is, consisting of discrete spheroidal particles with diameters 100–1000 Å [9], which are formed by even smaller particles with diameters of around 10 Å, where the coordination between these small particles is so high that there is no detectable micropore structure within the secondary globules [19].

Other forms of silica are mineral and artificial opals, materials characterized by the presence of SiO_2 microspheres obtained by self-assembly methods. Thereafter, the resulting material is periodically arranged forming close-packed arranges, where the enormous interest in artificial opals is mainly due to the interesting behavior imposed by the high order of the silica microspheres. In fact, these materials are dielectrics with periodic structures in which the refractive index varies in three dimensions (3D) [11–13].

In Figure 7.2 silica microspheres are shown obtained with the help of the Stobe–Fink–Bohn (SFB) synthesis method, which comprises two steps, that is, hydrolysis of tetraethyl orthosilicate (TEOS) in ethanol, methanol, n-propanol, or n-butanol in the presence of ammonia as a catalyst, together with the successive polymerization, in which the siloxane (Si–O–Si) bonds are created, provided small nuclei with diameters of ca. 100–150 Å are formed in this process that subsequently grows [2].

(a)   (b)

FIGURE 7.2   SEM micrographs of 68C: (a) bar = 10,000 Å and b-70bs2 (b) bar = 5,000 Å.

If the SFB process is modified [20–22], being in this case, the hydrolysis along with the condensation of TEOS is catalyzed, for example, by triethylamine (TEA). Hence, the amine will have a template effect producing the dispersion of silica particles through the aggregation of inorganic–organic composite species. Subsequently, the microsphere configuration will be lost (Figure 7.2b) and the surface of the obtained silica will be increased [22].

The common point linking the different forms of silica are the tetrahedral silicon–oxygen blocks [14], giving that one of the Pauling ionic radius-ratio rules asserts that since $R(Si^{4+})/R(O^{2-})$ is between 0.225 and 0.414. Thenceforward, silicon is tetrahedrically joined to oxygen in silica. If the tetrahedral units are regularly arranged, in that case, periodic structures materializing the different crystalline polymorphs are developed [23]. In the case when tetrahedra are randomly packed, producing a nonperiodic structure that gives rise to various forms of amorphous silica, in which this random association of tetrahedra shapes the complexity of the nanoscale and mesoscale morphology of amorphous silica pore systems [24].

To be definite, we can now state that every porous medium could be depicted as a tridimensional arrangement of matter together with empty space, in which matter along with empty space are divided by an interface. For amorphous silica the complexity of that interface is virtually unlimited, being classified as the morphological complexity of the interface falling in the following three categories [25]:

1. At atomic and molecular level it is common to imagine a surface roughness.
2. At larger scales the notion of pore size and pore shape is very convenient.
3. At even larger scales the concept of pore network topology is valuable.

However, it is the author's opinion that instead of a theoretical discussion of the possible morphologies, it is a better approach to experimentally study the morphology of the silica surface with the help of physical adsorption. Afterward, with the obtained data calculate some very well-defined parameters, such as surface area, pore volume, and pore size distribution (PSD), as was previously explained. Then complement this line of attack with a study of the morphology of these materials by scanning electron microscopy (SEM), transmission electron microscopy (TEM), scanning probe microscopy (SPM), or atomic force microscopy (AFM), and the characterization of its molecular and supramolecular structure by Fourier transform infrared (FTIR) spectrometry, nuclear magnetic resonance spectrometry (NMR), thermal methods, and other methods.

The molecular properties of silica are strongly affected by the nature of their surface sites [26–29]. In this regard, in sol-gel synthesized porous silica surfaces the unsaturated surface valence are saturated by surface hydroxyl functionalities, that depending on the calcination temperature exist as (a) vicinal (hydrogen-bonded silanols), (b) geminal (two hydroxyl groups attached to the same silicon atom), or (c) isolated (no hydrogen bonds possible) silanol sites (Figure 7.3) [23,30–32].

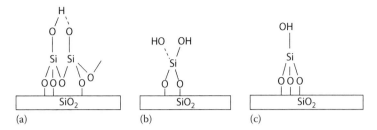

FIGURE 7.3 Representation of silanol group types present on silica surfaces: (a) vicinal, (b) eminal, and (c) isolated.

The hydrogen-bond interaction of OH groups at the surface is determined by the Si–O–Si ring size together with their opening degree, the number of hydroxyls per silicon site, and the surface curvature, where the concentration of OH groups at the surface is approximately $4–5 \times 10^{18}$ OH/m^2 been found to be almost independent of the synthesis conditions of porous silica [3,30].

While silanols or surface hydroxyls become adsorption sites for different molecules, surface hydroxyls are, indeed, particularly reactive with H_2O [4,5] together with other polar molecules, such as NH_3 [31] or molecules with developed quadrupolar moments such as CO_2 [32], which are likely to be physically adsorbed [33–37].

7.3 SYNTHESIS OF PRECIPITATED AMORPHOUS SILICA MATERIALS

Porous silica is one of the different forms of amorphous silica. It can be prepared by acidification of basic aqueous silicate solutions, and when reaction conditions are correctly adjusted porous silica gels are obtained [35,36]. In this regard, silica gel can be synthesized by acidification of basic aqueous silicate solutions as follows [1,2]:

$$Na_2SiO_3 + 2H^+ \overset{H_2O}{\Rightarrow} SiO_2 + H_2O + 2Na^+$$

in which once the reaction requirements are correctly regulated porous silica gels are obtained, having the reaction mechanism in two steps, that is, (1) silicate neutralization producing silicic acids reaction and (2) subsequently, condensation polymerization of the silicic acids [9]. If water is evaporated from the pores of silica hydrogels prepared in this fashion, porous xerogels are obtained. Consequently, with the help of this procedure, high surface area catalysts, catalyst supports, and chromatographic stationary phases together with adsorbents are industrially prepared [6].

Moreover, an alternative method for the synthesis of amorphous silica comprise the reaction of alkoxides with water, provided, in this case, silicic acid is initially generated by the hydrolysis of a silicon alkoxide, formally a silicic acid ether.

Then the silicic acids produced can experience self-condensation or condensation with the alkoxide [3]. Consequently, the global reaction continues as a condensation polymerization to form high molecular weight polysilicates that connects to generate a network whose pores are filled with solvent molecules, that is, a gel is formed [9,33,34].

In summary, sol-gel processing is a method of solid material synthesis, performed in a liquid at low temperature (typically $T < 100°C$), where the synthesized inorganic solids are formed by chemical transformation of chemical solutes termed as precursors. More specifically, for silica synthesis, sol-gel processing refers to the hydrolysis and condensation of alkoxide-based precursors, such as TEOS, that is, $Si(OC_2H_5)_4$ [2], dating the earliest examples of such reactions to the nineteenth century. However, sol-gel methods did not get broad attention until a methodology was created for preparing oxide films from sol-gel precursors in the late 1930s, which proved useful in the manufacturing of stained glass [16].

Meanwhile, other forms of silica, such as pyrogenic silica and opals are nonporous. In particular pyrogenic silica is composed of silica particles with a very narrow dispersion of their particle size, which are obtained by vaporizing SiO_2 in an arc or a plasma jet, or by the oxidation of silicon compounds [7,8]. On the other hand, opals are materials characterized by the presence of SiO_2 microspheres obtained by self-assembly methods, that is, they are arranged to form closely packed structures [10,38,39]. In this regard, synthetic silica microspheres (Figure 7.2a) for synthetic opal preparation can easily be obtained by using the Stobe–Fink–Bohn (SFB) method [2], where this method consists in the hydrolysis of TEOS ($Si(C_2H_5O)_4$) in ethanol, methanol, n-propanol, or n-butanol in the presence of ammonia as a catalyst, provided the batch composition for the synthesis of silica microspheres by means of TEOS in an alcohol (methanol or isopropanol) in the presence of ammonia as a catalyst with and without didistilled water (DDW) in the synthesis media is presented and reported in Table 7.1, where the source materials for the synthesis were TEOS, DDW, methanol (MeOH), isopropanol, and ammonium hydroxide (40 wt. % NH_4OH in water) [16].

TABLE 7.1

Composition for the Synthesis of the Silica Microspheres

Sample	TEOS (mL)	DDW (mL)	MeOH (mL)	Isoprop (mL)	NH$_4$OH (mL)
68F	0.75	0	30	0	3.0
80	1.5	0	30	0	6.0
81C	1.5	8.4	30	0	6.0
68C	1.5	4.5	30	0	6.0
69B	1.5	0.6	0	30	6.0
10	1.5	0.6	0	30	6.0
68E	2.4	0	30	0	6.0

Moreover, the layout to perform the synthesis following the recipes shown in Table 7.1 was as follows [2,20–23]:

1. Alcohol + catalysts (base) + DDW (if needed) (mixed with strong agitation)
2. TEOS is added to the reaction mixture
3. The mixture is stirred at room temperature for 1.5 h
4. After the synthetic procedure the product is heated at 70°C during 20 h

Moreover, in Tables 7.2 and 7.3, the microsphere diameter determined by SEM, labeled D_{SEM} for the materials synthesized followed by the above-accounted procedure, together with the Brunauer–Emmett–Teller (BET)-specific surface (S), DFT-pore volume (W), and DFT-pore-width mode (d) corresponding to the synthesized samples and the MCM-41 mesoporous molecular sieve are reported, where the relative error in the reported data is 20% [35].

Besides, some modifications to the SFB method were introduced to synthesize silica-based materials of remarkably high specific surface area [22,35]. To be precise, in some instances, isopropanol as a solvent and synthesis media was used. Besides, DDW in some cases, was eliminated [20,21]. In addition, amines dissolved

TABLE 7.2
SEM Sphere Diameters

Sample	D_{SEM} (Å)
68F	500
80	2000
81C	2250
68C	2750
10	3750
68E	4500

TABLE 7.3
BET-Area S, DFT-Pore Volume W, and DFT-d

Sample	S (m²/g)	W (cm³/g)	d (Å)
68F	625	1.18	81
80	440	0.49	39
81C	438	0.58	35
68C	320	0.46	21
69B	300	0.52	35
68E	18	0.04	61
MCM-41	820	1.69	35

TABLE 7.4

Synthesis by Hydrolysis of TEOS in Methanol or Ethanol in the Presence of an Amine as a Catalyst with and without DDW in the Synthesis Media

Sample	TEOS (mL)	DDW (mL)	NH$_4$OH (mL)	Amine (mL)	MeOH (mL)	EtOH (mL)	T (°K)
70bs2	0.25	0	0	2	0	10	300
68bs1F.	0.25	0	0	1	10	–	300
75bs1	0.35	0	0	2.5	10	–	300
79BS2	0.45	0	0	2.5	0	10	300
74bs5	0.35	2	0	2.5	–	–	300
68C	0.50	1.5	2	0	10	–	300

in water or in a strong base were used, instead of NH$_4$OH as catalysts, in order to get materials of particularly high specific surface area, provided in Table 7.4 the recipes used to synthesize these materials followed by the above-explained procedure are reported [26,35].

7.4 SILICA MODIFICATION

The large concentration of silanol groups contained in silica samples can be functionalized through elimination reactions, provided this postsynthesis procedure is capable to alter the pore size along with the variation of the hydrophobicity of the pore wall, where the incorporation of hydrophobic terminal groups improves substantially the interaction of the silica surface with nonpolar molecules [28]. In this regard, the attachment of an organic functional group to a silica surface through a covalent bond is the most reliable method of modification of the silica surface [40], provided the covalent bond is used for the binding of mostly Si–O–Si bond, in which one of the silicon atoms is on the silica surface, whereas the other comes from organosilicon compounds, whereas the Si–O–Si bond is formed by the reaction of an Si–OH group on the silica surface with the organosilicon compounds containing a leaving group of high reactivity on the silicon atom [41].

To be definite, it is necessary to state that the organosilicon compounds mainly used in the functionalization process are those containing an alkoxy leaving group [42]. In this regard, a feasible functionalization mechanism is shown in Figure 7.4, where the organic additives contain a reactive alkoxysilane group that perform the attachment of the molecule to the silanol group (–Si–OH) existent on the silica surface [43].

Moreover, many practical catalysts consist of one or several active components deposited on a high surface area support. In particular, highly dispersed transition metal clusters supported on oxides, such as silica have broad applications for reforming, abatement of automobile emissions, and oxidation and hydrogenation reactions [44–48]. But since the catalytic metal is often expensive they are applied in finely dispersed form as nanoparticles on a high specific surface area support.

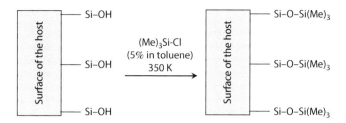

FIGURE 7.4 Attachment mechanism of an alkoxysilane group to a silica surface.

Subsequently, in these conditions a large fraction of the metal atoms are exposed to the reactant molecules [47], provided it is the usual method for the metal preparation in the impregnation of a preformed support with metal precursors along with the ensuing calcinations and reduction, having these well-dispersed transition metal catalysts have widespread industrial applications [44].

7.5 CARBON DIOXIDE ADSORPTION ON SILICA

7.5.1 INTRODUCTION

Adsorption is not simply an important industrial unitary operation, but also a formidable methodology for the characterization of the surface materials. At this point, both aspects will be emphasized [23]. In this regard, here it is considered the interaction of carbon dioxide with the surface of a group of well-characterized sphere and particle packing [20,21], amorphous [22] and crystalline microporous molecular sieve silica [32] with the help of different research methods along with an appropriate theoretical methodology [51–58].

Adsorption on nanoporous materials is an excellent method for the separation and recovery of gases [17,18]. In this regard, the separation and recovery of carbon dioxide are, in particular, a very important research problem because it is a greenhouse gas that contributes to global warming and a reactant in significant industrial processes, for example, in the Solvay process for the production of $NaHCO_3$ [59]. Hence, independent of the fact that carbonaceous adsorbents, molecular sieves, alumina, or silica gel can be applied for carbon dioxide separation, applying the pressure swing adsorption (PSA) process, activated alumina is currently considered the most suitable sorbent for removing carbon dioxide from air in a PSA process [60,61].

Another important application of carbon dioxide adsorption as a characterization tool is the measurement of the micropore volume and the PSD of nanoporous materials [62–64]. Usually, these measurements are carried out with the help of nitrogen adsorption isotherms measured at 77 K [23]. However, at such a low temperature, the transport kinetics of the nitrogen molecules through the micropores is very slow in pores smaller than 0.7 nm. However, in the case of carbon dioxide adsorption, at 273 K, the higher temperature of adsorption applied causes the carbon dioxide molecules to have a bigger kinetic energy [62]. Subsequently, they are capable of going into the microporosity for pores of sizes less than 0.7 nm. Therefore, an experimental solution for this difficulty is the use of carbon dioxide adsorption at 273 K, instead of nitrogen at 77 K [63], to apply the Dubinin adsorption isotherm to describe carbon dioxide adsorption and calculate the correspondent adsorption parameters [64].

Another significant application of carbon dioxide adsorption as a means for the characterization of the surface chemistry of adsorbents is the study of the infrared (IR) spectra of adsorbed carbon dioxide because it is a small weakly interacting probe molecule that is very useful for the study of the acids and basic properties of solid surfaces [65–70].

7.5.2 SYNTHESIS

As was explained earlier, all the consumed chemicals were of analytical grade without additional purification: water was bidistilled and the reagent was of analytical grade. The sphere packing (labeled SP), Stöber silica, and particle packing (labeled PP) silicas were synthesized with the help of the SFB method [1], provided they are the source materials for the synthesis of TEOS, DDW, MeOH, isopropanol, and ammonium hydroxide (40 wt. % NH_4OH in water) [23]. Moreover, two silica aerogel samples were synthesized to get materials by the supercritical drying procedure [16] as follows: gels were flushed three times with liquid CO_2 at 100 bar and 300 K during 3 h for each flush. Thereafter, the temperature was increased to 313 K and at the same pressure the samples were flushed with gaseous CO_2 at 100 bar during 6 h to get the AER-309 and 24 h in the case of the AER-909 aerogel, where the supercritical drying process was carried out in the HELIX supercritical dryer developed and supplied by Applied Separations Inc., Allentown, PA [32]. Besides, the MMS–MCM-41, the hexagonal phase of the family of mesoporous molecular sieves (MMSs) was used in the present study as a standard because it is a very well-studied silica surface. Concretely, the MCM-41 used in the present study was synthesized using the methods described in literature, whereas the molecular sieve (MS), MS–SSZ-24 was provided by R. Lobos. In addition, the tested MS–DAY was the DAY–20F (Si/Al = 20) were produced by Degussa AG, Germany, whereas the MS–ZSM-5, both 3020 and 5020, were provided by the PQ Corporation Malvern.

7.5.3 CHARACTERIZATION

The SEM study was performed with a JEOL CF 35 electron microscope in secondary electron mode at an accelerating voltage of 25 kV to image the surface of the studied samples, provided the samples are attached to the holder with an adhesive tape, then coated under vacuum by a cathode sputtering with a 30–40 nm gold film prior to observation. In Figure 7.5 the SEM images of the sphere packing silicas SP-80 and SP-FM3a-GEO are shown, where the spherical form of the Stöber silica along with a micrograph of the particle packing silica PP-68bs1e is evidently shown in which it is obviously revealed that the spherical geometry was broken.

Next, the nitrogen adsorption specific surface area S, measured together with carbon dioxide pore volume W, along with PSD, investigation of the produced samples were performed in a Quantachrome Autosorb-1 automatic physisorption analyzer, specifically S_{BET} was determined applying the BET method, whereas for the calculation of the micropore size distribution (MPSD) the Saito–Foley (SF) method [71]

SP-80   SP–FM3a-GEO   PP-68bs1e

FIGURE 7.5   SEM images of the silicas SP-80, SP-FM3a-GEO, and PP-68bs1e.

was used at the same time as the density functional theory (DFT) method [72–77] was employed for the calculation of the PSD in the mesopore region. In Table 7.5 the calculated parameters are reported [32].

Where S_{BET} is the BET-specific surface area, W_{Mic} is the micropore volume, W_{DFT} is the DFT pore volume, which include the mesopore region. Meanwhile, d_{Mic} and d_{Mes} are the SF and DFT pore width mode, respectively. Finally, Φ is the particle size. The previously reported morphologic data of the SP and PP silica samples indicate that these adsorbents have a complex pore structure, including micropores and mesopores, where this morphology is a result of the particular internal structure of the SP and PP silicas, which are composed of the secondary particles, evidenced by SEM where these agglomerates are composed of primary particles. That is, the observed morphology is explained by the agglomeration of the primary particles revealed with the help of the SAXS study [32] to form the secondary particles because in the agglomeration process void spaces between the primary particles are created. This morphology will be reflected in the adsorption of carbon dioxide as will be later explained.

TABLE 7.5
Morphological Data of the Different Tested Materials

Sample	S_{BET} (m²/g)	W_{Mic} (cm³/g)	W_{DFT} (cm³/g)	d_{Mic} (nm)	d_{Mes} (nm)	Φ (nm)	References
SP-80	400	0.16	0.49	0.57	3.90	250	[59,61]
SP-69	300	0.06	0.52	0.59	8.10	350	[59,61]
SP-FM3a	–	–	–	–	–	250	[59,61]
MCM-41	800	–	1.69	–	3.50	–	[60]
PP-70bs2	1600	0.18	3.00	0.55	6.50	–	[62,63]
PP-74bs5	1200	0.14	1.70	0.57	6.50	–	[62,63]
PP-68bs1	1500	0.25	2.40	0.55	8.10	–	[62,63]
PP-79bs1	1400	0.16	2.70	0.57	12.10	–	[62,63]
AER-309	500	0.16	4.20	0.58	5.00	–	[62,63]
AER-909	1000	0.40	2.00	0.57	7.3	–	[62,63]
DAY	–	0.30	–	–	0.60	–	[91]
ZSM(3020)	–	0.11	–	–	0.54	–	[5]
ZSM(5020)	–	0.12	–	–	0.54	–	[5]

7.5.4 CARBON DIOXIDE ADSORPTION IN SILICA

The adsorption isotherms of CO_2 at 273 K in samples degassed at 573 K during 3 h in high vacuum (10^{-6} Torr) were obtained in an upgraded Quantachrome Autosorb-1 automatic sorption analyzer. Since, the carbon dioxide vapor pressure at 273 K is $P_0 = 26, 141$ Torr, the adsorption process takes place in the following relative pressure range: $0.00003 < P/P_0 < 0.03$, where the adsorption process is very well described by the Dubinin–Radushkevitch (D–R) adsorption isotherm equation [54–58]:

$$n_a = N_a \exp\left(-\frac{RT}{E}\ln\left[\frac{P_0}{P}\right]\right)^n \tag{7.1}$$

which is expressed in linear form as follows:

$$\ln(n_a) = \ln(N_a) - \left(\frac{RT}{E}\right)^n \ln\left(\frac{P_0}{P}\right)^n \tag{7.2}$$

Is a very powerful tool for the description of the experimental data of adsorption in microporous material, where the fitting process of this equation to the experimental adsorption of carbon dioxide in the tested silica allowed us to calculate the best-fitting parameters, that is, N_m and E together with W for $n = 2$. The parameters calculated with the D–R adsorption isotherm equation (Table 7.6) allow us to evaluate not only the micropore volume of the sample, as was previously explained, but also the adsorption heat released during adsorption data, which allows the assessment of the adsorbate–adsorbent interaction. In this regard, we will show now that it is possible to calculate q_{iso} in an original way using only one isotherm as follows [56]:

$$q_{iso} = -\Delta G(\Theta) + EF(T,\Theta) \tag{7.3}$$

TABLE 7.6
Dubinin Equation Parameters for CO_2 Adsorption on Silica at 273 K

Sample	N_m (mmol/g)	W (cm³/g)	E (kJ/mol)	q_{iso} ($\Theta = 0.37$) (kJ/mol)
SP-69	4.31	0.21	24	28
SP-80	2.67	0.13	24	28
SP-FM3a	4.72	0.23	24	28
PP-75bs1	4.72	0.23	24	27
PP-68bs1	3.48	0.17	24	28
MMS	4.89	0.24	19	22
ZSM-5	3.48	0.17	33	38
DAY	6.77	0.33	17	20

where, $\theta = n_a/N_m$, then

$$\Delta G(\Theta) = RT \ln\left(\frac{P}{P_0}\right) \tag{7.4}$$

Along with

$$F(T,\Theta) = \left(\frac{\alpha T}{2}\right)\left[\ln\left(\frac{1}{\Theta}\right)^{(1/n)-1}\right] \tag{7.5}$$

It is also possible now to assert that since $E = \Delta G(1/e)$, where $\theta = 1/e$, in which $e \approx 2.71828183$ is the base of the n logarithm system. Now, with the help of Equations 7.3 through 7.5, for $\theta = 1/e \approx 0.37$, it is possible to get the following equation:

$$q_{iso}(0.37) = -\Delta G(0.37) + EF(T,0.37) = [1 - F(T,0.37)]E$$

To calculate $F(T, 0.37)$, we need trustworthy experimental calorimetric data reported in the literature. In this regard, the experimental heat of adsorption data that will be applied to compute $F(T, 0.37)$ is $q_{iso} = 22$ kJ/mol for the adsorption of carbon dioxide at 298 K, in the range of $0.1 < n_a < 0.7$ mmol/g in MCM-41. Accordingly, it is possible to estimate that $F(T, 0.37) \approx 1.16$. Finally, the equation used to calculate the isosteric heat of adsorption reported in Table 7.6 is

$$q_{iso}(0.37) = 1.16E \tag{7.6}$$

The reported data indicate that the tested SP and PP silicas show similar values for the isosteric heat of adsorption. This value is also identical to the experimental isosteric heat of adsorption. Concretely, $q_{iso} = 28$ kJ/mol, measured for the adsorption of carbon dioxide at 296–306 K, in the range of $0.1 < n_a < 1.5$ mmol/g in silicalite [78]. The similarity between our data and the values reported in literature for the isosteric heat of adsorption of carbon dioxide on silicalite is due to the similitude between silicalite and the tested silica in the micropore range, that is, the studied silica samples, in the micropore region exhibit pore sizes in the range of 0.56 nm $< d_{Mic} < 0.59$ nm, and the 10 MR channels of the MFI framework of silicalite display pore openings of 0.51×0.55 nm² and 0.53×0.56 nm². Accordingly, since, for porous materials with a similar surface chemistry, the pore size is the main parameter in the determination of the adsorption heat in porous materials. Then, our conclusion reasonably results.

The previously described adsorption data indicate that the interaction of the carbon dioxide molecule with silica is not strong. Specifically, we are in the presence of a physical adsorption process [79], where this physical adsorption process binds the carbon dioxide molecule inside the silica surface micropores by the influence of the dispersive forces and the attraction of the quadrupole interaction [80–82], given that when a molecule interacts with the surface of a solid adsorbent, it becomes subjected to diverse fields, such as the dispersion energy φ_D, repulsion energy φ_R, polarization energy φ_P, field dipole energy $\varphi_{E\mu}$, field gradient quadrupole energy φ_{EQ}, and some specific interactions, such as the acid–base interaction φ_{AB}, and the adsorbate–adsorbate interaction energy φ_{AA} [83,84], where dispersion and repulsion are the main forces during physical adsorption in all adsorbents. In this regard, for molecules such as H_2, Ar, CH_4, N_2, and O_2, these are the only forces present, given that the dipole moments of these molecules are zero, the quadrupole moment is very low or absent, and the polarization effect will be only noticeable in the case of adsorbents with high electric fields [23,85,86]. Meanwhile, the electrostatic interactions between the adsorbed molecule and the adsorbent framework, that is, φ_P, $\varphi_{E\mu}$, and φ_{EQ} depend on the structure and composition of the adsorbed molecule and the adsorbent itself, that is, the interaction between the adsorbent and molecules with, for example, a noticeable quadrupolar moment, such as the carbon dioxide molecule gives rise to specific interactions, where it is a combination of the dispersive and electrostatic forces [87,88].

7.6 ACTIVE CARBON

Activated carbons are amorphous solid adsorbents that can be produced from almost all carbon-rich materials, including wood, fruit stones, peat, lignite, anthracite, shells, and other raw materials, provided the properties of the obtained active carbon depend not only on the preparation method, but also on the starting raw material used for their production [89–100]. However, independent of the great variety of accessible carbonaceous materials, lignocellulosic materials account for 47% of the total raw materials used for active carbon production [92].

For the production of these materials, an organic material is converted into activated carbon by carbonization. Specifically, through pyrolysis under inert atmosphere, followed by activation explicitly by heat treatment with an oxidizing agent or by simultaneous carbonization and activation with a dehydrating compound [93,96], these processes create a very large adsorption capacity, in some cases up to 3000 m^2/g through the formation of an amorphous microporous structure, and so can yield high separation factors for some adsorbate mixtures [94], where their exceptional adsorption properties result from its high surface area, PSD, broad range of surface functional groups, together with mechanical strength. Consequently, porous carbonaceous materials are used regularly in numerous industrial processes for the removal of impurities from gases and liquids, gas separation, vehicle exhaust emission control, solvent recovery, environmental technology, and as catalyst support [92,93].

Along with their developed pore structure, small pore sizes, and large surface area, their surface hydrophobicity is a very important property [91], provided it is especially useful for the sorption of organic species [95–97]. On the other hand, active carbon contains heteroatoms such as oxygen, and to a smaller degree nitrogen and sulfur, provided these atoms were bound to their surface in the form of acidic or basic functional groups. Those that provide the activated carbon surface are acidic or basic in character [92].

7.7 MORPHOLOGY AND SURFACE CHEMISTRY OF ACTIVE CARBON

In Figure 7.6 graphite unit cell, which is the basic component of the activated carbon structure is displayed, consisting of carbon atoms ordered in parallel stacks of hexagonal layer planes of sp^2 carbon atoms forming regular hexagons, that is, each carbon is bonded to other three carbons by a σ bond, and the p_z orbitals containing one electron form a delocalized π bond, been the different layers are linked by van der Waals forces [101].

Factually, the pioneering work in the investigation of the surface chemistry of active carbon was performed by Garten and Weiss together with Boehm and coworkers, who found that the surface chemistry of active carbon is highly influenced by heteroatoms [102], such as oxygen, hydrogen, nitrogen, phosphorous, along with sulfur, which can be located in the carbon matrix conforming the surface in the form of single atoms and/or functional groups, where these atoms are chemically connected to the carbon atoms with unsaturated valences that are located at the edges of graphite basal planes [103]. Particularly, as it is shown in Figure 7.7, oxygen is the main heteroatom, generating carboxyl, carbonyl, phenols, enols, lactones, and quinones produced during the activation procedure or introduced after the preparation by an oxidation treatment, amination, or impregnation with various inorganic compounds [92,104].

A very important consequence of the presence of the previously enumerated groups is the fact that they determine the acidic and basic character of the carbon surfaces, provided the acidic character is associated with surface complexes, such as carboxyl, lactone along with phenol [105]. Meanwhile, the basic nature is generally assigned to surface groups, such as pyrone, ether, and carbonyl [106].

To quantify the amount, diversity, and strength of the different surface groups selective titration technique are applied. In this regard, in classical titration a solution of accurately known concentration called standard solution is added gradually to another solution of unknown concentration, until the chemical reaction between both the solutions is complete [107], that is, a known amount of the tested carbon is

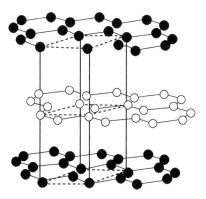

FIGURE 7.6 Structure of hexagonal graphite.

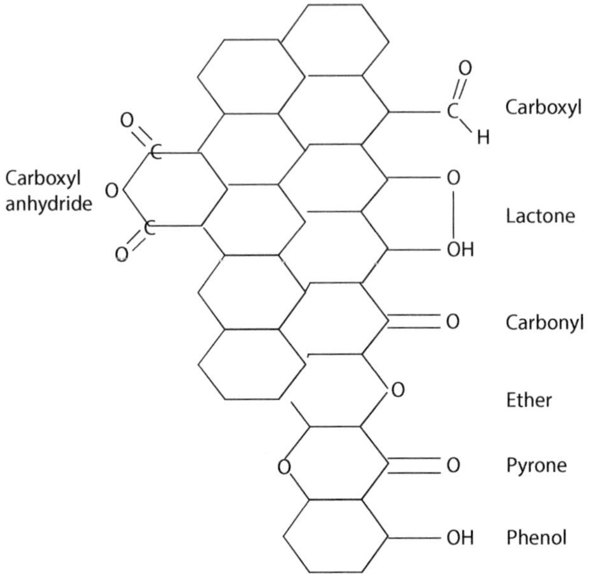

FIGURE 7.7 Oxygen surface groups on a carbon surface.

shaken in closed bottles holding 0.05–0.1 N solutions of different basic compounds, showing different strengths, as sodium bicarbonate ($NaHCO_3$), sodium carbonate (Na_2CO_3), sodium hydroxide ($NaOH$), and sodium ethoxide ($NaOC_2H_5$) [108]. This method allows a selective titration of the different surface oxygen groups, since a base with a specified pK_a will only neutralize the oxygen complexes with the same or lower pK_a [91]. Subsequently, considering the pK_a of these bases together with those of standard oxygen surface complexes it can be anticipated that $NaHCO_3$, which is the feeblest base in the chosen group will titrate solely carboxyl groups [109]. Furthermore, phenolic groups with a pK_a between 8 and 11, as well as carboxylic groups, will react with the stronger base $NaOH$. Moreover, the groups with intermediary acid strength among those of carboxylic groups along with phenolic groups can be tested by means of Na_2CO_3, given that this base can neutralize lactone groups together with carboxylic groups. Finally, $NaOC_2H_5$, which is the strongest base is used to get information about all acidic groups, including those with very low acidity [92].

As a conclusion, it is possible to state that the adsorption of organic compounds by activated carbon is controlled by two major interactions, that is, physical interactions, which include size exclusion together with microporosity effects, along with chemical interactions encompassing the chemical nature of the surface, the adsorbate, and the solvent [110,111]. For example, in liquid-phase adsorption, it has been established that the adsorption capacity of an activated carbon depends on the adsorbent–pore structure, ash content, and functional groups [112], the nature of the adsorbate, its pK_a, functional groups present, polarity, molecular weight, and size [113], and finally, the solution conditions such as pH, ionic strength, and the adsorbate concentration [114].

7.8 ACTIVE CARBON PRODUCTION

Generally, the raw materials used in commercial production of activated carbons are those with elevated carbon contents, such as peat, wood, and lignite together with coal of different ranks. Nevertheless, during the past years, growing attention has shifted to the use of other low-cost and profusely accessible agricultural byproducts, such as coconut shell, rockrose, eucalyptus kraft, lignin, apricot stone, cherry stone, and olive stone to be converted into activated carbons [115]. Their production is performed by applying fundamentally two principal procedures, that is, physical and chemical activation methods, where physical activation consists of a two-step methodology accomplished at high temperature (800°C–1000°C) [103], that is, carbonization under inert gas, usually nitrogen, together with activation under oxidizing agents, customarily, carbon dioxide or water vapor, whereas chemical activation comprises the treatment of the initial material with dehydrating means, for example, sulfuric acid, phosphoric acid, zinc chloride, potassium hydroxide, or another at temperatures varying from 400°C to 1000°C followed by the removal of the dehydrating agent by careful washing [102].

A concrete physical activation process is performed in the following manner [116]: the dried raw material is crushed and sieved to the desired size fraction. Thereafter, during carbonization, purified nitrogen at is used as the purge gas, been the furnace temperature increased from 25°C to 600°C at a rate of 10°C/min, then held at this temperature for circa 3 h. Hence, the resulting chars are activated at 500°C–900°C for 10–60 min under purified CO_2 flush. To illustrate the previously explained procedure in Figure 7.8, a flowchart is shown for the explanation of the physical activation method [106].

A standard chemical activation process is produced in this fashion: in addition, the dried raw material is crushed and sieved to the desired size fraction. Thenceforth, the produced powdered material is mixed with a concentrated solution of a dehydrating compound, such as sulfuric acid, been the mixture of powdered raw material and dehydrating agent dried afterward and heated under inert atmosphere, typically purified nitrogen in a furnace whose temperature is increased from room temperature to 400°C–700°C, and held at this temperature for several hours. After that, the ensuing

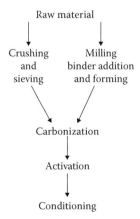

FIGURE 7.8 Schematic description of the physical activation method.

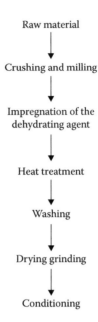

Raw material

↓

Crushing and milling

↓

Impregnation of the
dehydrating agent

↓

Heat treatment

↓

Washing

↓

Drying grinding

↓

Conditioning

FIGURE 7.9   Schematic description of the chemical activation method.

material is carefully washed in order to remove the activating agent. Finally, the activated carbon is separated from the slurry and dried and conditioned. In Figure 7.9 a flowchart is shown for the visual account of the chemical activation method [92].

7.9 ADDITIONAL APPLICATIONS OF PRECIPITATED SILICA IN ADSORPTION PROCESSES

7.9.1 Adsorption of NH_3, H_2O, CO, N_2O, CO_2, and SH_2 in Precipitated Silica

Sorption operations used for separation processes imply molecular transfer from a gas or a liquid to the adsorbent pore network [117–120]. As was formerly acknowledged, the silica gel surface is generally terminated with OH groups bonded with a silicon atom, Si–OH units, that is, silanol groups (Figure 7.3) [14,16], been the concentration of OH groups at the silica surface is approximately 4–5×10^{18} OH/m^2, almost independently of the synthesis conditions [3]. In particular, these silanols are reactive with polar molecules, such as H_2O, SH_2, CO, N_2O, CH_3Cl, CH_3F, HCl, NH_3, and others [20–22,35,121–123], which are prone to be physically adsorbed provided that when these molecules contact the silica solid surface it becomes subjected to the following energy fields [23,26,27]:

1. Dispersion (ϕ_D), repulsion (ϕ_R)
2. Polarization (ϕ_P), field dipole energy ($\phi_{E\mu}$), field gradient quadrupole energy (ϕ_{EQ})

3. Sorbate–sorbate interaction energy (ϕ_{AA})
4. Acid–base interactions (ϕ_{AB}) if the surface contains hydroxyl sites

In the case of NH_3 adsorption, experiments have shown that the principal interaction mechanism is bonding of the hydrogen included in the Si–OH group to the nitrogen atom contained in the NH_3 molecule [124,125], been the molar ammonia adsorption heats on a highly dehydrated silica measured with the help of an adsorption calorimeter, averaged at circa -40 kJ/mol [126]; a significant amount indicating a strong interaction between the ammonia molecule and the silica surface. Therefore, it is necessary to state that silica is an excellent ammonia adsorbent [35,121–126]. Consequently, this gas is extensively used in chemical industries, such as fertilizer and urea production plants together with the fact that their concentration in the gaseous effluents of these industries has to be less than 1 ppm, had the consequence that adsorption of ammonia on silica gel recently received significant attention, owing to their potential application in ammonia removal from gaseous effluents [127,128] together with solar energy cooling cycle refrigeration systems [129].

In the case of CO, H_2O, CO_2, and N_2O the experimental evidence demonstrated that the main interaction mechanism is hydrogen bonding of the hydrogen included in the Si–OH groups to the carbon end of the carbon monoxide and the oxygen end of water, carbon dioxide, and nitric oxide [32,104], owing to these properties, silica gel can behave as good adsorbents for water vapor together with contaminating gases. Particularly zeolites along with silica gels are the most common drying agents nowadays, displaying adsorption capacities of water vapor circa 0.3–0.4 kg of water/kg of adsorbent [130,131].

One more significant step in natural gas processing, which could be performed using silica as an adsorbent is the so-called *sweetening* process, that is, the removal of acidic gases, such as carbon dioxide and hydrogen sulfide from a natural gas stream [132,133]. Usually, the *sweetening* process is performed by gas–liquid absorption-stripping procedures by means of aqueous solutions of alkanolamines [134]. Nevertheless, this procedure is exceptionally energy intensive for regeneration of the solvents together with the presence of serious corrosion problems. On the other hand, chemical adsorption of SH_2 on ferrous oxide is also utilized, but the adsorbent cannot be regenerated as a result of circumstances, which cause environmental difficulties [133]. However, if the *sweetening* process is performed by physical adsorption methods, using silica as adsorbent, the regeneration of the adsorbent is possible by sweeping of the adsorbent bed with a hot stream, the circumstances increasing the cost of energy along with the investment on heating facilities [132]. Nowadays, as long as natural gas consumption is increased, the requirements for the development of efficient techniques for acid-gas elimination in natural gas streams with lower economic costs is also increased, as for example, by the elaboration of solid adsorbents that could be regenerated, also showing high selectivity for acidic gases, together with high adsorption capacity became a need [134].

Supposedly, many solid materials are potential sorbents for carbon dioxide together with hydrogen sulfide. But activated carbon and zeolites have been so far the most widely used solid sorption materials. Nevertheless, the industrial application of these materials as acidic gas adsorbents is restricted, owing to the low selectivity toward acidic gases for activated carbon together with the strong water inhibition effect for zeolite and silica materials [3,23,26,28–30]. Conversely, silica are ideal solid supports for active functionalities supported on their surface, owing to their large surface areas along with the well-defined pore structures. More significantly, the hydroxyl groups on their surfaces are significant for many surface phenomena, such as gas adsorption, surface modification, wetting, and other applications [122,123]. Owing to the high concentrations of surface silanol groups (Si–OH), silica and ordered mesoporous silicas are the most widely used materials for surface modification, that is, grafting of functional groups onto the pore walls of the silica through silylation reactions between the surface silanol groups and the grafting material [4–16].

7.9.2 APPLICATION OF PRECIPITATED SILICA IN HYDROGEN STORAGE

It is the author's opinion that another possible application of NH_3 adsorption in silica is hydrogen storage, where this process is understood as the confinement of hydrogen in appropriate containers for their delivery [135]. In this regard, within the past three decades, much attention has been raised in the assembly of transportable reversible systems for high capacity hydrogen storage [136], a circumstance that is critical to the large-scale use of hydrogen fuel cells, especially for mobile applications. Moreover, novel methods for storing hydrogen more efficiently can quicken the pace of the transition to the hydrogen economy dramatically. Consequently, new methods of hydrogen storage must be considered, being a very promising method of storage is by physical adsorption of molecular hydrogen on different materials, that is, carbon, silica, alumina, or zeolites [137].

To date the emphasis in hydrogen storage has generally been on liquid hydrogen, together with metal-hydride systems. However, liquid hydrogen is very dangerous, energy intensive, and requires a big tank volume, whereas the interstitial hydride method suffers from high cost and weight, also having instability along with reversibility control to regenerate the material problems. Moreover, both the liquid hydrogen and hydride methods have low energy efficiency [136]. However, higher energy efficiency could be obtained in systems where hydrogen is concentrated by physical adsorption above 70 K using an adsorbent. Hence, following this objective, activated carbons have been studied with promising results. In addition, zeolites, silica, and alumina have also been investigated. However, the adsorption results are far to be near the target figure of 6.5 wt. % [137]. Consequently, the criteria for ideal solid hydrogen storage systems are the following [136]:

1. High storage capacity, that is, a minimum of 6.5 wt. %
2. Reversibility of uptake and release
3. Low cost
4. Low toxicity
5. Nonexplosive
6. Inertness

In addition, an interesting alternative for hydrogen storage is ammonia adsorption on an extremely high specific surface silica materials [35], given that, since a key principle for the development of an ideal solid hydrogen system is the use of light elements, such as Li, Be, B, C, N, O, F, Na, Mg, Al, Si, and P for solid storage system. Consequently, silica (SiO_2) could be considered for these purposes. Moreover, silica is nontoxic, non-explosive, inert, inexpensive, and ammonia adsorption on silica is highly reversible [127]. Accordingly, as an interesting alternative for hydrogen storage the author proposed the ammonia physical adsorption of micro/mesoporous silica particle packing materials, followed by their decomposition into H_2 and N_2, applying Ir catalysts [138]. In this regard, silica particle-packing materials showing extremely high specific surface area, specifically, up to 2200 m^2/g [139], given that the hydroxyl concentration in the surface of silica adsorbents is $4.5 \pm 0.5 \times 10^{18}$ OH/m^2 [3]. So 9.9×10^{22} OH/g will be present on the surface of such silica material, if an adsorption proportion of 1 NH_3 per OH is predicted. Thus, 0.44 gram of H per gram of silica will be adsorbed, that is, 31 wt. %; a huge amount by far larger than the goal of 6.5 wt. %. Nevertheless, in practice, this will not be the case; since, taking into consideration real data reported for NH_3 adsorption in silica materials [127], it is feasible to anticipate that at least 0.09 g or 8.3 wt. % of H per gram of silica will be stored in the extremely high specific surface area materials. In addition, a significant quantity above the present estimates of H storage for transportation purposes, which as was previously stated is 6.5 wt. %.

7.9.3 Adsorption of Volatile Organic Compounds in Precipitated Silica

Volatile organic compounds (VOCs) are some of the most extensively released compounds by chemical, petrochemical, and related industrial air pollutants [92,140–147]. These compounds can be defined as those possessing a vapor pressure greater than 133.3 Pa (1 mm Hg) at room temperature [141]. Examples of these chemicals are benzene, toluene, xylenes, hexane, cyclohexane, thiophene, diethylamine, acetone, and acetaldehyde, which are crucial reactants for photochemical reactions [142].

VOCs of oxygenated hydrocarbons are significantly released in the printing, painting, and solvent industries, provided with an applied adsorption process for their removal from the air. In this regard, activated carbon, silica gel, and zeolite are common adsorbents used to attain VOC reduction in industry [143,145].

Benzene derivatives are particularly dangerous to the environment and human health between VOCs, then strict control of VOC emissions is one of the chief purposes of government regulations introduced under, for example, the 1990 U.S. Clean Air Act Amendment [146]. Some VOCs polluting sources [147] are as follows:

1. Automobile exhaust emissions.
2. Vapor released, that is, associated to the paint, dye, lacquer, and varnish industries.
3. Gas fumes given off from storage tanks.
4. Solvent vapors emanating from paints and from liquids utilized in cleaning or degreasing operations.
5. Adhesives.
6. Discharges originated by plastic industries.

In this regard, it is frequent to remove these compounds from air adsorption in silica, owing to their benefits in contrast to other adsorbents such as activated carbons, zeolites, and microporous silica, where the elimination of VOCs from a bulk gas stream by an adsorption process is an ordinary engineering procedure that has been getting growing consideration during the past few years [8,26,95]. Between adsorbents potentially helpful for an effective VOC removal, microporous inorganic materials represent one of the best alternatives to carry out this job [96]. In this sense, active carbon is the most extensively used adsorbent for environmental cleaning, given that the generally activated carbon has a big affinity for volatile organic compounds, owing to the nonpolar character of its surface, in contrast to other solid sorbents [92]. However, this material presents certain drawbacks, including the fact that the adsorbed molecules are not very often destroyed or decomposed. However, instead they are merely weakly held at the surface [109]. Owing to the polar surface of silica gel, it adsorbs molecules from the gas phase, and keeps them on the surface. Subsequently, a systematic evaluation of the adsorption capacities of organic vapors on silica gel was started in the late 1980s [141]. For this reason, it is worth trying to investigate the performance of silica microporous adsorbents, which can have stronger interactions toward the adsorbed molecules [146].

7.10   SOME APPLICATIONS OF ACTIVATED CARBONS, AND OTHER CARBONACEOUS MATERIALS IN GAS PHASE ADSORPTION PROCESSES

7.10.1   ADSORPTION AND REMOVAL OF H_2O AND CO_2, SH_2 AND SO_2 WITH ACTIVE CARBON

Activated carbons are the most extensively applied adsorbents for the elimination of contaminants from gaseous, aqueous, and nonaqueous streams, provided the adsorptive properties of activated carbons are regulated by their porous structure and surface chemical properties, given that the largely nonpolar character of the carbon surface causes activated carbon to selectively adsorb nonpolar molecules rather than polar molecules [106–109], whereas adsorption isotherms are applied to estimate the optimum size of the adsorbent beds along with their operating conditions to properly apply these adsorbents [23,85–87]. In particular, for gas-phase applications, carbon adsorbents are usually used in the shape of hard granules, hard pellets, fiber, cloths, and monoliths because these prevent an extreme pressure drop [105–109].

Moreover, a research subject whose interest has been growing in recent years is the interaction between the water vapor and the activated carbon, if the water molecule's adsorption on carbon surfaces is a phenomenon relevant to a wide variety of commercial processes, as for example, VOC elimination from humid air streams and steam regeneration of activated carbon, adsorption of water vapor has been found [148,149].

Furthermore, carbon-supported platinum catalysts used in fuel cells [150] are affected by the presence of water vapor [151–154], where the adsorption of water on the carbon augments platinum use, therefore permitting enlarged fuel cell efficiency [155].

Notwithstanding the significance of these commercial applications, the fundamental understanding of water adsorption in carbon is yet basic. In this sense, the current theories link surface chemistry together with the structure of carbon to water adsorption affinity [156], provided that carbon adsorbents are composed of misaligned aromatic platelets that form a nanoporous network in which oxygenated functional groups are contained. Then the oxygenated functional groups form primary sites for the adsorption of water [157], where the polar character of the water molecule allow it to bond with the individual oxygenated functional groups, if at low pressures followed by the Henry's law. Meanwhile, at greater pressures hydrogen bonding among free and adsorbed water molecules takes place, whereas the development of clusters begins. Further, at greater pressures, water clusters increase and hydrogen bonding between clusters takes place [151], in which the characteristics of the resultant equilibrium isotherm are dependent, not merely on the total amount of primary adsorption sites, but also on their surface density [158]. As a conclusion, it is possible to affirm that the presence of water in activated carbons has a large effect on the performance of industrial adsorbers, often reducing their useful lifetime by as much as 50% [159].

Furthermore, carbon dioxide capture technology on active carbon has been relatively widely tested. however, it is expensive and energy intensive [160,161]. Hence, to make it possible, some technologies have been developed, for example, pressure swing adsorption (PSA), which is one of the possible methods that could be pertinent for the elimination of CO_2 from gas streams [162,163]. PSA procedures are founded on favored adsorption of the desired gas on a porous adsorbent at high pressure and recuperation of the gas at low pressure. Consequently, the porous sorbent can be reused for subsequent adsorption [164]. In this regard, activated carbon showed preferential adsorption of CO_2 at 25°C and pressures up to 300 psi, and the adsorption of CO_2 was found to be reversible. Hence, excellent separation of CO_2 was obtained from gas mixtures containing both CO_2/N_2 and $CO_2/H_2/He$ utilizing activated carbon during competitive gas adsorption studies performed with a microreactor [162].

In the case of hydrogen sulfide removal from air using activated carbons, it is not only adsorbed but is also oxidized owing to the catalytic effect of the carbon surface. Hence, it is elemental if sulfur along with sulfuric acid are the principal products of this surface reactions. Consequently, the spent carbons have to be substituted by a fresh material. Consequently, a regeneration process, if feasible, has to be done *in situ*, flowing steam to the carbon bed [166,167].

For the elimination of SO_2 with active carbon, analogously, as the case of hydrogen sulfide, SO_2 also reacts in the presence of O_2 and H_2O at relatively low temperatures (20°C–150°C) in the surface of carbon, a process involving a series of reactions directing to the formation of sulfuric acid as the final product, provided the overall reaction [168,169] is

$$SO_2 + \frac{1}{2}O_2 + H_2O + C \rightarrow C - H_2SO_4$$

7.10.2 Hydrogen Storage with Active Carbon and Other Carbonaceous Materials

As was previously commented, a secure, effective, and inexpensive storage system is vital for the upcoming utilization of hydrogen as an energy carrier. The established techniques, for example, liquid hydrogen, compressed gas, and the storage in a solid-state matrix as a metal hydride [96] has the advantage of being safe, but fails because of the total weight of the tank system [135]. Consequently, carbon with its low atomic weight could assist to surmount these drawbacks [97,135–140].

On the contrary to the chemisorption in metal hydrides, microporous carbon adsorbs the undissociated hydrogen molecules by van der Waals forces at its pore network and surface. However, as these binding forces are weak, the physisorption process at room temperature is nearly impeded by the thermal motion. Consequently, to store large amounts of hydrogen the carbon samples have to be cooled [135]. However, a cryogenic hydrogen storage method in the majority of cases is economically unsuccessful [137].

In recent years, the storage of hydrogen in carbon has involved much consideration due to the accessibility of novel carbon nanomaterials, such as fullerenes, nanotubes, and nanofibers [97,135]. The cylindrical structure of carbon nanotubes can increase the adsorption potential in the tube core leading to capillary forces and to enhanced storage capability [99,135]. However, notwithstanding the optimism raised with the reports on the use of carbon nanotubes [99] and carbon nanofibers [138], owing to further research, the obtained results have become questionable [97]. Then, with the study of hydrogen storage in carbonaceous materials with the current type of carbonaceous adsorbents and sorption conditions a maximum uptake of 2.1 wt. % has been observed, amply below the DOE figure target of 6.5 wt. % for mobile applications [97].

A new type of carbonaceous porous materials called mesocarbon microbeads (MCMBs) are microcarbon spheres produced by mesophase pitches. These materials have been principally used as filler in paints, elastomers, and plastics to change the mechanical and electrical properties of the materials [53]. The activated MCMBs with high specific surface area greater than 3000 m²/g can be prepared. These types of MCMBs are called super high surface area carbons or super surface carbons [53]. Many investigations show that a-MCMBs are expected to be more ordered in structure than the activated carbon fibers (ACF), and the calculations show that the adsorption amount of hydrogen at 10 MPa can reach 3.2 wt. % and 15 wt. % at 298 K and 77 K, respectively, which are also higher scores as compared to the other carbon materials [53].

7.10.3 METHANE STORAGE IN ACTIVATED CARBON AND OTHER CARBONACEOUS MATERIALS

During the past years, the interest in natural gas as a vehicular fuel has increased significantly. Since then, natural gas has important benefits over standard fuels, both from an environmental point of view and for its natural abundance [48]. However, it is recognized that its greater inconvenience is its lower heat of combustion per unit of volume when compared with usual fuels [141]. Another motive for this growing interest is for the reason that natural gas is much cheaper than usual petroleum-based gasoline and diesel fuel [142]. Consequently, the attention of different countries in diminishing their dependence on imported oil has directed to the creation of new technologies to facilitate the employment of other fuels, as for instance, natural gas to meet their transportation energy needs [142,143].

Regrettably, methane cannot be stored at a density as high as other fuels. As a result, methane has an energy density about one-third that of gasoline for compressed natural gas (CNG) at 24.8 MPa [48,143]. Besides, the use of CNG has some disadvantages, for example, the high cost of the used cylinders [144].

Between other options, the employment of adsorbents (adsorbed natural gasor ANG) has been thought to be a way for obtaining methane densities comparable to CNG, but at much lower pressure (3.5–4 MPa) [48,142,144–147]. This lower storage pressure reduces the cost of the storage vessel that allows the use of single-stage compressors, and represents a lesser safety hazard than the higher pressures used for CNG [48,143,144]. The U.S. Department of Energy storage goal for ANG has been set at 150 V/V, this means 150 STP liters of gas stored per liter vessel volume at pressure $P = 101.325$ kPa, and temperature $T = 298$ K [142]. Among the available adsorbents, activated carbons exhibit the largest adsorptive capacity for methane storage [48,145,148]. The recent growth of high-capacity carbonaceous adsorbents has generated a great amount of literature, showing the fast increase and significance of the area under discussion [48,142–153], showing that activated carbons are very good adsorbents, presenting the highest ANG energy densities [142–144,146].

High pressure adsorption data were attained for methane, ethane, ethylene, propane, carbon dioxide, and nitrogen on activated carbons [145]. Adsorption isotherms that were achieved up to 6 MPa and at various temperatures were measured. In addition, the isosteric heats of adsorption were also calculated [149].

Methane storage in activated carbon fibers (ACFs) [142,146] and chemically activated carbons prepared from an anthracite [142] have shown that these adsorbents can meet the above-mentioned DOE target presenting quite high storage capacities. In addition, the amount of methane adsorbed in MCMBs at 298 K and 4 MPa could reach 36 wt. %, which is a huge figure [53,154].

7.10.4 Adsorption of Volatile Organic Compounds in Activated Carbon

As stated earlier, VOCs are among the most widespread air contaminants released by chemical, petrochemical, and other industries. Benzene, toluene, xylenes, hexane, cyclohexane, thiophene, diethylamine, acetone, and acetaldehyde are examples of VOCs. The control of air emissions of organic vapors is one of the primary objectives of the stringent regulations introduced under the Clean Air Act amendments [92].

In this sense, one of the most formidable challenges posed by the gradually stringent regulations on air pollution is the search for efficient and economical control strategies for VOCs existing several methods, such as condensation, absorption, adsorption, contact oxidation, and incineration for removal and/or recovery of organic vapors. Nevertheless, the most presently relevant technology for VOC control is adsorption on an activated carbon. It is a well-known technology broadly used in industrial processes for the elimination and recovery of hydrocarbon vapors from gaseous streams. In addition, it offers several benefits over the others, that is, the opportunity of pure product retrieval for reuse, high removal efficiency at low inlet concentrations, and low fuel/energy costs [96–100].

One of the major uses of activated carbon is in the recovery of solvents from industrial process effluents: dry cleaning, paints, adhesives, and polymer manufacturing, and printing are some examples. Since, as a result of the highly volatile character of many solvents, they cannot be emitted directly to the atmosphere [170], if the typical solvents recovered by active carbon are acetone, benzene, ethanol, ethyl ether, pentane, methylene chloride, tetrahydrofuran, toluene, xylene, chlorinated hydrocarbons, and other aromatic compounds [92]. In addition, automotive emissions make a large contribution to urban and global air pollution [170].

7.10.5 AIR-CONDITIONING WITH ACTIVATED CARBON

Growing public alertness and a general alarm over the air quality have produced a need for enhanced treatment of the air that is supplied to public spaces such as airports, hospitals, theaters, and office blocks. Then granular activated carbon filters have been developed in combination with air-conditioning systems for the elimination of harmful trace contaminants from air. On the other hand, the construction of adsorption refrigerating machines constitutes another application of these materials, which is currently undergoing a rapid development [172]. Both the operational principle and the first commercial applications of the adsorption refrigerating cycle were developed several years ago. But this technique of cold production was abandoned for the advantage of the gas compression cycle. Nevertheless, the study of adsorption refrigerating cycles has been encouraged by new regulations on the use of chlorofluorocarbons (CFCs). These can be found in the Montreal Protocol edited in 1987, and in the new amendments made during the London Conference in 1990 [173].

There are many possibilities of utilization of an adsorption cycle covering a great range of temperatures, for instance, refrigeration and ice making from solar energy. From the fundamental discontinuous cycle, numerous procedures were projected to permit for pseudocontinuous cold production on one hand and, on the other hand, enhancement in the efficiency of the adsorption systems [172,174].

REFERENCES

1. R.K. Iler, *The Chemistry of Silica*, John Wiley & Sons, New York, 1979.
2. W. Stobe, A. Fink, and E. Bohn, *J. Colloid. Interf. Sci.*, 26 (1968) 62.
3. C.J. Brinker and G.W. Scherer, *Sol-Gel Science*, Academic Press, New York, 1990.
4. K. Unger and D. Kumar, in *Adsorption on Silica Surfaces* (E. Papirer, Ed.), Marcel Dekker, New York, 2000, p. 1.
5. C. Burda, X. Chen, R. Narayanan, and M.A. El-Sayed, *Chem. Rev.*, 105 (2005) 1025.
6. A.C. Pierre and G.M. Pajonk, *Chem. Rev.*, 102 (2002) 4243.
7. M. Pagliaro, *Silica Based Materials for Advanced Chemical Applications* (1st ed.), Royal Society of Chemistry Publications, London, UK, 2009.
8. L.L. Hench, *Sol-Gel Silica: Properties, Processing and Technology Transfer* (1st ed.), Noyes Publications, Upper Saddle River, NJ, 1998.
9. T.J. Barton, L. Bull, G. Klemperer, D. Loy, B. McEnaney, M. Misono, P.A. Monson et al., *Chem. Mater.*, 11 (1999) 2633.
10. S.M. Yang, H. Miguez, and G.F. Ozin, *Adv. Funct. Mater.*, 11 (2002) 425.

11. D. Wang, V. Salgueiriño-Maceira, L.M. Liz-Marzán, and F. Caruso, *Adv. Mater.*, 14 (2002) 908.
12. R.C. Schroden, M. Al-Daous, C.F. Blanford, and A. Stein, *Chem. Mater.*, 14 (2002) 3305.
13. C. Armellini, A. Chiappini, A. Chiassera, and I. Cacciari, *Adv. Sci. Technol*, 55 (2008) 118.
14. J. Persello, in *Adsorption on Silica Surfaces* (E. Papirer, Ed.), Marcel Dekker, New York, 2000, p. 297.
15. G.J.A.A. Soler-Illia, C. Sanchez, B. Lebeau, and J. Patarin, *Chem. Rev.*, 102 (2002) 4093.
16. B.L. Cushing, V.L. Kolesnichenko, and C.J. O'Connor, *Chem. Rev.*, 104 (2004) 3893.
17. S. Sircar, Adsorption, in *The Engineering Handbook* (R.C. Dorf, Ed.), CRC Press, Boca Raton, FL, 1996, Chapter 59.
18. D.W. Green and R.H. Perry, *Chemical Engineering Handbook* (8th ed.), McGraw-Hill, New York, 2008.
19. G.M.S. El Shaffey, in *Adsorption on Silica Surfaces* (E. Papirer, Ed.), Marcel Dekker, New York, 2000, p. 35.
20. R. Roque-Malherbe and F. Marquez-Linares, *Mat. Sci. Semicond. Proc.*, 7 (2004) 467.
21. R. Roque-Malherbe and F. Marquez-Linares, *Surf. Interf. Anal.*, 37 (2005) 393.
22. F. Marquez-Linares and R. Roque-Malherbe, *J. Nanosci. Nanotech.*, 6 (2006) 1114.
23. R. Roque-Malherbe, *Physical Chemistry of Materials: Energy and Environmental Applications*, CRC Press, Boca Raton, FL, 2009.
24. A. Cotton, G. Wilkinson, C.A. Murillo, and M. Bechman, *Advanced Inorganic Chemistry* (6th ed.), Wiley-Intersciences, New York, 2007.
25. H. van Damme, in *Adsorption on Silica Surfaces* (E. Papirer Ed.), Marcel Dekker, New York, 2000, p. 119.
26. J. Rouquerol, F. Rouquerol, P. Llewellyn, G. Maurin, and K.S.W. Sing, *Adsorption by Powder and Porous Solids* (2nd ed.), Elsevier, Amsterdam, the Netherlands, 2012.
27. B.A. Morrow and I.D. Gay, in *Adsorption on Silica Surfaces* (E. Papirer, Ed.), Marcel Dekker, New York, 2000, p. 9.
28. J. Goworek, in *Adsorption on Silica Surfaces* (E. Papirer, Ed.), Marcel Dekker, New York, 2000, p. 167.
29. R. Duchateau, *Chem. Rev.*, 102 (2002) 3525.
30. I.I. Hinic, G.M. Stanisik, and Z.V. Popovic, in *Advances in the Science and Technology of Sintering* (B.D. Stojenovics, V.V. Shorokhod, and N.V. Nikolic, Eds.), Kluwer Academic Publications, New York, 1999, p. 277.
31. R. Roque-Malherbe and F. Diaz-Castro, *J. Mol. Catal. A*, 280 (2008) 194.
32. R. Roque-Malherbe, R. Polanco, and F. Marquez-Linares, *J. Phys. Chem. C*, 114 (2010) 17773.
33. L.T. Zhuraliov, in *Encyclopedia of Surface and Colloid Science* (3rd ed.), (P. Somasundaran, Ed.), CRC Press, Boca Raton, FL, 2015, p. 398.
34. M. Arkan and M. Dogan, in *Encyclopedia of Surface and Colloid Science* (3rd ed.), (P. Somasundaran, ed.), CRC Press, Boca Raton, FL, 2015, p. 6623.
35. R. Roque-Malherbe, F. Marquez, W. del Valle, and M. Thommes, *J. Nanosci. Nanotech.*, 8 (2008) 5993.
36. J.M. Thomas and W.J. Thomas, *Principles and Practice of Heterogeneous Catalysis*, Wiley-VCH, Weinheim, Germany, 1996.
37. B.C. Dunn, D.J. Covington, P. Cole, R.J. Pugmire, H.L.C. Meuzelaar, R.D. Ernst, E.C. Heider, and E.M. Eyring, *Energy Fuels*, 18 (2004) 1519.
38. A. Chiappini, C. Armellini, A. Chiasera, Y. Jestin, M. Ferrari, M. Mattarelli, M. Montagna et al., *Optoelectronics Lett.*, 3 (2007) 184.

39. S.-H. Kim, S.Y. Lee, S.-M. Yang, and G.-R. Yi, *NPG Asia Mater.*, 3 (2011) 25.
40. E.P. Plueddemann, *Silane Coupling Agents* (1st ed., reprint), Springer-Verlag, Berlin, Germany, 2013.
41. K. Flodström, V. Alfredsson, and N. Källrot, *J. Am. Chem. Soc.*, 125 (2003) 4402.
42. N.Y. Turova, *Russian Chem. Rev.*, 73 (2004) 1041.
43. X.S. Zhao and G.Q. Lu, *J. Phys. Chem. B*, 102 (1998) 1556.
44. B.C. Gates, in *Handbook of Heterogeneous Catalysis* (G. Ertl, H. Knozinger, and J. Weitkamp, Eds.), Wiley-VCH, Weinheim, Germany, 1997, p. 793.
45. H.F. Rase, *Handbook of Commercial Catalysts, Heterogeneous Catalysts* (1st ed.), CRC Press, Boca Raton, FL, 2000.
46. J.K. Nkrskov, F. Studt, F. Abild-Pedersen, and T. Bligaard, *Fundamental Concepts in Heterogeneous Catalysis* (1st ed.), John Wiley & Sons, New York, 2014.
47. J.M. Thomas and W.J. Thomas, *Principles and Practice of Heterogeneous Catalysis*, Wiley-VCH, Weinheim, Germany, 1996.
48. I.I. Salame and T.J. Bandosz, *Langmuir*, 16 (2000) 5435.
49. K.S.W. Sing, D.H. Everett, R.A.W. Haul, L. Moscou, R.A. Pirotti, R.J. Rouquerol, and T. Siemieniewska, *Pure Appl. Chem.*, 57 (1985) 603.
50. S.U. Rege and R.T. Yang, in *Adsorption Theory, Modeling and Analysis* (J. Toth, Ed.), Marcel Dekker, New York, 2002, p. 175.
51. D.H. Everet and R.H. Ottewill, *Surface are Determination, International Union of Pure and Applied Chemistry*, Butterworth, London, UK, 2013.
52. S. Lowell, J.E. Shields, M.A. Thomas, and M. Thommes, *Characterization of Porous Solids and Powders: Surface Area, Pore Size and Density*, Kluwer Academic Press, Dordrecht, the Netherlands, 2004.
53. J.B. Condon, *Surface Area and Porosity Determination by Physisorption*, Elsevier, Amsterdam, the Netherlands, 2006.
54. M.M. Dubinin, *Prog. Surf. Membr. Sci.*, 9 (1975) 1.
55. M.M. Dubinin, *American Chemical Society Symposium Series*, ACS, Washington DC, 40 (1977) 1.
56. B.P. Bering, M.M. Dubinin, and V.V. Serpinskii, *J. Coll. Int. Sci.*, 38 (1972) 185.
57. B.P. Bering and V.V. Serpinskii, *Izv. Akad. Nauk, SSSR, Ser. Xim.*, (1974) 2427.
58. R. Roque-Malherbe, *Mic. Mes. Mat.*, 41 (2000) 227.
59. M. Walss, *The Solvay Process* (1st ed.), Tiger Bark Press, New York, 2009.
60. H. Scott-Fogler, *Elements of Chemical Reaction Engineering*, Prentice Hall, NJ, 1999.
61. D.W. Ruthven, *Principles of Adsorption and Adsorption Processes*, John Wiley & Sons, New York, 1984.
62. D. Cazorla-Amoros, J. Alcaniz-Monge, and A. Linares-Solano, *Langmuir*, 12 (1996) 2820.
63. D. Cazorla-Amoros, J. Alcaniz-Monge, J. de la Casa-Lillo, and A. Linares-Solano, *Langmuir*, 14 (1998) 4589.
64. Powder Tech Note 35, Quantachrome Instruments, Boynton Beach, FL.
65. B. Bonelli, B. Civalleri, B. Fubini, P. Ugliengo, C.O. Arean, and E. Garrone, *J. Phys. Chem. B*, 104 (2000) 10978.
66. F.X. Llabrés i Xamena and A. Zecchina, *Phys. Chem. Chem. Phys.*, 4 (2002) 1978.
67. H.A. Al-Abadleh and V.H. Grassian, *Langmuir*, 19 (2003) 341.
68. P. Galhotra, J.G. Navea, S.C. Larsen, and V.H. Grassian, *Energy Environ. Sci.*, 2 (2009) 401.
69. X. Liu, in *Zeolite Characterization and Catalysis* (A.W. Chester and E.G. Derouane, Eds.), Springer-Verlag, Berlin, Germany, 2009.
70. J. Baltrusaitis, J. Schuttlefield, E. Zeitler, and V.H. Grassian, *Chem. Eng. J.*, 170 (2011) 471.
71. A. Saito and H.C. Foley, *AIChE J.*, 37 (1991) 429.

72. R. Evans, in *Fundamentals of Inhomogeneous Fluids* (D. Henderson, Ed.), Marcel Dekker, New York, 1992, p. 85.
73. Y. Rosenfeld, M. Schmidt, H. Lowen, and P. Tarazona, *Phys. Rev. E*, 55 (1997) 4245.
74. A.V. Neimark, P.I. Ravikovitch, and A. Vishnyakov, *Phys. Rev. E*, 64 (2001) 011602.
75. D. Scholl and J.A. Steckel, *Density Functional Theory: A Practical Introduction* (1st ed.), John Wiley & Sons, New York, 2009.
76. E. Engel and R.M. Dreizler, *Density Functional Theory: An Advanced Course: Theoretical and Mathematical Physics*, Springer-Verlag, Berlin, Germany, 2011.
77. K. Koshelev, About density functional theory interpretation, *arXiv*:0812.2919 (2015).
78. J.A. Dunne, R. Mariwala, M. Rao, S. Sircar, R. Gorte, and A.L. Myers, *Langmuir*, 12 (1996) 5888.
79. A.L. Goodman, *Energy Fuels* 23 (2009) 1101.
80. R. Bai, J. Deng, and R.T. Yang, *Langmir*, 19 (2003) 2776.
81. R.M. Barrer and R. Gibbons, *Trans. Faraday Soc.*, 61 (1965) 948.
82. R.M. Barrer, *Zeolites and Clay Minerals as Sorbents and Molecular Sieves*, Academic Press, London, UK, 1978.
83. A. Corma, *Chem. Rev.*, 95 (1995) 559.
84. R. Roque-Malherbe and F. Diaz-Castro, *J. Mol. Catal. A*, 280 (2008) 194.
85. F. Rouquerol, J. Rouquerol, K. Sing, P. Llewellyn, and G. Maurin, *Adsorption by Powders and Porous Solids* (2nd ed.), Academic Press, New York, 2013.
86. D.H. Everet and R.H. Ottewill, *Surface are Determination, International Union of Pure and Applied Chemistry*, Butterworth, London, UK, 2013.
87. S. Lowell, J.E. Shields, M.A. Thomas, and M. Thommes, *Characterization of Porous Solids and Powders: Surface Area, Pore Size and Density*, Kluwer Academic Press, Dordrecht, the Netherlands, 2004.
88. J.B. Condon, *Surface Area and Porosity Determination by Physisorption*, Elsevier, Amsterdam, the Netherlands, 2006.
89. N.N. Avgul and A.V. Kiselev, in *Chemistry and Physics of Carbon*, Vol. 6, (P.J. Walker, Jr., Ed.), Marcel Dekker, New York, 1970, p. 1.
90. M.M. Dubinin, *Carbon*, 18 (1980) 355.
91. H.P. Boehm, *Carbon*, 32(1994) 759.
92. F. Rodriguez-Reinoso and A. Sepulveda-Escribano, in *Handbook of Surfaces and Interfaces of Materials*, Vol. 5 (H.S. Nalwa, Ed.), Academic Press, New York, 2001, p. 309.
93. D. Lozano-Castello, D. Cazorla-Amoros, A. Linares-Solano, and D.F. Quinn, *J. Phys. Chem. B*, 106 (2002) 9372.
94. X. Shao, W. Wang, R. Xue, and Z. Shen, *J. Phys. Chem. B*, 108 (2004) 2970.
95. F. Cecen and A. Ozgur, *Activated Carbon for Water and Wastewater Treatment: Integration of Adsorption and Biological Treatment* (1st ed.), Wiley-VCH, Mannheim, Germany, 2011.
96. S. Alam, *Preparation of Activated Carbon from Low Cost Precursors: And its Use for the Wastewater Treatment*, Lambert Academic Publishing, Germany, 2012.
97. T. Bandosz (Ed.), *Activated Carbon Surfaces in Environmental Remediation, Interface Science and Technology*, Vol. 7, Academic Press, New York, 2006.
98. C. Avellaneda (Ed.), *Proceedings of Nano Carbon 2011*, Springer-Verlag, Berlin, Germany, 2012.
99. T.E. Rufford, J. Zhu, and D. Hulicova-Jurcakova (Eds.), *Green Carbon Materials: Advances and Applications* (1st ed.), Pan Stanford Publishing, Singapore, 2014.
100. T.E. Long, B. Voit, and O. Okay (Eds.), *Porous Carbons*, Springer-Verlag, Berlin, Germany, 2014.

101. C.E. Housecroft and A.G. Sharpe, *Inorganic Chemistry* (4th ed.), Pearson-Prentice Hall, Harlow, UK, 2012.

102. V.V. Turov and R. Leboda, in *Chemistry and Physics of Carbon*, Vol. 27 (L.R. Radovic, Ed.), Marcel Dekker, New York, 2001, p. 67.

103. Z. Vukcevich, Z.M. Jovanovich, Z.V. Lauševich, and M.D. Lauševich, *J. Serb. Chem. Soc.*, 76 (2011) 757.

104. S. Ege, *Organic Chemistry* (5th ed.), Houghton-Mifflin, Boston, MA, 2003.

105. I.I. Salame and T.J. Bandosz, *Langmuir*, 16 (2000) 5435.

106. R.C. Bansal, J.B. Donnet, and F. Stoeckli, *Active Carbon*, Marcel Dekker, New York, 1988.

107. Z.W. Wang, M.D. Shirley, S.T. Meikle, R.L.D. Whitby, and S.V. Mikhalovsky, *Carbon*, 47 (2009) 73.

108. S.L. Goertzen, K.D. Thériault, A.M. Oickle, A.C. Tarasuk, and H.A. Andreas, *Carbon*, 48 (2010) 1252.

109. C.A. Leon-Leon and L.R. Radovic, in *Chemistry and Physics of Carbon*, Vol. 24 (P.A. Thrower, Ed.), Marcel Dekker, New York, 1992, p. 213.

110. T. Karanfil and J.E. Kilduff, *Environ. Sci. Technol.*, 33 (1999) 3217.

111. Z. Yue and J. Economy, *J. Nanopart. Res.*, 7 (2005) 477.

112. C. Hsieh and H. Teng, *J. Col. Interf. Sci.*, 230 (2000) 171.

113. M. Franz, H.A. Arafat, and N.G. Pinto, *Carbon*, 38 (2000) 1807.

114. F. Haghseresht, S. Nouri, J.J. Finnerty, and G.Q. Lu, *J. Phys. Chem. B*, 106 (2002) 10935.

115. J.M. Dias, M.C.M. Alvim-Ferraz, M.F. Almeida, J. Rivera-Utrilla, and M. Sánchez-Polo, *J. Environ. Manag.*, 85 (2007) 833.

116. A.C. Lua and J. Guo, *Langmuir*, 17 (2001) 7112.

117. J.D. Seader, E.J. Henley, and D.K. Roper, *Separation Process Principles* (3rd ed.), John Wiley & Sons, New York, 2011.

118. H. Scott-Fogler, *Elements of Chemical Reaction Engineering* (5th ed.), Prentice Hall, Boston, MA, 2016.

119. C.J. Geankoplis, *Transport Processes and Separation Process Principles* (4th ed.), Prentice Hall, New York, 2003.

120. B.K. Dutta, *Principles of Mass Transfer and Separation Process* (1st ed.), Prentice Hall, Delhi, India, 2007.

121. B. Fubini, V. Bolis, A. Cavenago, E. Garrone, and P. Ugliengo, *Langmuir*, 15 (1999) 5829.

122. J. Zhang and D. Grischkowsky, *J. Phys. Chem. B*, 108 (2004) 18590.

123. E. Garrone and F. Fajula, Acidity and basicity of ordered silica-based mesoporous materials, in *Molecular Sieves Science and Technology: Acidity and Basicity* (A. Auroux, A. Brait, E. Brunner, F. Fajula, E. Garrone, A. Jentys, J.A. Lercher and H. Pfeifer, Eds.), Springer-Verlag, Berlin, Germany, p. 213.

124. E. Garrone and P. Ugliengo, *Langmuir*, 7 (1991) 1409.

125. M. Armandi, V. Bolis, C. Otero-Arean, P. Ugliengo, B. Bonelli, and E. Garrone, *J. Phys. Chem. C*, 115 (2011) 23344.

126. B. Fubini, V. Bolis, A. Cavenago, E. Garrone, and P. Ugliengo, *Langmuir*, 15 (1999) 5829.

127. J. Helmenin, J. Helenius, and E. Paatero, *J. Chem. Eng. Data*, 46 (2001) 391.

128. T.J. Bandosz and C. Petit, *J. Coll. Int. Sci.*, 328 (2009) 329.

129. M. Prasad, Multi-stage water and ammonia refrigeration systems in the light of ozone hole problem, *International Refrigeration and Air Conditioning Conference*, http://docs.lib.purdue.edu/iracc/167, 1992.

130. Z. Knez and Z. Novak, *J. Chem. Eng. Data*, 46 (2001) 858.

131. Y.I. Aristov, M.M. Tokarev, A. Freni, I.S. Glaznev, and G. Restuccia, *Mic. Mes. Mat.*, 96 (2006) 65.

132. L. Zhou, L. Zhong, M. Yu, and Y. Zhou, *Ind. Eng. Chem. Res.*, 43 (2004) 1765.

133. H.Y. Huang, R.T. Yang, D. Chinn, and C.L. Munson, *Ind. Eng. Chem. Res.*, 42 (2003) 2427.

134. V. Smil, *Natural Gas: Fuel for the 21st Century* (1st ed.), John Wiley & Sons, New York, 2015.

135. W. Grochala and P.E. Edwards, *Chem. Rev.*, 104 (2004) 1283.

136. G.W. Crabtree, M.S. Dresselhaus, and M.V. Buchanan, *Phys. Today*, 57 (12) (2004) 39.

137. L. Klebanoff (Ed.), *Hydrogen Storage Technology: Materials and Applications* (1st ed.), CRC Press, Boca Raton, FL, 2016.

138. W. Chen, I. Ermanoskii, and T. Madey, *J. Amer. Chem. Soc.*, 127 (2005) 5014.

139. R. Roque-Malherbe and F. Marquez-Linares, US Provisional Patent Application No. 10/982,798, filed November 8, 2004.

140. J.H. Choung, Y.W. Lee, and D.K. Choi, *J. Chem. Eng. Data*, 46 (2001) 954.

141. C.-M. Wang, T.-W. Chung, C.-M. Huang, and H. Wu, *J. Chem. Eng. Data*, 50 (2005) 811.

142. R. Koppmann, *Volatile Organic Compounds in the Atmosphere*, Wiley-Blackwell, New York, 2007.

143. M. Ncube and Y. Su, *Int. J. Sust. Built Environ*, 1 (2012) 259.

144. T. Dobre, O.C. Pârvulescu, G. Iavorschi, M. Stroescu, and A. Stoica, *Ind. Eng. Chem. Res.*, 53 (2014) 3622.

145. D. Won, R.L. Corsi, and M. Rynes, *Environ. Sci. Technol.*, 34 (2000) 4193.

146. M.A. Hernandez, J.A. Velasco, M. Asomoza, S. Solis, F. Rojas, and V.H. Lara, *Ind. Eng. Chem. Res.*, 43 (2004) 1779.

147. M. Lordgooei, M.J. Rood, and M.R. Abadi, *Environ. Sci. Technol.*, 35 (2001) 613.

148. R.C. Bansal and M. Goyal, *Activated Carbon Adsorption* (1st ed.), CRC Press, Boca Raton, FL, 2005.

149. R.T. Yang, *Adsorbents: Fundamentals and Applications*, John Wiley & Sons, New York, 2003.

150. A. Suleiman, C. Cabrera, R. Polanco, and R. Roque-Malherbe, *RSC Advances*, 5 (2015) 7637.

151. S.W. Rutherford and J.E. Coons, *Langmuir*, 20 (2004) 8681.

152. P. Mani, R. Srivastava, and P. Strasser, *J. Phys. Chem. C*, 112 (2008) 2770.

153. P. Patel, *MIT Technology Review*, May 5, 2010.

154. P. Strasser, S. Koh, T. Anniyev, J. Greeley, K. More, C. Yu, Z. Liu et al., *Nat. Chem.*, 2 (2010) 454.

155. J. Maruyama and I. Abe, *J. Electroanal. Chem.*, 545 (2003) 109.

156. E.F. Vansant, P. van der Voort, and K.C. Vranken, *Stud. Surf. Sci. Catal.*, 93 (1995) 59.

157. S.J. Gregg and K.S.W. Sing, *Adsorption Surface Area and Porosity*, Academic Press, London, UK, 1991.

158. J.K. Brennan, K.T. Thomsom, and K.E. Gubbins, *Langmuir*, 18 (2002) 5438.

159. A. Striolo, in *Nanomaterials Design and Simulations* (P.B. Balbuena and J.M. Seminaris, Eds.), Elsevier, Amsterdam, the Netherlands, 2007.

160. B. Guo, L. Cheng, and K. Xie, *J. Nat. Gas Chem.*, 15 (2006) 223.

161. P.-H. Huang, H.-H. Cheng, and S.-H. Lin, *J. Chem.*, 10 (2015) 10.

162. R.V. Siriwardane, M.S. Shen, E.P. Fisher, and J.A. Poston, *Energy Fuels*, 15 (2001) 279.

163. S. Cavenati, C.A. Grande, and A.E. Rodrigues, *Energy Fuels*, 20 (2006) 2648.

164. S. Sircar, *Ind. Eng. Chem. Res.*, 41 (2002) 1389.

165. K. Liu, C. Song, and V. Subramani, in *Pressure Swing Adsorption in Hydrogen Production and Purification technology*, Wiley, New York, 2009, Chapter 10.

166. F. Adib, A. Bagreev, and T.J. Bandosz, *Environ. Sci. Technol.*, 34 (2000) 686.

167. A. Bouzaza, A. Leplanche, and S. Monteau, *Chemosphere*, 54 (2004) 481.

168. A.A. Lizzio and J.A. DeBarr, *Energy Fuels*, 11 (1997) 284.

169. L.-C. Wu, T.-H. Chang, and Y.-C. Chung, *J. Air Waste Manag. Assoc.*, 57 (2007) 1461.

170. F. Derbyshire, M. Jagtoyen, R. Andrews, A. Rao, I. Martin-Guillon, and E.A. Grulke, in *Chemistry and Physics of Carbon*, Vol. 27, (L.R. Radovic, Ed.), Marcel Dekker, New York, 2001, p. 1.

171. X.S. Zhao, Q. Ma, and G.Q. Lu, *Energy Fuels*, 12 (1998) 1051.

172. S. Follin, V. Goetz, and A. Guillot, *Ind. Eng. Chem. Res.*, 35 (1996) 2632.

173. Y. Zhao and E. Hu, *Adsorption Refrigeration*, Lambert Academic Publishing, Germany, 2012.

174. R.E. Critoph, *Carbon*, 27 (1989) 63.

8 Microporous and Mesoporous Molecular Sieves

8.1 NANOPOROUS MATERIALS

8.1.1 INTRODUCTION

The development of new materials is an essential objective of materials science, since the electronic industry initiated the advancement of minuscule components, ranging dimensions in the region between 1 and 100 nm. Consequently, within this dominion materials properties normally differ from those of the bulk material [1–10].

In the particular case of adsorbents, the most important group is the so-called porous materials, that is, those with pore diameters between 3 and 500 Å, where this collection of adsorbents could be classified as microporous materials, specifically those with pores between 3 and 20 Å, whereas those possessing pore diameters in the range between 20 and 500 Å are designated as mesoporous, whereas those with pores bigger than 500 Å are macroporous adsorbents [11]. These materials have gotten the attention of chemists and materials scientists, on account of their commercial applications together with the scientific curiosity in the challenges posed by their synthesis, processing, characterization, investigation of their sizes, shapes, and adsorption space in relation to the capacity of these materials to perform different functions in certain uses [12–31].

Chemical process at relatively low temperatures and pressures from molecular or colloidal precursors obviously offers an original approach to obtain customized nanostructured materials [12–17,32]. For instance, the mild conditions of sol-gel chemistry offer reacting systems, generally, under kinetic control [32]. Consequently, small modification of the experimental parameters, specifically pH, concentrations, temperatures, nature of the solvent, counter ions, and structure-directing agents, can lead to considerable alteration of the resulting supramolecular assemblies [33]. In particular, zeolites and mesoporous molecular sieves can be incorporated in a collection of materials: crystalline and ordered nanoporous materials, correspondingly, because synthetic zeolites are crystalline materials comprising pores together with cavities of molecular dimensions, specifically 3–15 Å, hence generating a microscale framework, which can be filled with water or other guest molecules [34]. On the other hand, mesoporous ordered silicas and aluminosilicates containing pores of mesoporous dimensions, specifically from 20 to 100 Å, which display different ordered structures but not crystalline structures building up a mesoscale framework, which could also be occupied with water or other molecules [17].

8.1.2 Microporous Molecular Sieves

Zeolites and related materials are molecular sieves, which can be applied for the development of selective separation processes [2]. Heterogeneous acid catalysts [35,36] and adsorbents [2,24,30,38–40] are being produced by the majority of world's gasoline by the fluidized catalytic cracking (FCC) of petroleum using zeolite catalysts, because of their pore size and shape selectivity together with the potential for strong acidity [36].

Furthermore, some minerals are characterized by a negatively charged framework bearing cavities, cages, or tunnels in which water molecules along with inorganic cations are occluded as charge-compensating ions. For instance, natural zeolites define a family of crystalline microporous aluminosilicates, presenting pore sizes of $d < 1$ nm [24,38], having these materials applications in wastewater cleaning, agriculture, fertilizers, aquaculture, animal health, animal nourishment, gas separation, solar refrigeration, gas cleaning, deodorization, solid electrolytes, construction materials, and cleaning of radioactive wastes [40].

In reply to the developing needs in both industry and fundamental research, it has been an increasingly growing attention in increasing the pore sizes of materials from the micropore region to the mesopore subdivision of porosity. In 1988, the first description of a crystalline microporous material with regular pores larger than 10 Å appeared. To be exact, the aluminophosphate VPI-5 was synthesized opening the area of extralarge pore crystalline materials [16]. Later more extralarge pore crystalline materials were developed, that is, ALPO4-8, Cloverite, JDF-20, ULM-5, UTD-1, ULM-16, CIT-5, ND-1, FDU-4, and NTHU-1 [15].

8.1.3 Mesoporous Molecular Sieves

In parallel to the work on extra-large pore crystalline materials, scientists working at Mobil Corporation developing the concepts used in zeolite synthesis have taken advantage of this methodology to generate novel inorganic materials. This resulted in the discovery of the M41S family of mesoporous-ordered silicas with hexagonal and cubic symmetry and pores sizes ranging from 20 to 100 Å through the use of surfactants as organizing agents [17]. The mesoporous structure can be regulated by a complex choice of templates, that is, surfactants, adding auxiliary organic chemicals, for example, mesitylene and changing reaction parameters, for example, temperature and compositions [33,42].

For the synthesis of mesoporous molecular sieves (MMS), surfactant liquid crystals with long *n*-alkyl chains are applied as structure-directing agents, allowing a thorough control of the pore size on this methodology, provided the pore diameters are tunable from 15 to 300 Å producing narrow pore size distribution (PSD), high surface area, and pore volumes [21], particularly a family of a hexagonal phase known as MCM-41, a cubic phase known as MCM-48, and a lamellar phase known as MCM-50 were developed [43]. The comprehension of the mechanism for the arrangement of these mesostructured materials placed the bases for the synthesis of silicas with a defined PSD, as silica thin films [44], fibers [45], and spheres [46].

8.2 STRUCTURE OF MICROPOROUS AND MESOPOROUS MOLECULAR SIEVES

8.2.1 MICROPOROUS MOLECULAR SIEVES

Natural zeolites were the first example of molecular sieves. In this regard, the initially found zeolite, that is, stilbite was discovered in 1756 by the Swedish scientist Freiherr Axel Fredrick Cronsted throughout the gathering of minerals in a copper mine in Lappmark, Sweden. He baptized the new mineral with the name zeolite, since the mineral during the blowpipe test showed a characteristic intumescence. Thereafter, the expression zeolite is derived from two Greek roots *zeo* to boil and *lithos* a stone [12]. Later, in 1932, McBain introduced the term molecular sieve when he discovered that the zeolitic mineral, chabazite, selectively adsorbed molecules smaller than 5 Å in diameter [47].

Natural and synthetic zeolites belong to the group of molecular sieves existing in more than 1000 reported materials [34] that are three-dimensional microporous crystalline solids. In this regard, as an exemplification of how the zeolite structures are assembled, here, precisely, three of the most important framework types are shown: the LTA framework type (Figure 8.1a) corresponding to the following materials A, LZ-215, SAPO-42, ZK-4, ZK-21, ZK-22, and alpha, along with the framework-type FAU (Figure 8.1b) related to the natural zeolite faujasite and the synthetic ones: X, Y, EMC-2, EMT, ZSM-3, and ZSM-20. Finally, the HEU framework-type (Figure 8.1c) compatible with the natural zeolites heulandite and clinoptilolite and the synthetic zeolite LZ-219 is shown [2,24,34].

The LTA framework structure is composed of sodalite cages placed in the corners of a cube with axis, $a \approx 11.9$ Å, linked by 4–4 secondary building units forming 8-member rings (MR) windows leading to an α-type supercage (Figure 8.1a). Meanwhile, the FAU framework involves sodalite cages connected 6–6 secondary building units in a face centered cubic lattice (FCC) producing a cube with $a \approx 7.24$Å, whereas in diamond only half of the tetrahedral sites of the FCC lattice are occupied producing a cube with an axis, $a \approx 24.3$ Å (Figure 8.1b), containing the generated framework 12-MR windows leading to a β-type supercage. Finally, the structure of clinoptilolite (HEU type) shows 3 channels, that is, one 8-MR channel along [100] with an access of 2.6 A × 4.7 A, together with two parallel channels along [001] one 8-MR with a window of access with 3.3 A × 4.6 A and a 10-MR with an access of 3.0 A × 7.6 A (Figure 8.1c) [12,13,24].

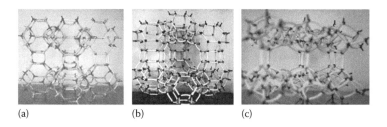

(a)　　　　　　　(b)　　　　　　　(c)

FIGURE 8.1 Representation of the LTA (a), FAU (b), and HEU (c) framework types.

The previously reported frameworks are related to aluminosilicate zeolites whose frameworks are built from AlO_2^- and SiO_2 tetrahedrons linked in the corners sharing all oxygen atoms, provided it is necessary to recognize at this point that each of the framework tetrahedrons really contains two oxygen atoms each, since the oxygen atom is shared between two tetrahedrons. Hence, it must be described as TO_2 and not TO_4. Further, the presence of tetracoordinated Al(III) generates a negative charge, which is balanced by extra-framework cations, that is, one per Al(III) [34,48]. Consequently, the chemical composition of aluminosilicate zeolites can be expressed as $M_{x/n}[(AlO_2)_x(SiO_2)]zH_2O$, where M denotes the balancing cations, which compensate the charge from the *Al(III)*, and z is the water contained in the voids of the zeolite, provided the balancing cation is a metal or another species. For example, ammonium, which is located within the zeolite channel or cavity can be exchanged giving the zeolite its ion-exchange property [49,50].

Si and Al are the T atoms in aluminosilicate zeolites. Nevertheless, other elements, such as P, Ge, Ga, Fe, B, Be, Cr, V, Zn, Zr, Co, Mn, and others can also be T atoms, provided these elements in tetrahedral coordination form a crystalline microporous material similar to a zeolite with or without charge-compensating cations, such that the electroneutrality principle is satisfied [2,35–37]. Figure 8.2 shows a framework-type FAU. However, the most important group of molecular sieves is microporous solids with high Si/T ratio, in which T is trivalent Al, Fe, B, Ga, or tetravalent Ti, Ge.

High together with all silica zeolites are essentially impure silica or silica polymorphs that are intrinsically hydrophobic and organophilic, provided it is established that the following categories classify them, that is, high silica zeolites for those where Si/T < 500 and zeosils or all-silica zeolites when Si/T > 500, where the last mentioned compounds are basically Si-based molecular sieves. However, opposing to clathrasils, the porosity of these materials is accessible, provided both zeosils and clathrasils are the family of porosil silica-based materials [51–58]. The majority of the members of the ZSM molecular sieves group along with some of the members of the SSZ family are high silica zeolites. From this ZSM-5 with MFI

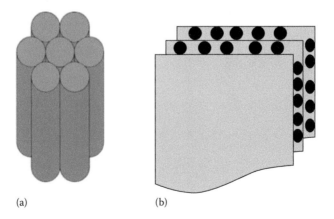

(a)   (b)

FIGURE 8.2 Hexagonal phase (a) and lamellar configuration (b).

framework-type and ZSM-11 with MEL-type, there are some of the most important materials in zeolite science and technology [34].

The Si/Al relation of aluminosilicate zeolites is also increased by dealumination. This process can be made by hydrothermal treatments, such as steaming, acid leaching, or treatments in flowing $SiCl_4$ at 200°C–300°C or hexafluorosilicates [59,60]. Dealuminated materials have different uses in catalysis and adsorption, specifically USY zeolite (ultra stable zeolite Y) is widely used in catalytic cracking [30]. Dealuminated Y (DAY) zeolite is a FAU-type zeolite. However, unlike Na–Y, DAY ideally has no Al in the framework, and hence no Na^+ ions to balance the charge [61,62].

All silica zeolites are three-dimensional microporous crystalline solids, which are built from SiO_2 tetrahedra linked in the corners, sharing all oxygen atoms. In this regard, in 1978 a new polymorph of silica, labeled silicalite displaying the MFI framework was synthesized [62], being constituted as the framework of this material by a three-dimensional system of intersecting 10-MR channels as in the case of the ZSM-5 zeolite whose Si/Al ratio varies between 10 and 500 [12,34]. To date, all silica zeolites with the following framework types: MEL [63], MTW [64], AFI [65] BEA [66], IFR [67], ITE [68], AST [66], CFI [67], CHA [68], MWW [69], STF, STT [67], ISV [56], CON [56], MTT [57], RUT [58], ITW [59], and LTA [60] have been synthesized, mostly in fluoride media [61].

8.2.2 MESOPOROUS MOLECULAR SIEVES

Notwithstanding with the considerable amount of work dedicated to zeolites and related materials, their dimensions together with the ease of access in the pores of these materials are confined to the microporous scale. This reality reduces the use of zeolites to small molecules. Therefore, a noteworthy effort has been made to produce materials exhibiting larger pore size [17,23,35], being a significant result of this endeavor the discovery in 1992 by researchers at Mobil Corporation of the M41S family of mesoporous molecular sieves [43].

This new family of mesoporous silica together with aluminosilicate compounds were obtained by the introduction of supramolecular micellar aggregates, rather than molecular template species were used as structure-directing agents so that the growth of inorganic or hybrid networks templated by structured surfactant assemblies permitted the construction of a novel type of nanostructured materials in the mesoscopic scale (2–100 nm) [42,70–75], leading this supramolecular directing concept to a family of materials whose structure, composition, and pore size can be tailored during synthesis by variation of the reactant stoichiometry, nature of the surfactant molecule, or by postsynthesis functionalization techniques [17]. Thereafter, the obtained solid phases, which are not ordered crystalline mesoporous materials because of the lack of precise atomic positioning in the pore wall structure as shown by MAS–NMR and Raman spectroscopy [15,76] exhibiting sharp pore size dispersions, giving rise to inorganic solids with enormous differences in morphology and structure [23].

It is required now to clarify that the assemblage mechanism of this mesoporous type of molecular sieves is regulated by two features: first, the dynamics of the surfactant molecules to develop molecular structures conducting to micelles, and ultimately, liquid crystal formation [17]. Second, it is the ability of the inorganic oxide to experience condensation reactions to structure the extended thermally stable configurations [42].

The initially produced materials implicated the formation of silicates using alkyl-trimethylammonium cationic surfactants in a basic medium. Later efforts have demonstrated that these structures can also be formed in acid media [77,78] or by using neutral normal amines, [79] nonionic surfactants [80], and dialkyldimethylammo-nium cationic surfactants [81]. Moreover, several mechanistic studies have expanded the initial pathway studies to a more generalized view of an organic/inorganic charge balance driving force for the formation of these structures [76–78,82,86]. In this regard, the representation of one of the original members of the M41S family, that is, the hexagonal phase-labeled MCM-41 is shown in Figure 8.2 (a) together with the stabilized lamellar phase named MCM-50 (b) [17,42].

The MCM-41 shows an X-ray diffraction pattern, including three or more low-angle peaks, that is, below 10° in 2θ, which can be indexed to a hexagonal lattice [85]. The structure is proposed to have a hexagonal stacking of uniform diameter porous tubes whose size can be varied from about 15 to more than 100 Å [24]. In this regard, an example of the characteristic X-ray diffraction pattern of the MCM-41 mesoporous molecular sieve is shown in Figure 8.3 [85,86].

MCM-48, the cubic material, exhibits an X-ray diffraction pattern consisting of several peaks that can be assigned to the Ia3d space group [87]. The structure of MCM-48 has been proposed to be bicontinuous with a simplified representation of two infinite three-dimensional, mutually intertwined, unconnected network of rods [88]. Meanwhile, the MCM-50, that is, the stabilized lamellar structure shows an X-ray diffraction pattern consisting of several low-angle peaks that can be indexed to (h00) reflections. Hence, this material can be a pillared layered material with inorganic oxide pillars separating a two-dimensional sheet similar to layered silicates, such as magadiite or kenyaite [42]. Alternatively, the lamellar phase could be represented by a variation in the stacking of surfactant rods such that the pores of the inorganic oxide product could be arranged in a layered form.

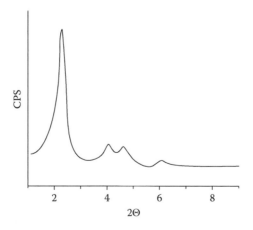

FIGURE 8.3   X-ray diffraction pattern corresponding to MCM-41.

Another M41S-type mesoporous materials are the materials labeled SBA-1, which are phases whose framework is cubic satisfying the *Pm3n* group symmetry [76] together with the material named SBA-2 whose framework is also cubic but fulfilling the P63/mmc group symmetry [89]. Further, synthesis efforts have produced new materials such as the SBA-15 mesoporous silica, presenting a well-defined hexagonal structure, large surface area, and high hydrothermal stability, which give them high potential for a variety of technological applications [85,91,92].

Between the diversity of silica mesophases, the utmost widely studied have been the MCM-41 together with the SBA-15. Nevertheless, independent of the fact that both materials display two-dimensional hexagonal P6mm frameworks exhibit some differences, that is [42,43,85,89–94]

- SBA-15 has larger pores and thicker pore walls than MCM-41.
- MCM-41 is simply mesoporous, whereas SBA-15 show micropores.
- While MCM-41 channels are not connected those of SBA-15 are interconnected.

In addition to the nature of the pore system, specifically pore size, shape, together with connectivity, depending on the morphology of the mesophase may be significant. In this regard, simple morphologies with short, unconstrained path lengths such as small spheres, crystal-like, and straight rods are useful for applications controlled by intraparticle diffusion processes, such as catalysis, separation, guest molecule encapsulation, and internal surface modification [81]. Consequently, not unexpectedly, extensive work has been devoted to the morphological control of mesoporous silicas [95–98], provided most of the approaches are based on changes in synthesis conditions, including the silica source, the nature of the surfactants, cosurfactants, cosolvents, and additives, and the overall composition of the synthesis mixture [81], where the understanding of the formation mechanism of these mesostructured materials laid the foundations for the synthesis of silica mesophase with a defined PSD, such as thin films [44], fibers, spheres [45], short rod-like particles, along with more exotic particles having doughnut, discoid-like, and other shapes have also been obtained using organic additives and inorganic salts [98].

8.3 SYNTHESIS

8.3.1 ALUMINOSILICATE ZEOLITE SYNTHESIS

Aluminosilicate zeolites are as a rule synthesized in hydrothermal conditions using solutions containing sodium hydroxide, sodium silicate, or sodium aluminate, where the particular zeolite produced is evidently determined by the reactants along with the synthesis specifications used, such as temperature, pH, and time [12–16]. In particular, aluminosilicate zeolites are synthesized in three steps: (1) induction, (2) nucleation, and (3) crystallization, where nucleation is the stage where germ nuclei are formed from very small aggregates of precursors becoming larger with time. Then, crystallization starts comprising the germ nuclei from the nucleation step, and other components of the reaction mixture being in this process are affected by several factors that can be modified during the synthesis procedure, specifically the presence of cations in the reaction mixture, OH$^-$, concentration, SiO_2/Al_2O_3 ratio, H_2O, content, temperature, pH, time, aging, stirring of the reaction mixture, order of mixing, and other factors [13–16].

The first hypothesis [12] to explain the crystallization of aluminosilicate zeolites stated that it is produced through the formation of the aluminosilicate gel or reaction mixture along with the nucleation and growth of zeolite crystals from the reaction mixture. This pioneering model has been practically abandoned and is displaced by another hypothesis [13]. Supposing that the formation of zeolite crystals takes place in solution. Consequently, in this model the nucleation and growth of crystalline nuclei are a result of condensation reactions between soluble species, where the gel perform a limited role as a reservoir of matter.

Summarizing, aluminosilicate zeolites are obtained by hydrothermal crystallization of a heterogeneous gel, which consists of a liquid and a solid phase, where the reaction media contain the following sources: T atoms as Si, Al, P,..., that form the framework, charge compensating cations, that is, Na, K, along with the sources of mineralizing agents, specifically HO$^-$ and in some cases organic, cationic, or neutral molecules along with the solvent in general water.

8.3.2 HIGH SILICA ZEOLITES

The synthesis techniques in the cases of high silica, all silica, and other nonaluminosilicate zeolites are similar to the methods used in aluminosilicate zeolites. Nevertheless, in this case the initial gel composition is different [99–121], provided some aspects are changed that are involved in the process. Concretely, organic substances known as structure directing agents (SDA), normally organic cations are included in the synthesis batch [33], where these SDAs stabilize structures with cavities and shapes of similar dimensions to the ones with the organic cations [35,99]. So far, amines and related compounds, that is, quaternary ammonium cations along with linear or cyclic ethers and coordination compounds, that is, organometallic complexes have been the most frequently applied organic templates [35,62]. The first high silica zeolites, that is, EU, NU, and the ZSM series were patented in the late 1960s or early 1980s [100,123], whereas all silica zeolite, silicalite, as was previously stated, was obtained in 1978. In this sense, pure silica zeolites have been synthesized fundamentally in fluoride media [58] using different SDA, the following framework types are being produced: MFI [39], MEL [40], MTW [41], AFI [42] BEA [43], IFR [44], ITE [45], AST [46], CFI [47], CHA [48], MWW [49], STF [50], STT [51], ISV [52], CON [53], MTT [54], RUT [55], ITW [56], and LTA [57] are represented as members of this series of molecular sieves.

8.3.3 ALUMINOPHOSPHATES

Microporous aluminophosphate molecular sieves AlPO, SAPO, and MeAPO were developed by Union Carbide, that is, SAPO molecular sieves were obtained by incorporation of Si in the AlPO framework [62], whereas MeAPO molecular sieves were obtained by the inclusion of Me (Me = Co, Fe, Mg, Mn, Zn) in the AlPO framework

[101,121–125], growing the crystalline microporous phosphates considerably with the synthesis of beryllophosphates [98], vanadophosphates [99], beryllophosphates [102], gallophosphates [103], zincophosphates, and ferrophosphates [104].

In the case of aluminophosphate families and derived compounds (SAPO, MeAPO, etc.), the reaction pH is between 3 and 10 [16,124]. Anions such as hydroxide or fluoride collaborate in the dissolution of the reactive silica moieties in the gel and their transport to the developing crystals [33]. In addition, F^- anions can play the role of a costructuring agent by stabilizing certain building blocks of the inorganic network [105]. Moreover, nonaqueous routes [106,107] or dry synthesis methods [109] have also been explored.

The SDAs are often occluded in the microporous cavities and channels of the synthesized materials, contributing to the stability of the obtained zeolite. The guest-framework stabilizing interactions can be of coulombic, H-bonding, or van der Waals nature. Nevertheless, guest–guest interactions can also contribute to the total energy [10,14–16]. Different factors concerning SDA must also be considered, for instance, [10,16,109] size. The size of the SDA is in direct relation with the cavity or pore size of the zeolite, even though this effect clearly depends on the temperature.

Besides the selection of the SDAs and the batch temperature of other components and/or factors in the synthesis are of relevant importance, for example, the hydroxide ion that acts by adjusting the degree of polymerization, hence increasing the crystal growth. In addition, the OH^-/SiO_2 ratio has been correlated with the pore size [10,16].

On the other hand, many synthesis procedures are carried out using fluoride anions, instead of hydroxide ions, resulting in zeolites or related material with higher crystal sizes and lower structural defects with respect to conventional procedures using OH^- [51–60], temperature also being a significant factor. In this regard, the temperatures applied are below 350°C, giving that high values of temperature yield more condensed phases. Furthermore, the pH of the synthesis mixture is of crucial importance, happening that are in general alkaline, that is, pH $>$ 10. Finally, the reaction time is another parameter to consider. It should be optimized because in different reaction times, different zeolites or phases can be obtained at the same reaction mixture, whereas stirring the reaction mixture affects the zeolitic structure and clearly the particle size [61–69].

All silica zeolite phases can be prepared hydrothermally using either hydroxide or fluoride as mineralizers. Both methods will be designated here according to the mineralizer as the OH^- and F^- methods, where, usually, tetraethyl orthosilicate (TEOS) is hydrolyzed in an aqueous solution of the suitable structure-directing agent in its hydroxide form. Then, the resultant mixture is left under stirring until complete evaporation of the ethanol is produced, that is, within the detection limit of 1H MAS NMR no ethanol could remain in the synthesis mixture after evaporation [110]. This indicates that these water/alcohol/silica/SDA mixtures do not form azeotropic compositions as opposed to the known behavior of water/ethanol mixtures. After ethanol evaporation, HF (48% aqueous solution) is added and the mixture is homogenized by stirring. The resulting mixture is always a slurry with a viscosity that depends on the specific SDA used and the final water content [118].

Concretely, for the synthesis of pure silica phases, it is possible to use different SDAs in fluoride media. The SDAs are chosen primarily according to criteria defined as significant in determining a high structure-directing ability, rigidity, size, shape, along with the C/N^+. Consequently, SDAs with polycyclic moieties (giving rise to SDAs with relatively large and rigid portions) predominate, where after the synthesis procedure, the zeolite should be calcined to remove the organic compounds that are blocking the pores, where calcination entails heating the as-synthesized zeolite in an air flow to a temperature normally ranging from 350°C to 400°C [51–69].

In general, zeolite synthesis requires considering many factors, where the relation between these experimental conditions together with the nature of the synthesized product is not trivial. Hence, to overpower these difficulties different strategies have been proposed, among them the combinatorial approach is a hopeful methodology consisting of the use of miniaturized multiautoclave systems to finally explore a very high number of possibilities.

8.3.4 MESOPOROUS MOLECULAR SIEVES

In the past two decades, an important effort has been made on obtaining molecular sieves showing larger pore size. The introduction of supramolecular assemblies, that is, micellar aggregates, rather than molecular species, as SDAs allowed a new family of mesoporous silica and aluminosilicate compounds (M41S) to be obtained [16,17,70–75]. The novel group of MMS, M41S was discovered by expanding the concept of zeolite synthesis with small organic molecules as SDAs to longer-chain surfactant molecules. Then, rather than individual molecular directing means participating in the ordering of the reagents to form the porous material, assemblies of molecules are responsible for the formation of these pore systems [43]. This supramolecular-directing notion has led to a type of materials whose structure, composition, and pore size can be tailored during the synthesis by variation of the reactant stoichiometry, nature of the surfactant molecule, or by postsynthesis functionalization techniques [42].

The characteristic methodology for the development of mesostructured materials is the use of the next standard synthesis conditions: low temperatures, coexistence of inorganic and organic moieties, and extensive choice of precursors, where the chemical, spatial, and structural properties of the texturing agent must be cautiously adjusted by controlling the rates of chemical reactions, the nature of the interfaces, and the encapsulation of the growing inorganic phase, provided the judicious tailoring of the organomineral interface the most important fact to obtain well-defined textured phases [33].

In this regard, the formation mechanism of this mesoporous group of molecular sieves is determined by two features: (1) the dynamics of surfactant molecules to form molecular assemblies, which lead to micelle and (2) ultimately liquid crystal formation together with the ability of the inorganic oxide to undergo condensation reactions to form extended and thermally stable structures. Thereafter, the reaction gel chemistry is considered to perform a significant role in MMS synthesis, provided the understanding of the chemistry of surfactant/silicate solution is a prerequisite for the synthesis and mechanisms that are responsible for the formation of MCM-41 and other MMS from its precursors [33,42].

In a simple binary system of water–surfactant, surfactant molecules show themselves as very active constituents with changeable structures in agreement with concentrations, where they exist as monomolecules at low concentrations, whereas with growing concentration, surfactant molecules combine to produce micelles to reduce the system entropy [84], provided the initial concentration limit at which monatomic molecules accumulate to form isotropic micelles named critical micellization concentration (CMC). Moreover, as the concentration process persists, hexagonal closely packed arrays emerge producing hexagonal phases. Next, as the coalescence persists, parallel cylinders to produce the lamellar phase or the cubic phase also appear prior to the lamellar phase [127]. Further, the specific phase present in a surfactant solution at a particular concentration depends also on the length of the hydrophobic carbon chain, hydrophilic head group, along with the counter ion, pH, temperature, the ionic strength, and other additives, the fact reflected in the effect of the aforesaid matters on the CMC, which decreases with the growth of the chain length of a surfactant, valence of the counter ions together with the ion strength in a solution. However, it increases with the growing counter ion radius, pH, and temperature [128].

For instance, if in an aqueous solution at 25°C, the CMC for the surfactant $C_{16}H_{33}(CH_3)_3N^+Br^-$ is about 0.83 mM been present in 11 wt. % of small spherical micelles. While in the concentration range of 11–20.5 wt. %, elongated flexible rod-like micelles are formed [129], whereas hexagonal liquid crystal phases appear in the concentration region between 26 and 65 wt. % followed by the formation of cubic, lamellar, and reverse phases with increasing concentration [84]. Besides, at 90°C, the hexagonal phase is observed at a surfactant concentration of more than 65% [130], being the significance of this new *organized matter soft chemistry synthesis* endlessly growing [131]. These materials are possible candidates for a diversity of applications, in the fields of catalysis [132], optics, photonics, sensors, separation, drug delivery, and sorption, acoustic, or electrical insulation, and ultra-light structural materials [133–135].

Silica-based materials are the most studied systems for several reasons: a great variety of possible structures, a precise control of the hydrolysis–condensation reactions, enhanced thermal stability of the obtained amorphous networks, and strong grafting of organic functions [33]. In this regard, the M41S family was originally obtained by hydrothermal synthesis in basic media from inorganic gels containing silicate (or aluminosilicate) in the presence of quaternary trimethylammonium cations [17]. Moreover, a pseudomorphic synthesis based on the dissolution–reprecipitation of silica microspheres in alkaline media in the presence of C_{16}TMABr allowed the production of morphologically controlled MCM-41 [136]. Hexagonal mesoporous silica (HMS) compounds are obtained at neutral pH, according to a new synthesis path using primary amines $C_nH_{2n+1}NH_2$ ($n = 8 - 18$) as amphiphilic molecules [137,138].

A great deal of work has been dedicated to the pore size control, being possible to tailor the pore size from 15 to 45 Å by varying the chain length of C_nTMA$^+$ cations between 8 and 18 carbon atoms, whereas the addition of organic molecules, such as 1,3,5-trimethylbenzene [42] or alkanes [139] permitted increased pore size up to 100 Å, where these swelling agents are soluble in the hydrophobic part of the micelle growing the volume of the template [33]. Nevertheless, this method, although simple

in form, is difficult to put into practice, lacks reproducibility, and produces less organized mesophases. Then, as an option to the swelling agent, an efficient method relies
on an extended hydrothermal treatment in TMA^+ solutions. This process makes the
pore organization better, as well as growing the pore size [140]. Nevertheless, the
pore size of MCM-41 and related materials is limited by the size of the micellar templates. A natural expansion to increase the pore size consists of making use of bigger
molecules, such as polymers or more complex texturing agents.

Amphiphilic block copolymers (ABC) could be utilized for ordered mesoporous
materials synthesis. ABCs represents a new class of functional polymers with a
strong application potential mainly due to the high energetic and structural control that can be exerted on the material interfaces [150]. The chemical structure
of ABCs can be programmed to tailor interfaces between materials of totally different chemical natures, polarities, and cohesion energies [150]. The *traditional*
surfactants polymer organized systems (POS) formed by ABC polymers are excellent templates for the structuring of inorganic networks [151]. They have also been
used for growth control of discrete mineral particles [151], diblock (AB), or triblock
(ABA). Block copolymers are generally used, in which A represents a hydrophilic
block (polyethylene oxide [PEO] or polyacrylic acid [PAA]) and B, a hydrophobic
block (polystyrene [PS], polypropylene oxide [PPO], polyisoprene [PI], or polyvi-
nylpyridine [PVP]) [150].

8.3.5 MODIFICATION OF ORDERED SILICA MESOPOROUS MATERIALS

The enlarged pore size of ordered mesoporous silicas contrasted with microporous zeolites offer us with many opportunities to include diverse organic guest
species within extremely porous structures in a very methodical and planned way
[10,152–165]. Organic moieties are able to easily functionalize. This process creates
active sites in the solid state for catalysis, ion exchange, or adsorption, while taking
advantage of the exclusive textural properties of the mesoporous materials [165].
Such porous composite materials are supposed to bring new possibilities for the
investigation of exceptional physical and chemical behaviors of molecules confined
inside the nanospace, which will be significant for the design of innovative materials for chromatography, sensing, electronic, and optoelectric devices, or recyclable
stable heterogeneous catalysts [165].

So far, mesoporous organic-silica composite materials were, as a rule, prepared
both by cocondensation reactions of organosilanes directly during the synthesis of
the mesoporous material, or by grafting organosilanes onto preprepared mesoporous
silica surfaces [10,165]. Significant advantage of both approaches is on the chemical
and thermal stability of the organic moieties taking place from the strong covalent
bonding between the organics and the silica walls. The great order of MCM-41 and
related phases makes these materials particularly interesting as support materials,
since the framework ensures an ordered structure, and thermal and mechanical stability [10].

Postsynthesis methods, such as functionalization of pore walls can affect the
pore size and the surface chemistry in mesoporous molecular sieves materials. For
example, MCM-41 samples contain a large concentration of silanols, which can be

functionalized through simple elimination reactions. This postsynthesis technique can be used to alter the pore size or affect the hydrophobicity of the pore wall [24]. Otherwise, other species can be used to anchor moieties having specific catalytic or adsorptive properties. That is, the organic species incorporated to the material allow fine control of the interfacial and bulk properties, such as hydrophobicity, porosity, accessibility, optical, electrical, or magnetic properties [10].

Organic functions can be attached onto the oxide walls leading to hybrid meso-structured materials with tunable surfaces. This is a certainly promising subject in the design of sophisticated materials, such as catalysts, membranes, sensors, and nanoreactors [153]. A considerable quantity of methods have been developed or modified to add organic functions to the walls of the mesoporous silica [154] combining the properties of a mesoporous inorganic structure with the surface organic groups [10].

The inclusion of the organic functions can, in principle, be carried out in two modes: (1) covalent binding on the inorganic walls of the material and (2) by direct inclusion of the organic functions on the synthesis process [10]. In the first method, organochlorosilanes or organoalkoxysilanes have been extensively utilized to connect particular organic groups by condensation reactions with silanol or Si–O–Si groups of the silica framework [155,156]. The mesoporous hosts must be carefully dried before adding the organosilane precursors to elude their autocondensation in the presence of water [10]. The concentration and delivery control of the organic functions are limited by the surface silanols and their ease of access. The grafting ratio depends on the precursor reactivity, being also restricted by diffusion and steric features. A substitute line of attack for pore functionalization relies on a direct synthesis based on the cocondensation of siloxane and organosiloxane precursors *in situ* to give up modified MCM-41 in one step [157]. Although siloxane precursors guarantee the formation of the mineral network, organosiloxane moieties perform a twofold role. That is, contribute as building blocks of the inorganic structure and provide organic groups [10]. This one-pot pathway presents several advantages, such as high modification ratios, homogeneous incorporation, and short preparation times [158].

On the other hand, the periodic nanostructures of MCM-41 mesoporous oxides and other mesoporous molecular sieve materials recommend that they could operate as model hosts for reasonable nanomanufacturing [159]. The purpose of numerous studies has been to develop a methodology of using ordered nanoporous oxide materials as generic nanoscale reactors for making and replicating technically important nanomaterials [150–153].

Two approaches have been elaborated to transport precursor molecules or ions for the construction of nanoparticles or nanorods inside the channels of MCM-41 and related mesoporous oxides. The first technique requires the direct impregnation of mesoporous materials with precursor molecules or ions [160–162]. The next method use functional ligands to, at random, functionalize together the internal and the external surfaces of mesoporous oxides followed by inserting the precursor compounds through affinity interaction between the functional ligands and metal ions [163].

The construction of silver nanorods inside SBA-15 through the direct impregnation and evaporation of silver ion sources has been reported [161]. A new gas-phase transport method to load MCM-41 with cluster compounds have also been developed [162]. The main problem associated with these methodologies is the complication in controlling the site of the growth of nanoparticles and preventing their uncontrolled aggregation on external surfaces of the MCM-41 [159]. An accurately regulated technique of building nanoparticles within the mesopores of externally functionalized MCM-41 materials through the controlled transport of metal ions through ion-exchange reactions have also been developed [159]. This new methodology makes use of the single mesoscopic arrangement of as-synthesized MCM-41 whose external surfaces can be selectively passivated with inert organic groups to remove the external nucleation sites but still retain the ion exchangeability of the internal pore surfaces [159].

Copper–tellurolate cluster [(Cu$_6$(TePh)$_6$(PPh$_2$Et)$_5$] has been loaded into the pores of MCM-41 by solid-state impregnation techniques [164].

Independently, numerous studies were reported on the encapsulation of organic polymers within the channels of mesoporous silica materials [165–171]. For example, Bein and coworker reported the polymerization of polyaniline and other monomers inside the mesopore system of MCM-41 [166]. The polyaniline system with conjugated polymers with mobile charge carriers within nanometer-size galleries can be considered as a noteworthy pace in the direction of the design of nanoscale electronic devices [166]. Afterward, Moller et al. studied the polymerization of methyl methacrylate within microporous and mesoporous silicas with the resultant polymer possessing different physical properties as compared with the bulk polymer [166,167]. Tolbert and coworkers demonstrated that semiconducting polymers aligned within mesopores showed exclusive energy transfer and photophysical properties that are promising for the preparation of electronic and optoelectric devices [169,170].

In all these studies, the polymers filled the entire volume of the silica mesopores, resulting in nonporous materials for most of the cases. However, certain notable aspects of the nanoscale chemistry and physics of polymers confined within mesoporous channels were revealed. Very recently, Shantz and coworkers reported the synthesis of dendrimers inside the mesoporous silica [171].

8.3.6 ZEOLITE MODIFICATION

The majority of the methods developed for zeolite modification are related to aluminosilicate zeolites [59]. For example, one very common modification is the ion exchanging of the zeolitic structures. Synthetic zeolites are normally obtained in the sodic form, and natural zeolites contain Na$^+$, K$^+$, Ca^{2+}, and Mg^{2+} in its mineral form [31] between other cations. Therefore, sometimes it is necessary to exchange these cations for other cation, or also to obtain the zeolites in their acid form by exchanging the sodium for ammonium, and heating in an air flow [12,59].

Alumino silicate zeolites are also dealuminated. The dealumination process can be done by hydrothermal treatments (steaming), acid leaching, or treatments in flowing SiCl$_4$ at 200°C–300°C or hexafluorsilicates [59]. These treatments produce high silica zeolites. The dealumination treatment also creates mesopores by extraction of Al from the zeolite lattice, so causing partial collapse of the framework. These dealuminated materials have different uses: for example, in catalysis, USY zeolite, that is, ultra stable zeolite Y is widely used in catalytic cracking [30]. In addition, hydrophobic zeolites, such as DAY, that is, a zeolite with a very low aluminum content possess a high adsorption capacity of organic compounds dissolved in water [62,120–123].

The incorporation of molecules inside the zeolite cavities is a very interesting tool for all silica zeolite modification. Neutral molecules can be adsorbed by stirring the compound with the zeolite in an appropriate solvent. This process is normally very efficient with small molecules. The incorporation of large molecules in the zeolite can be accomplished in solution and it depends on the dimensions of the molecule, and also on the diffusion properties across the zeolite channels [36,59]. For example, iridium carbonyl can be deposited on zeolites by adsorbing the complex $[Ir(CO)_2$ (acac)] from a solution in hexane. Rhodium neutral clusters: $[Rh_6(CO)_{16}]$ can also be obtained when $[Rh(CO)_2Cl]_2$ is introduced in the cages of NaY or even acidic HY zeolites, and treated at room temperature with 1 atm of CO in the presence of water, then $[Rh_6(CO)_{16}]$ is formed [127].

However, sometimes the molecules to incorporate have dimensions larger than the zeolite channels. For instance, faujasite has supercages of 13 Å diameter with windows giving access to these cages of ca. 7.4 Å [36]. Larger molecules of 7.4 Å diameter could not be incorporated by conventional procedures. Nevertheless, to overcome this difficulty the large molecules can be synthesized within the zeolite supercages, this procedure is called ship-in-a-bottle synthesis [128]. The examples published so far are the incorporation of tetra-butyl substituted iron phtalocyanine [129], perfluorphthalocyanines of iron [130], cobalt [134,135] copper [131,132], and manganese [132], and iron–tetranitrophthalocyanine. However, in this case the complex was formed in the outer surface of the zeolite [133]. In the case of porphyrin-type ligands the encapsulation of iron and manganese tetramethylporphyrins in the supergage of zeolite Y was claimed [134].

It is also possible to incorporate transition metal complexes by synthesizing the molecular sieve structure around the performed complexes [128]. To carry out this procedure the complex must be stable under the conditions of zeolite synthesis, that is, pH, temperature, and hydrothermal conditions. Besides, it should show solubility in the synthesis medium [128]. The synthesis of mordenite from gels containing bipyridine, phenantonitrile, or phthalocyaninen complexes was reported. However, it was not clearly shown the encapsulation of the complexes [128]. It was also claimed that homogeneous encapsulation of transition metal complexes in zeolites is possible in zeolite X [136,137].

8.4 SOME APPLICATIONS OF CRYSTALLINE AND ORDERED NANOPOROUS MATERIALS IN GAS SEPARATION AND ADSORPTION PROCESSES

8.4.1 GAS CLEANING

8.4.1.1 Zeolites

After the degasification process, gas or vapor molecules can go through the pore structure of crystalline and ordered nonporous materials through a series of channels and/or cavities. Each layer of these channels and cavities is separated by a dense, gas-impermeable division, and within this adsorption space the molecules are subjected to force fields. The interaction with this adsorption field within the adsorption space is the base for the use of these materials in adsorption process.

Zeolites A, X, ZSM-5, chabazite, clinoptilolite, mordenite, and other nanoporous materials are used for removing H_2O, NH_3, NO, NO_2, SO_2, SH_2, CO_2, and other impurities from gas streams [31,122,171–178]. For example, cleaning zeolites in gas are normally used for the removal of H_2O, SH_2, and CO_2 from sour natural gas streams [31,122,172–177]. They could also be used for drying of the CO_2 used in cryosurgery, selective removal of NH_3 produced during the gasification of coal, removal of NH_3, SO_2, NO_x, and CO_2 from air, and for the selective adsorption of SH_2 from methane (CH_4) streams [31,122,172,173].

Between natural zeolites, clinoptilolite is the more profusely distributed in the earth core, and consequently they are mainly applied in adsorption applications [31]. Clinoptilolite is a member of the heulandite family with a molar ratio Si/Al > 4 of about 22 water molecules that are composed of the unit cell, where Na, K, Ca, and Mg are the most common charge-balancing cations [31].

In Figure 8.4 the breakthrough curve following from the dynamic adsorption of H_2O from a CO_2–H_2O mixture by a Mg–CMT natural zeolite is shown [175]. The water concentration in the mixture prior to breakthrough was $C_0 = 0.32$ mg/L, which was reduced after the passage of the gas flow through the adsorption reactor to 0.01–0.03 mg/L [175,176]. The reactor had a cross-sectional area of $S = 10.2$ cm^2, column length, $D = 6.9$ cm, adsorbent mass in the bed, $M = 30$ g, the volume of the bed was 70 cm^3, the volume free of adsorbent in the bed was about 35 cm^3, and the volumetric flow rate was $F = 7.7$ (cm^3/s) [175,176]. The Mg–CMT natural zeolite is an homoionic magnesium natural zeolite sample labeled CMT, which is a mixture of clinoptilolite (42 wt. %) and mordenite (39 wt. %) and other phases (15 wt. %). Whereas others are montmorillonite (2–10 wt. %), quartz (1–5 wt. %), calcite (1–6 wt. %), feldspars (0–1 wt. %), and volcanic glass [31,176]. The reactor had a cross-sectional area $S = 10.2$ cm^2, column length $D = 6.9$ cm, adsorbent mass in the bed $M = 30$ g, the volume of the bed was 70 cm^3, the volume free of adsorbent in the bed was about 35 cm^3, and the volumetric flow rate was $F = 7$ (cm^3/s) [175,176].

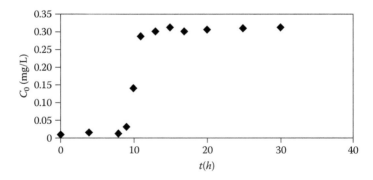

FIGURE 8.4 Breakthrough curve resulting from the dynamic adsorption of H_2O from a CO_2–H_2O mixture by an Mg–CMT natural zeolite [175,176].

TABLE 8.1

Breakthrough Mass (B.M.) for Air Drying in the Adsorption Reactor Filled with Natural (HC Sample) and Homoionic Sodium Clinoptilolite (Na–HC) and Homoinic Calcium Clinoptilite (Ca–HC)

Sample	B.M. (mg of H_2O/cm^3)	Sample	B.M. (mg of H_2O/cm^3)
HC	39	MP	32
Na–HC	44	Na–MP	42
Ca–HC	67	Ca–MP	49
Na–X	96	Alumina	28

Source: [156]

In Table 8.1 [176] the breakthrough mass (B.M.) for air drying in an adsorption bed reactor filled with natural clinoptilolite (HC sample), mordenite (MP sample), and ionic exchanged clinoptilolite and mordenite, that is, homoionic sodium clinoptilolite (Na–HC sample), mordenite (Na–MP sample), and homoinic calcium clinoptilite (Ca–HC sample) and mordenite (Ca–MP sample) are shown. The B.M. was measured in mg of water/cm³ of adsorption bed, where the volume of the bed was 70 cm³ and the volume free of adsorbent in the bed was 35 cm³ [176].

The natural zeolite sample labeled as HC is relatively pure clinoptilolite containing 85 wt. % of clinoptilolite and 15 wt. % of other phases, where the other phases are montmorillonite (2–10 wt. %), quartz (1–5 wt. %), calcite (1–6 wt. %), feldspars (0–1 wt. %), and volcanic glass [31,176]. The natural zeolite sample labeled MP is a reasonably pure mordenite containing 80 wt. % of mordenite and 20 wt. % of other phases, where the other phases are montmorillonite (2–10 wt. %), clinoptilolite (1–5 wt. %), quartz (1–5 wt. %), calcite (1–6 wt. %), feldspars (0–1 wt. %), and volcanic glass [31,176]. The adsorbents used for comparison in the present experiment was a synthetic Na–X zeolite provided by Laporte and alumina (Al_2O_3) provided by Neobor [176]. The water concentration in the tested air before breakthrough was 1.2 mg/L and it was reduced after the pass of the air flow through the adsorption reactor to 0.01–0.03 mg/L [176].

The HEU framework of clinoptilolite is a two-dimensional micropore channel system, where the 10-MR channel A and the 8-MR channel B run parallel to each other and to the c axis of the unit cell, whereas channel C (8-MR) is placed along the a axis intersecting both the A and B channels (Figure 8.3) [37]. The elliptical shaped 8-MR and 10-MR that make up the channel system are nonplanar, and consequently, cannot be simply dimensioned. The structure of mordenite (MOR framework type [37]) is characterized by the presence of a two-dimensional micropore channel system, one of the channels is the principal system because the other channel is normally blocked, which is a 12-MR channel running along the [001] axis with free access of 6.5 A × 7 A, the other channel is a 8-MR side pocket channel running along [010] with a window of access of 2.6 A × 5.7 A connecting the 12-MR channels [12,38].

The selectivity and uptake rate of gases by clinoptilolite and mordenite natural zeolites, and also by other zeolites are influenced by the type, number, and location of the charge-balancing cations residing in the A–C channels [12,31,38,173–181]. Therefore, the positions adopted by exchangeable cations and the adsorbed molecules are interdependent, then variations in the cation composition cause changes in the amount of adsorbed molecules as it is evident from Table 8.1 [176].

In the case of the synthetic zeolite Na–A, it is possible to cation exchange it with potassium and calcium in order to get the 3A and 5A molecular sieves, respectively. Different studies [173,174,178,182,183] have shown that the pore-opening size of zeolite and other molecular sieves can be controlled to fit desired applications by postsynthesis modification techniques such as internal or external surface modification by chemical reactions, preadsorption of polar molecules [173], chemical vapor deposition [181] or similar coating processes, and thermal treatment [182].

It has also been reported that by calcination of NaA zeolites at 953–1033 K after water vapor adsorption, the adsorption predilection toward oxygen over nitrogen is improved, a fact which can be credited fundamentally to pore-size shrinkage [178]. It has also been shown [174] that the pore size and the affinity of a zeolite structure can be modified by chemical treatment of the zeolite structure, the silane, borane, or disilane molecules are chemisorbed on the zeolite surface by reacting with the silanol groups of the zeolite [178]. Polar molecules, for example, water and amines presorbed in the zeolite can also be applied to modify the molecular sieve performance and the interaction toward adsorbate molecules of the zeolite [178].

8.4.1.2 Mesoporous Molecular Sieves

The attention paid to this new family of adsorbents is mainly due to the unique mesopore structure, which is not shared by any other families of adsorbent. Consequently, mesoporous ordered silica has been proposed as a reference material for the study of adsorption processes in mesopores [184–190]. In addition, one of the most striking features of these novel materials to the pore structure are the large BET surface area and pore volume [191,192]. Then, the surface properties of MMS could be of major importance both for the preparation of active and stable catalysts [156] and for the modification of their sorption properties.

The interactions of the mesoporous molecular sieves surface and test molecules can provide useful information on the nature and local arrangement of the surface groups, which cause either hydrophobic or hydrophilic behavior characterized by a heat of adsorption of smaller or larger than the heat of liquefaction of water, respectively [184]. Accordingly, adsorption of water on M41 systems has been studied using microcalorimetry and IR spectroscopy. It was shown that after thermal treatment at 423 K, the samples exhibit two types of surface patches: one hydrophobic distinguished by isolated silanols not interacting with water, and the other highly hydrophilic [184]. However, it is generally recognized that the internal surface of these materials is hydrophobic notwithstanding the presence of silanol groups [193–196]. Consequently, the hydrophobic surface nature reveal these materials as selective adsorbents for the removal of volatile organic compounds (VOCs) and other organic compounds present in high humidity gas streams or wastewater [193]. Consequently, MCM-41 is a possible adsorbent to substitute activated carbon for controlling VOCs [185,193,195]. However, the adsorption equilibrium of VOCs on MCM-41 regularly shows very low adsorption capacity in the low-concentration region, because of its mesoporous structure, this fact considerably limits the application of MCM-41 as an adsorbent for low-concentration VOC abatement [185].

The pore sizes of MCM-41 materials can be modified during the synthesis by selecting various surfactants with different carbon chain lengths, the minimum pore size that can be obtained seems to be about 2 nm [21–24]. Simultaneously, as the pore size is lessened, the pore volume is also diminished, then the modification of the pore diameter of MCM-41 by postmodification is attractive and essential to attain a shape selective adsorption property [185]. Consequently, the pore-opening size of MCM-41 could be reduced to be in the microporous region using a chemical vapor deposition technique for a selective modification in order to improve the adsorption properties of the modified MCM-41, which significantly improves the adsorption performance of MCM-41 for low-concentration VOCs [193].

The adsorption characteristics of MCM-41 for polar molecules greatly depend on the concentrations of surface silanol groups (SiOH) [184,193–196]. It has been demonstrated that several types of SiOH groups exist over MCM-41 surfaces, which can be qualitatively and quantitatively determined by a number of techniques [184,193–196]. Those SiOH groups allow various accomplishments in the modification of MCM-41 for catalysis, adsorption, and novel composites [196]. For example, a sorption separation process using modified MCM-41 for purification of water has been proposed [197].

Besides, the large pore volume, pore size flexibility, and structural variety of MCM-41 can be extensively used for the selective adsorption of a diversity of gases and liquids [185,186,193–196]. It has been shown, for example, an extremely high sorption capacity for benzene [186,193,195]. Widespread work has been carried out on sorption properties of some adsorbates, such as nitrogen, argon, oxygen, water, benzene, cyclopentane, toluene, and carbon tetrachloride, and certain lower hydrocarbons and alcohols on MCM-41 [194].

It was also shown that the adsorptive capacity of the mesoporous materials is in excess of an order of magnitude superior than that of conventional porous adsorbent materials. Consequently, MCM-41 therefore is promising as a selective adsorbent in separation techniques, for example, high-performance liquid chromatography and supercritical fluid chromatography [194].

In addition, the substitution of the surface hydroxyl groups in the pore wall with trimethylchlorosilane groups creates a more hydrophobic environment that substantially reduces the sorption capacity of polar molecules [196]. The surface chemistry of MCM-41 and other ordered mesopores molecular sieves can be efficiently modified to be more hydrophobic by chemical addition of organic species, a process called silylation. Besides, the surface modification of ordered mesoporous molecular sieves can be conducted in various ways, such as esterification and chemical depositions [196]. A few types of silanol groups exist on MCM-41 surfaces, among which both free and geminal ones (Chapter 7, Section 7.2) are responsible for active silylation.

Consequently, free and germinal silanol groups over MCM-41 surfaces are responsible for such active modification [196]. Siliceous MCM-41 samples were modified by silylation using trimethylchlorosilane (TMCS), the degree of silylation was found to linearly increase with increasing preoutgassing temperature before silylation [196]. It was finally shown that surface modification of MCM-41 by silylation is an effective technique in the development of selective adsorbents for the removal of organic compounds from streams or wastewater [196].

In summary, the outstanding characteristics of the majority of ordered mesoporous materials are thus well-defined pore shapes, fine distribution of pore sizes, insignificant pore networking or pore blocking effects, especially high degree of pore ordering over micrometer length scales, tailoring and modification of the pore dimensions, large pore volumes, excellent sorption capacity. Therefore, as a result of the large pore volume, very high surface area (700–1500 m^2/g), large amount of internal hydroxyl (silanol) groups (40%–60%), high surface reactivity, ease of modification of the surface properties, enhanced catalytic selectivity in certain reactions, and excellent thermal, hydrothermal, and mechanical properties [194] are some characteristics.

8.4.2 PRESSURE SWING ADSORPTION

In the present monograph, we have not discussed the adsorption process in multicomponent gas systems because we are studying adsorption here, fundamentally, from the point of view of materials science, that is, we are interested in the methods for the use of single-component adsorption in the characterization of the adsorbent surface and pore volume, the study of the parameters characterizing single-component transport processes in porous systems, and in less extent the adsorption energetic and dynamic adsorption in bed reactors. However, from the point of view of the application of adsorbent materials it is necessary to discuss gas mixture separation, and this is the aim of the present section.

Pressure swing adsorption (PSA) is a cyclic process used to selectively adsorb and separate components of a feed gas mixture, thereby producing partially purified gas products. Since the first patent of a PSA process [198], a great variety of complicated PSA processes have been developed and commercialized, largely taking advantage of PSA's low-energy requirement and low capital investment [199]. It is a very multipurpose technology for separation and purification of gas mixtures, which offers an additional level of thermodynamic freedom for describing the adsorption process in comparison with other standard separation methods, for example, distillation, extraction, or absorption [200].

Some of the main industrial applications comprise [200–212]:

1. Gas drying
2. Solvent vapor recovery
3. Fractionation of air
4. Production of hydrogen from steam methane reformer and petroleum refinery of gases
5. Separation of carbon dioxide and methane from landfill gas

6. Carbon monoxide–hydrogen separation
7. Normal isoparaffin separation
8. Alcohol dehydration

The concept of PSA for gas separation is fairly simple, that is, some components of a gas mixture are selectively adsorbed on a microporous–mesoporous solid adsorbent at a comparatively high pressure by contacting the gas with the solid in a packed column of the adsorbent in order to yield a gas stream enriched in the less strongly adsorbed components of the feed gas [200–208]. The adsorbed components are, after that, desorbed from the solid by reducing their superincumbent gas-phase partial pressures in the interior of the column in order that the adsorbent can be reused. Then, the desorbed gases are enriched in the more strongly adsorbed components of the feed gas [200–208]. During the process, no external heat is generally used for desorption [200]. On the other hand, a vacuum swing adsorption (VSA) process undergoes the adsorption step at a near-ambient pressure level and desorption is achieved under vacuum [200].

In PSA processes presently available, as a rule, only one component is the chosen product, this is regularly the most weakly adsorbed one. Nevertheless, for a better economics, it is imperative to recuperate as many of the components from the feed as is possible [200].

Many microporous–mesoporous types of adsorbents, as-synthesized or modified, such as activated carbons, zeolites, aluminas, silica gels, and polymeric sorbents exhibiting different adsorptive properties for separation of gas mixtures, that is, equilibria, kinetics, and heats are accessible for application in PSA or VSA processes [200–212].

For example, many different zeolites having diverse thermodynamic selectivities and capacities for adsorption are employed to develop PSA process for gas separation [200,212]. More concretely, zeolites A and X, chabazite, clinoptilolite, mordenite, and other zeolites can be used for PSA separation processes, that is, H_2–N_2, N_2–CH_4, and other couples can be separated with the assistance of zeolites [31]. The zeolite related processes for N_2–O_2 separation are based, in general, on nitrogen selective zeolite adsorbents [200,201], such as natural chabazite, clinoptilolite, and mordenite, which could be used for this purposes with a performance comparable or better than those reported for synthetic mordenite or zeolites type A [198,199]. The PSA method with natural zeolite was also employed for methane purification [200].

The working principle of a PSA installation where vacuum is used for desorption, more exactly a VSA equipment for N_2–O_2 separation with zeolites [209] (Figure 8.5) is as follows: the pretreatment columns are filled with silica gel desiccant followed by a granulated zeolite, the main column is filled with a zeolite. The main (3a and 3b) and pretreatment columns (4a and 4b) are activated at 500°C. Filtered air is sent to the pretreatment column (e.g., 4a) where H_2O, CO_2, and SO_2 are eliminated. The purified gas is sent to the main column (3a). Nitrogen is mainly adsorbed in the columns and O_2 with a small amount of Ar and N_2 in the gas tank (1). Before the column is exhausted the air supply is sent to the pretreatment column (4b), and main column (4b), and the columns (3a and 4a) are connected automatically with the vacuum pump exhausting the N_2 [209].

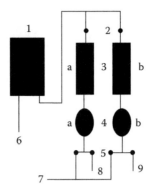

FIGURE 8.5 The vacuum swing adsorption (VSA) equipment for oxygen generation from air include the oxygen gas tank (1), automatic stopcocks (2 and 5), multiple zeolite adsorption beds (3a and 3b), pretreatment columns (4a and 4b), vacuum pump (7), and the air supply (8 and 9).

8.4.3 OTHER SEPARATION APPLICATIONS

Gas separation with zeolites is also applied as an analytical tool using zeolites as adsorbent columns for analysis in a gas chromatograph [213–215].

In addition, we have previously discussed gas separation with the help of pressure swing methods, which imply cycles of favored adsorption of one component over the other and succeeding desorption [200]. The replacement of swing adsorption methods with a steady-state membrane process could be useful because of the reduction of operating costs and energy consumption. In this sense, microporous inorganic membranes could be very useful in gas separation [216]. These membranes could be manufactured with zeolites and are potentially useful in gas separation for cleaning processes and catalytic reactors [217]. The zeolite in these membranes should be placed in the form of a thin film supported on a macroporous substrate [216,218].

In the Chapter 5 of the present book, we have discussed the gas transport mechanism through microporous inorganic membranes, and it was shown that the transport of molecules through zeolite cavities and channels determines the molecular sieving nature of these materials. Then zeolite-based membranes will be molecular sieves, and consequently it may be used for gas separation processes. The pore size distribution of these materials can be controlled by synthesis procedures so that their molecular sieving properties can be tailored to selectively separate gases and purify it [216].

8.4.4 AIR-CONDITIONING

Solar energy storage [31,219] and solar cooling [31,220–222] applications of natural zeolites are closely connected with the adsorption properties of these minerals. In solar cooling installations (Figure 8.6), solar heating is utilized to induce water desorption, and consequently the dehydration of a zeolite contained in a solar panel (1). The desorbed water is later condensed (2) and allowed to pass into a deposit (6) water

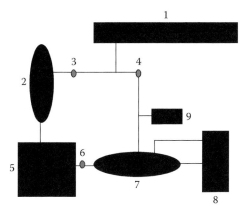

FIGURE 8.6 Solar cooling installation, zeolite-filled solar panel (1), water condenser (2), stopcocks (3, 4, and 6), water deposit (5), heat exchanger (7), evaporator (8), and buffer storage (9).

through a heat exchanger (7) to advance to an evaporator system (8) where later it is left to diffuse to the previously dehydrated zeolite (1). The cooling is provided by water evaporation to the dehydrated zeolite. The process is cyclic, this mean that zeolite desorption is carried out during the day and water adsorption in the zeolite is carried out at night [53].

REFERENCES

1. C. Delerue and M. Lannoo, *Nanostructures Theory and Modeling*, Springer-Verlag, Berlin, Germany, 2004.
2. R. Roque-Malherbe, *The Physical Chemistry of Materials: Applications in Pollution Abatement and Sustainable Energy*, CRC Press, Boca Raton, FL, 2009.
3. E.L. Wolf, *Nanophysics and Nanotechnology: An Introduction to Modern Concepts in Nanoscience* (3rd ed.), Wiley-VCH, Mannheim, Germany, 2015.
4. R. Roque-Malherbe, *Electrochemistry: Energy and Environmental Applications*, Lambert Academic Publishing, Saarbrücken, Germany, 2017.
5. K. Lu, *Nanoparticulate Materials Synthesis, Characterization and Processing*, John Wiley & Sons, Hoboken, NJ, 2012.
6. K. Lu, *Materials in Energy Conversion, Harvesting and Storage*, John Wiley & Sons, Hoboken, NJ, 2014.
7. W.-N. Wang, I.W. Lenggoro, and K. Okuyama, Preparation of nanoparticles by spray routes, in *Encyclopedia and Nanoscience and Nanotechnology*, Vol. 21 (H.S. Nalwa, Ed.), American Scientific Publishers, Stevenson Ranch, CA, 2011.
8. A. Suleiman, C. Cabrera, R. Polanco, and R. Roque-Malherbe, *RSC Ad.*, 5 (2015) 7637.
9. S.E. Lyshevskii (Ed.), *Dekker Encyclopedia of Nanoscience and Nanotechnology* (3rd ed.), Seven Volume Set, CRC Press, Boca Raton, FL, 2014.
10. B. Brushan (Ed.), *Encyclopedia of Nanotechnology* (2nd ed.), Springer-Verlag, Berlin, Germany, 2016.
11. K.S.W. Sing, D.H. Everett, R.A.W. Haul, L. Moscou, R.A. Pirotti, J. Rouquerol, and T. Siemieniewska, *Pure App. Chem.*, 57 (1985) 603.
12. D.W. Breck, *Zeolite Molecular Sieves*, John Wiley & Sons, New York, 1974.

13. R.M. Barrer, *Zeolites and Clay Minerals as Sorbents and Molecular Sieves*, Academic Press, London, UK, 1978.
14. J.W. Steed and P.A. Gale (Eds.), *Supramolecular Chemistry: From Molecules to Nanomaterials*, Eight Volume Set, John Wiley & Sons, New York, 2012.
15. H.K. Hunt, C.M. Lew, M. Sun, Y. Yan, and M.E. Davis, *Mic. Mes. Mat.*, 130 (2010) 49.
16. J.M. Garces, A. Kuperman, D.M. Millar, M. Olken, A. Pyzik, and W. Rafaniello, *Adv. Mater.*, 12 (2000) 1725.
17. C.T. Kresge and W.J. Roth, *Chem. Soc. Rev.*, 42 (2013) 3663.
18. R.K. Iler, *The Chemistry of Silica*, John Wiley & Sons, New York, 1979.
19. W. Stobe, A. Fink, and E. Bohn, *J. Colloid. Interf. Sci.*, 26 (1968) 62.
20. C.J. Brinker and G.W. Scherer, *Sol-Gel Science*, Academic Press, New York, 1990.
21. K. Unger and D. Kumar, in *Adsorption on Silica Surfaces* (E. Papirer, Ed.), Marcel Dekker, New York, 2000, p. 1.
22. C. Burda, X. Chen, R. Narayanan, and M.A. El-Sayed, *Chem. Rev.*, 105 (2005) 1025.
23. A. Corma, *Chem. Rev.*, 95 (1995) 559.
24. R. Roque-Malherbe, in *Handbook of Surfaces and Interfaces of Materials*, Vol. 5 (H.S. Nalwa, Ed.), Academic Press, New York, 2001, p. 495.
25. F. Rouquerol, J. Rouquerol, K. Sing, P. Llewellyn, and G. Maurin, *Adsorption by Powders and Porous Solids* (2nd ed.), Academic Press, New York, 2013.
26. R.T. Yang, *Adsorbents: Fundamentals, and Applications*, John Wiley & Sons, New York, 2003.
27. D.H. Everet and R.H. Ottewill, *Surface Area Determination, International Union of Pure and Applied Chemistry*, Butterworth, London, UK, 2013.
28. S. Lowell, J.E. Shields, M.A. Thomas, and M. Thommes, *Characterization of Porous Solids and Powders: Surface Area, Pore Size and Density*, Kluwer Academic Press, Dordrecht, the Netherlands, 2004.
29. J.B. Condon, *Surface Area and Porosity Determination by Physisorption*, Elsevier, Amsterdam, the Netherlands, 2006.
30. R. Roque-Malherbe, *Mic. Mes. Mat.*, 41 (2000) 227.
31. F. Marquez-Linares and R. Roque-Malherbe, *J. Nanosci. Nanotech.*, 6 (2006) 1114.
32. B.L. Cushing, V.L. Kolesnichenko, and C.J. O'Connor, *Chem. Rev.*, 104 (2004) 3893.
33. G.J.A.A. Soler-Illia, C. Sanchez, B. Lebeau, and J. Patarin, *Chem. Rev.*, 102 (2002) 4093.
34. C. Baerlocher, W.M. Meier, and D.M. Olson, *Atlas of Zeolite Framework Types* (5th ed.), Elsevier, Amsterdam, the Netherlands, 2001.
35. C.S. Cundy and P.A. Cox, *Chem. Rev.*, 103 (2003) 663.
36. J. Cejka, A. Corma, and Z. Sones (Eds.), in *Zeolites and Catalysts Synthesis and Applications*, Wiley-VCH, Mannheim, Germany, 2010.
37. R. Millini, *Science*, 355 (2017) 6329.
38. R. Roque-Malherbe, L. Lemes, L. López-Colado, and A. Montes, in *Zeolites'93*, Full Papers Volume (D. Ming and F.A. Mumpton, Eds.), International Committee on Natural Zeolites Press, Brockport, NY, 1995, p. 299.
39. M.M. Dubinin, *Prog. Surf. Memb. Sci.*, 9 (1975) 1.
40. M.M. Dubinin, E.F. Zhukovskaya, V.M. Lukianovich, K.O. Murrdmaia, E.F. Polstiakov, and E.E. Senderov, *Izv. Akad. Nauk SSSR*, (1965) 1500.
41. F.A. Mumpton, *Proc. Nat. Acad. Sci.*, 96 (1999) 3463.
42. T.J. Barton, L.M. Bull, G. Klemperer, D.A. Loy, B. McEnaney, M. Misono, P.A. Monson et al., *Chem. Mater.*, 11 (1999) 2633.
43. C.T. Kresge, M.E. Leonowicz, W.J. Roth, J.C. Vartuli, and J.S. Beck, *Nature*, 359 (1992) 710.
44. J.E. Martin, M.T. Anderson, J. Odinek, and P. Newcomer, *Langmuir*, 13 (1997) 4133.
45. S.H. Tolbert, T.E. Schäffer, J. Feng, P.K. Hansma, and G.D. Stucky, *Chem. Mater.*, 9 (1997) 1962.

46. P.J. Bruinsma, A.Y. Kim, J. Liu, and S. Baskaran, *Chem. Mater.*, 9 (1997) 2507.
47. G.T. Kerr, *ACS Symp. Ser.*, 368 (1988) 13.
48. F. Marquez and R. Roque-Malherbe, *Facets-IUMRS J.*, 2 (2003) 14.
49. S. Sherry, Ion exchange, in *Handbook of Zeolite Science and Technology* (1st ed.), (S.M. Auerbach, K.A. Carrado, and P.K. Dutta, Eds.), CRC Press, Boca Raton, FL, 2013, Chapter 21.
50. A. Zorpas and V.J. Inglezakis, Natural zeolites: Industrial and environmental applications, in *Handbook of Zeolites: Structure, Properties and Applications* (T.W. Wong, Ed.), Nova Science Publishers, New York, Incorporated, 2009, Chapter 2.
51. P.A. Barrett, M.A. Camblor, A. Corma, R.H. Jones, and L.A. Villaescusa, *Chem. Mater.*, 9 (1997) 1713.
52. L.A. Villaescusa, P.A. Barrett, and M.A. Camblor, *Chem. Mater.*, 10 (1998) 3966.
53. M.A. Camblor, A. Corma, M.J. Diaz-Cabanas, and C. Baerlocher, *J. Phys. Chem. B*, 102 (1998) 44.
54. L.A. Villaescusa, P.A. Barrett, and M.A. Camblor, *Chem. Commun.*, (Cambridge), 2329 (1998).
55. L.A. Villaescusa, P.A. Barrett, and M.A. Camblor, *Angew. Chem. Int. Ed.*, 38 (1999) 1997.
56. C. Jones, S. Hwang, T. Okubo, and M. Davis, *Chem. Mater.*, 13 (2001) 1041.
57. P.M. Piccione, B.F. Woodfield, J. Boerio-Goates, A. Navrotsky, and M. Davis, *J. Phys. Chem. B*, 105 (2001) 6025.
58. B. Marler, U. Werthmann, and H. Gies, *Mic. Mes. Mat.*, 43 (2001) 29.
59. P.A. Barrett, T. Boix, M. Puche, D.H. Olson, E. Jordan, H. Koller, and M.A. Camblor, *Chem. Commun.*, (Cambridge), 2114 (2003).
60. A. Corma, F. Rey, J. Rius, M.J. Sabater, and S. Valencia, *Nature*, 431 (2001) 287.
61. Z. Li, C.M. Lew, S. Li, D.I. Medina, and Y. Yan, *J. Phys. Chem. B*, 109 (2005) 8652.
62. E.M. Flanigen, J.M. Bennett, R.W. Grose, J.P. Cohen, R.L. Patton, R.L. Kirchner, and J.V. Smith, *Nature*, 271 (1978) 512.
63. D.M. Bibby, N.B. Milestone, and L.P. Aldridge, *Nature*, 280 (1979) 664.
64. C.A. Fyfe, H. Gies, G.T. Kokotailo, B. Marler, and D.E. Cox, *J. Phys. Chem.*, 94 (1990) 3718.
65. R. Bialek, W.M. Meier, M. Davis, and M.J. Annen, *Zeolites*, 11 (1991) 438.
66. M.A. Camblor, A. Corma, and S. Valencia, *J. Chem. Soc. Chem. Commun.*, 2365 (1996).
67. P.A. Barrett, M.A. Camblor, A. Corma, R.H. Jones, and L.A. Villaescusa, *Chem. Mater.*, 9 (1997) 1713.
68. M.A. Camblor, A. Corma, P. Lightfoot, L.A. Villaescusa, and P.A. Wright, *Angew. Chem. Int. Ed. Engl.*, 36 (1997) 2659.
69. M.A. Camblor, A. Corma, M.J. Diaz-Cabanas, and C. Baerlocher, *J. Phys. Chem. B*, 102 (1998) 44.
70. S. Samanta, S. Giri, P.U. Sastry, N.K. Mal, A. Manna, and A. Bhaumik, *Ind. Eng. Chem. Res.*, 42 (2001) 3012.
71. Q. Cai, Z.-S. Luo, W.-Q. Pang, Y.-W. Fan, X.-H. Chen, and F.-Z. Cui, *Chem. Mater.*, 13 (2001) 258.
72. M. Widenmeyer and R. Anwander, *Chem. Mater.*, 14 (2002) 1827.
73. S. Han, W. Hou, X.Y. Zhengmin, L.P. Zhang, and D. Li, *Langmuir*, 19 (2003) 4269.
74. J. Fan, C. Yu, L. Wang, B. Tu, D. Zhao, Y. Sakamoto, and O. Terasaki, *J. Am. Chem. Soc.*, 123 (2001) 12113.
75. C.C. Pantazis and P.J. Pomonis, *Chem. Mater.*, 15 (2001) 2299.
76. M.E. Davies, *Nature*, 417 (2002) 813.
77. A. Monnier, F. Schüth, Q. Huo, D. Kumar, D. Margolese, R.S. Maxwell, G.D. Stucky et al., *Science*, 261 (1993) 1299.

78. Q. Huo, D. Margolese, U. Ciesia, P. Feng, T.E. Gier, P. Sieger, R. Leon, P.M. Petroff, F. Schuth, and G.D. Stucky, *Nature*, 368 (1994) 317.

79. P.T. Tanev and T.J. Pinnavaia, *Science*, 267 (1995) 865.

80. S. Bagshaw, E. Prouzet, and T.J. Pinnavaia, *Science*, 269 (995) 1242.

81. V.R. Karra, I.L. Moudrakovski, and A. Sayari, *J. Porous Mater.*, 3 (1996) 77.

82. Q. Huo, D.I. Margolese, U. Ciesla, D.K. Demuth, P. Feng, T.E. Gier, P. Sieger et al., *Chem. Mater.*, 6 (1994) 1176.

83. J.S. Beck, J.C. Vartuli, G.J. Kennedy, C.T. Kresge, W.J. Roth, and S.E. Schramm, *Chem. Mater.*, 6 (1994) 1816.

84. X.S. Zhao, G.Q. Lu, and G.J. Millar, *Ind. Eng. Chem. Res.*, 35 (1996) 2075.

85. J. Fan, Y. Chengzhong, Y.L. Wang, B. Tu, D. Zhao, Y. Sakamoto, and O. Terasaki, *J. Am. Chem. Soc.*, 123 (2001) 12113.

86. R. Roque-Malherbe and F. Marquez, *Mat. Sci. Semicond. Proc.*, 7 (2004) 467.

87. R. Roque-Malherbe and F. Marquez, *Surf. Interf. Anal.*, 37 (2005) 393.

88. A. Firouzi, D. Kumar, L.M. Bull, T. Besler, P. Sieger, Q. Huo, S.A. Walker et al., *Science*, 267 (1995) 1138.

89. Q. Huo, R. Leon, P. Petroff, and G.D. Stucky, *Science*, 268 (1995) 1324.

90. D. Zhao, J. Feng, Q. Huo, N. Melosh, G.H. Fredrickson, B.F. Chmelka, and G.D. Stucky, *Science*, 279 (1998) 548.

91. A. Nossov, E. Haddad, F. Guenneau, A. Galarneau, F. Di Renzo, F. Fajula, and A. Gedeon, *J. Phys. Chem. B*, 107 (2003) 12456.

92. W.W. Lukens, Jr., P. Schmidt-Winkel, D. Zhao, J. Feng, and G.D. Stucky, *Langmuir*, 15 (2001) 5403.

93. R. Ryoo, C.H. Ko, M. Kruk, V. Antochshuk, and M. Jaroniec, *J. Phys. Chem. B*, 104 (2000) 11465.

94. A. Galarneau, H. Cambon, F. Di Renzo, and F. Fajula, *Langmuir*, 17, 8328 (2001).

95. A. Sayari, B.-H. Han, and Y. Yang, *J. Am. Chem. Soc.*, 126 (2001) 14384.

96. D. Zhao, J. Sun, Q. Li, and G.D. Stucky, *Chem. Mater.*, 12 (200) 275.

97. D. Zhao, P. Yang, B.F. Chmelka, and G.D. Stucky, *Chem. Mater.*, 11 (1999) 1174.

98. Q. Huo, J. Feng, F. Schuth, and G.D. Stucky, *Chem. Mater.*, 9 (1997) 14.

99. H. Robson, *Verified Synthesis of Zeolitic Materials* (2nd ed.), Elsevier, Amsterdam, the Netherlands, 2001.

100. R.J. Argauer and G.R. Landolt, U.S. Patent 3,702,886 (1972).

101. B.M. Lok, C.A. Messina, R.L. Patton, R.T. Gajek, T.R. Cannan, and E.M. Flanigen, *J. Am. Chem. Soc.*, 106 (1984) 6092.

102. A. Merrouche, J. Patarin, H. Kessler, M. Soulard, L. Delmotte, J.L. Guth, and J.F. Joly, *Zeolites*, 12 (1992) 22.

103. G. Ferey, *J. Fluorine Chem.*, 72 (1995) 187.

104. J.R.D. Debord, W.M. Reiff, C.J. Warren, R.C. Haushalter, and J. Zubieta, *Chem. Mater.*, 9 (1997) 1994.

105. H. Kessler, J. Patarin, and C. Schott-Darie, *Stud. Surf. Sci. Catal.*, 85 (1994) 75.

106. D.M. Bibby and M.P. Dale, *Nature*, 317 (1985).

107. Q. Huo, R. Xu, S. Li, Z. Ma, Z.J.M. Thomas, R.H. Jones, and A.M. Chippindale, *Chem. Commun.*, 875 (1992).

108. R. Althoff, K. Unger, and F. Schüth, *Micro. Mater.*, 2 (1994) 557.

109. R.F. Lobo, S.I. Zones, and M.E. Davis, *J. Inclusion Phenom. Mol. Recogn. Chem.*, 21 (1995) 47.

110. M.A. Camblor, L.A. Villaescusa, and M.J. Diaz-Cabañas, *Top. Catal.*, 9 (1999) 59.

111. E.M. Flanigen and R.L. Patton, U.S. Patent 4,073,865 (1978).

112. J.L. Guth, H. Kessler, and R. Wey, in *New Developments in Zeolite Science and Technology* (Y. Murakami, A. Iijima, and J.W. Ward, Eds.), Elsevier, Amsterdam, the Netherlands, 1986, p. 121.

113. J.P. Gilson, in *Zeolite Microporous Solids: Synthesis, Structure and Reactivity* (E.G. Derouane, F. Lemos, C. Naccache, and F.R. Ribeiro, Eds.), NATO ASI Series, No. C352, Kluwer, Dordrecht, the Netherlands, 1992, p. 19.

114. R. Szostak, *Molecular Sieves Principles of Synthesis, and Identification* (2nd ed.), Blackie, London, UK, 1998.

115. M.A. Camblor, P.A. Barrett, M.J. Diaz-Cabañas, L.A. Villaescusa, M. Puche, T. Boix, E. Perez, and H. Koller, *Mic. Mes. Mat.*, 48 (2001) 11.

116. L.A. Villaescusa and M.A. Camblor, *Recent. Res. Dev. Chem.*, 1 (2001) 93.

117. S.I. Zones, R.J. Darton, R. Morris, and S.-J. Hwang, *J. Phys. Chem. B*, 109 (2005) 652.

118. Y. Kubota, M.M. Helmkamp, S.I. Zones, and M.E. Davis, *Mic. Mat.*, 6 (1996) 213.

119. S.M. Auerbach, L.M. Bull, N.J. Henson, H.I. Metiu, and A.K. Cheetham, *J. Phys. Chem.*, 100 (1996) 5923.

120. Y.-K. Ryu, J.W. Chang, S.-Y. Jung, and C.-H. Lee, *J. Chem. Eng. Data*, 47 (2002) 363.

121. N.A. Briscoe, D.W. Johnson, M.D. Shannon, G.T. Kokotailo, and L.B. McCusker, *Zeolites*, 8 (1988) 74.

122. M.D. Shannon, J.I. Casci, P.A. Cox, and S.J. Andrews, *Nature*, 353 (1991) 417.

123. R. Roque-Malherbe and F. Marquez-Linares, *Facets-IUMRS J.*, 3 (2004) 8.

124. H. Lermer, M. Draeger, J. Steffen, and K.K. Unger, *Zeolites*, 5 (1985) 131.

125. A. Dyer, *An Introduction to Zeolite Molecular Sieves*, John Wiley & Sons, New York, 1988.

126. E.M. Flanigen, R.L. Patton, and S.T. Wilson, *Stud. Surf. Sci. Cat.*, 37 (1988) 13.

127. R. Roque-Malherbe, R. López-Cordero, J.A. González-Morales, J. Onate and M. Carreras, *Zeolites*, 13 (1993) 481.

128. R. Szostak, *Molecular Sieves Principles of Synthesis and Identification* (2nd ed.), Blackie, London, UK, 1998.

129. D.O. Hummel, *Handbook of Surfactant Analysis: Chemical, Physico-Chemical, and Physical Methods* (1st ed.), John Wiley & Sons, New York, 2000.

130. D. Myers, *Surfactant Science, and Technology* (4th ed.), John Wiley & Sons, New York, 2012.

131. C.Y. Chen, H.Y. Li, and M.E. Davis, *Mic. Mat.*, 2 (1993) 27.

132. A. Steel, S.W. Carr, and M.W. Anderson, *J. Chem. Soc. Chem. Com.*, 1571 (1994).

133. S. Mann, S.L. Burkett, S.A. Davis, C.E. Fowler, N.H. Mendelson, S.D. Sims, D. Walsh, and N.T. Whilton, *Chem. Mater.*, 9 (1997) 2300.

134. A. Corma, *Chem. Rev.*, 97 (1997) 2372.

135. A. Imhof and D.J. Pine, *Nature*, 389 (1997) 948.

136. C. Blanco and S.M. Auerbach, *J. Phys. Chem. B*, 107 (2003) 2490.

137. Y.-W. Chan and R.B. Wilson, Preprints division of petroleum chemistry, *Amer. Chem. Soc.*, 33 (1988) 453.

138. T. Martin, A. Galarneau, F. Di Renzo, F. Fajula, and D. Plee, *Angew. Chem. Int. Ed.*, 41 (2002) 2590.

139. P.T. Tanev, M. Chibwe, and T. Pinnavaia, *Nature*, 368 (1994) 321.

140. P.T. Tanev and T. Pinnavaia, *Science*, 267 (1995) 865.

9 Adsorption from Liquid Solution

9.1 GENERAL INTRODUCTION

Multicomponent adsorption at air/water, oil/water, gas/solid, and liquid/solid interfaces is a significant practical problem in foams, emulsions, detergents, catalysis, pollution abatement, and separation methods [1–35]. In general, it is determined by the adsorption equilibrium of the individual components, and after that the adsorption of the mixtures is described [1,2,7–16].

For the correlation of the liquid-phase adsorption equilibrium of a single component, several isotherm equations for gas-phase adsorption can, in principle, be extended to liquid-phase adsorption by the simple replacement of adsorbate pressure by its concentration [29]. These equations are the Langmuir, Freundlich, Sips, Toth, and Dubinin–Radushkevich equations [27–29]. However, the Langmuir and Freudlich equations are most widely used to correlate liquid-phase adsorption data [21,31–35].

Adsorption from liquid solution is practically another planet with respect to gas-phase adsorption, because the fundamental principles and methodology are different in almost all aspects [11,12]. In the simplest case, for example, in a binary solution the composition of the adsorbed phase is usually unknown. Besides, adsorption in the liquid phase is influenced by many factors, such as pH, type of adsorbent, solubility of adsorbate in the solvent, and temperature as well as the adsorptive concentration [10,11,13,14]. This is the cause that despite the industrial importance of adsorption from liquid phase, it is less studied than the adsorption from the gas phase. However, for the design and liquid-phase adsorptive separation processes, adsorption equilibrium data are required. These data have to be experimentally measured or calculated using various predictive multicomponent adsorption models or empirical correlations.

Fundamentally, liquid-phase adsorption [10–23] on activated carbon [10,13,14,22], silica [20,21], zeolites [15,17,18], and resins [16] provides a feasible technique and is one of the most extensively used technologies for the removal of organic pollutants from industrial wastewater [10,13–23]. In adsorption on activated carbon, the type of activated carbon plays an important role, because the carbon has a complex porous structure with associated energetic and chemical heterogeneities [10] (see Chapter 7, Sections 7.5 and 7.6). The energetic and chemical heterogeneities are determined by the variety of surface functional groups, irregularities, and strongly bound impurities, as well as structural nonuniformity. This heterogeneity considerably influences the process of physical adsorption [10].

On the other hand, the adsorption literature gives an account of numerous pure gas adsorption equilibrium data on various porous adsorbents of practical interest, such as activated carbons, silica and alumina gels, zeolites, and polymeric sorbents [24–26]. In comparison, the number of published data on binary gas equilibrium adsorption is less, and the data for three or more component gas mixtures are infrequent [25]. Mathematical models for design and optimization of adsorptive separation processes, such as pressure swing and thermal swing adsorption, on the other hand, require accurate multicomponent adsorption equilibrium data as was previously stated (see Section 8.4.2) [26]. These data have to be experimentally measured in the complete range of pressure, temperature, and gas composition met by the adsorbent during the separation processes or calculated from the corresponding pure gas equilibrium data using various predictive multicomponent adsorption models or empirical correlations [26].

In the present book we will not discuss the adsorption process in multicomponent gas systems, since as formerly acknowledged, we are studying adsorption, fundamentally, from the point of view of materials science, that is, we are interested in the methods for the use of single-component adsorption in the characterization of the adsorbent surface and pore volume, the study of the parameters characterizing single-component transport processes in porous systems, and in less extent the adsorption energetic and dynamic adsorption in bed reactors. However, the adsorption process from the liquid phase, notwithstanding the fact that it is a multicomponent process will be discussed here, because of its fundamental and practical importance.

9.2   SURFACE EXCESS AMOUNT AND AMOUNT OF ADSORPTION FOR LIQUID SOLID ADSORPTION SYSTEMS

The interfacial layer is the inhomogeneous space region that is intermediate between the two bulk phases in contact, and where properties are significantly different from, but related to, the properties of the bulk phases. Examples of such properties are compositions, molecular density, orientation or conformation, charge density, pressure tensor, electron density, and so on. The interfacial properties vary in the direction normal to the surface. Complex profiles of interfacial properties occur in the case of multicomponent systems with coexisting bulk phases where attractive/repulsive molecular interactions involve adsorption or depletion of one or several components.

The surface excess amount or Gibbs adsorption (see Section 2.2) of the component i, that is, n_i^σ is defined as the excess amount of this component actually present in the system over that present in a reference ideal system of the same volume as the real system, and in which the bulk concentrations in the two phases remain uniform up to the Gibbs dividing surface (GDS). That is, the surface excess amount, n_i^σ, for adsorption from liquid phase is

$$n_i^\sigma = n_i - V_o^l c_i^l \tag{9.1}$$

where:
   n_i is the total amount of component, i, in the liquid
   c_i^l is its concentration in the liquid phase after adsorption
   V_o^l is the volume of the liquid phase up to the GDS

This expression follows directly from the defining equation (Figure 9.1)

$$n_i^\sigma = n_i - V_o^\alpha c_\alpha + V_o^\beta c_\beta$$

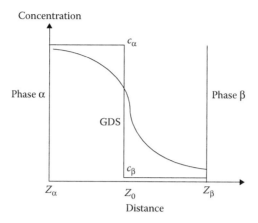

FIGURE 9.1 Gibbs dividing surface (GDS). In the reference system, the concentration remains constant up to the GDS. In the real system, the concentration vary across the interface of thickness $\gamma = Z_\beta - Z_\alpha$ from that of phase α, concentration c_α to that of phase β, concentration c_β.

However, the value of n_i^σ is dependent on the position of the GDS through the value of V_o^l (see Section 2.2) [11]. This is caused by the fact that some authors prefer to adopt alternative approaches for the calculation of the surface excess amount.

One of the most meaningful presentations of the liquid-phase adsorption equilibrium is the reduced surface excess [4,18,27], a function invariant with respect to the position of the GDS (see Section 2.2), which could be defined as a result of the following procedure [11]: first from Equation 9.1 it is possible to write

$$n_2^\sigma = n_2 - V_o^l c_2^l \tag{9.2}$$

and

$$n^\sigma = n^0 - V_o^l c^l \tag{9.3}$$

where $n^\sigma = n_1^\sigma + n_2^\sigma$ and $n^0 = n_1 + n_2$, and $c^l = c_1^l + c_2^l$, n_i^σ, where $i = 1,2$, is the surface excess amount of component i, total amount of component i in the liquid, c_i^l is its concentration in the liquid phase after adsorption. Then, the reduced excess amount for the component 1 is defined by the elimination of V_o^l from Equations 9.2 and 9.3 [4,11,18,27]:

$$n_1^e = n^0 \left(x_1^0 - x_1^e \right) \tag{9.4}$$

or referred to an amount of adsorbent:

$$\Gamma_1^e = \frac{n^0}{m_a} \left(x_1^0 - x_1^e \right) \tag{9.5}$$

where

$$x_1^0 = \frac{n_1^0}{n_1^0 + n_2^0} = \frac{n_1^0}{n^0}$$

and

$$x_1^e = \frac{n_1^0 - n_1^s}{n_1^0 - n_1^s + n_2^0 - n_2^s}$$

where:

n^0 is the total number of moles in contact with the adsorbent that describes the amount of substance of the liquid mixture

m_a is the adsorbent mass

x_1^0 and x_1^e are the mole fractions of component 1 in the initial and equilibrium solution, respectively

n_1^s and n_2^s are the moles adsorbed into the Gibbs phase

That is, the amount disappeared from the bulk phase solution [4]. Consequently, for components $i = 1,2$, $\Delta x_i^l = x_i^0 - x_i^e$ is the change in the mole fraction of component i, resulting from bringing a specified mass of adsorbent m_a, into contact with a specified amount of solution n^0 [1,2].

The use of reduced surface excess amounts is recommended by the IUPAC [2] as a convenient way of reporting the experimental results. For many years, this description of the experimental adsorption data was intuitively selected as a method of plotting the experimental data without any reference to the Gibbs approach. The obtained isotherm was, in general, named a *composite isotherm* [11]. Consequently, adsorption isotherms are usually reported in the form of reduced surface excess amounts, Γ_1^e versus x_1^l, which is the mole fraction of the solute component [1,2,4,9,11]. In Figure 9.2 five types of composite isotherms according to the classification of

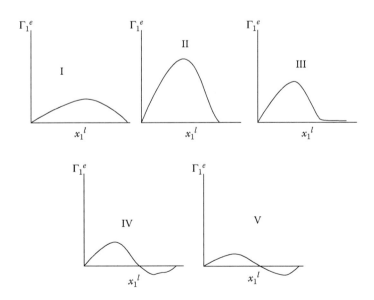

FIGURE 9.2   Composite isotherm according to the classification of Schay and Nagy.

Schay and Nagy are represented [9]. A negative value of Γ_1^e means that the solvent is preferentially adsorbed.

In practice, the amount of solute adsorbed from the liquid phase is also calculated by subtracting the remaining concentration after adsorption to the concentration at the beginning of the adsorption process [10]. First, it is necessary to determine the particle size of the adsorbent, adsorption temperature, pH of the solution, and contact time necessary to achieve the adsorption equilibrium [10,13,28]. Then, different amounts of the solid adsorbent, generally few milligrams, that is, 1–10 mg are contacted with equivalent volumes of solution, in general, 25–50 mL of the solution containing an established concentration of the substance to be adsorbed [10,15,28]. Thereafter, all solutions are equilibrated normally for 0.5–5 h, while being stirred. Once contact time has elapsed the adsorbent is centrifuged and filtered. The resulting solutions are poured into vials, which are completely filled, so the headspace is very low. Then, the lids are sealed with a tape until the vials are analyzed by, for instance, a spectrophotometer [28]. The amount of solute adsorbed from the liquid phase is calculated by subtracting the remaining analyzed concentration to the starting concentration [10,28]. Then, the amount adsorbed on the adsorbent is determined from the initial liquid-phase concentration and equilibrium concentration [28]:

$$q_1^e = \frac{V}{m_a}\left(C_1^0 - C_1^e\right)$$ (9.6)

that is, C_1^0 and C_1^e are the initial and equilibrium concentrations, respectively, V is the volume of solution, and m_a is the mass of the adsorbent.

9.3 EMPIRICAL ADSORPTION ISOTHERMS APPLIED FOR THE CORRELATION OF LIQUID–SOLID ADSORPTION EQUILIBRIA IN SYSTEMS COMPRISING ONE DISSOLVED COMPONENT

For the correlation of the liquid-phase adsorption equilibria of a single component, several isotherm equations for gas-phase adsorption can, in principle, be extended to the liquid-phase adsorption by the simple replacement of adsorbate pressure by its concentration [29]. These equations are the Langmuir, Freundlich, Sips, Toth, and Dubinin–Radushkevich equations [27–29]. The Langmuir equation is

$$q_1^e = \frac{q_m b_0 C_1^e}{1 + b_0 C_1^e}$$ (9.7)

and the Freundlich equation is

$$q_1^e = k\left(C_1^e\right)^{1/n}$$ (9.8)

equations are most widely used to correlate liquid-phase adsorption data [21,31–35]. Where the experimental equilibrium data express as q_1^e, which is the amount of the solute component 1 adsorbed per mass of adsorbent, this parameter is expressed in (mg/g) or (mmol/g), C_1^e is the final solute concentration in solution, or the equilibrium concentration in (mg/L) or (mmol/L). Besides, q_m and b_0 are the Langmuir's parameters, and k and n are the Freundlich parameters [28].

The Sips equation is [28,36]

$$q_1^e = \frac{q_0 (dC_1^e)^{1/S}}{1 + (dC_1^e)^{1/S}} \tag{9.9}$$

where q_0, d, and S are parameters of the Sips equation. The other three parameter equation is the Toth equation [4,28]:

$$q_1^e = \frac{q_0 b C_1^e}{1 + (bC_1^e)^{1/t}} \tag{9.10}$$

in which q_0, b, and t are parameters of the Toth equation. The calculation of the parameters for the Langmuir equation is, generally, carried out by linearly regressing each set of experimental data using the following equation:

$$y = C_1^e = q_m \left(\frac{C_1^e}{q_e} \right) + \frac{1}{b_0} = mx + b \tag{9.11}$$

which is a linear form of the Langmuir equation.

A standard regression analysis could be performed on each set of data to calculate the correlation coefficient for each regression as well as the standard error of the estimate and the values of the standard errors for the parameters applying a commercial regression software [37,38]. The calculation could be also carried out by nonlinearly regressing each set of experimental data using Equation 9.7 directly [37,38].

In the case of the Freundlich equation, the calculation of the parameters is obtained by linearly regressing each set of experimental data using the following equation [21]:

$$y = \log(q_e) = \log(k) + \left(\frac{1}{n} \right) \log(C_1^e) = mx + b \tag{9.12}$$

In addition, the linear regression analysis could be carried out by applying commercial regression software, which could also be used to nonlinearly regress each set of experimental data using Equation 9.8 directly [37,38].

It has been also shown that the preferential adsorption of various compounds from aqueous solutions, by activated carbons, follows an equation of the Dubinin–Radushkevihs–Kaganer type [39,40]:

$$q_1^e = q_m^{DRK} \exp \left\{ - \left[\frac{RT}{E_S} \ln \left(\frac{C_1^0}{C_1^e} \right) \right]^2 \right\} \tag{9.13}$$

In this expression, the vapor–solid adsorption potential $RT \ln(P_0/P)$ (see Section 3.2) is replaced by a new thermodynamic potential $RT \ln(C_1^0/C_1^e)$, where C_1^0 is the saturation concentration of the adsorbed species, C_1^e is the concentration at equilibrium to which corresponds the amount adsorbed, q_e and E_S are similar to the characteristic energy [40]. A generalized form of Equation 9.13 is [40–42]

$$q_1^e = q_m^{DRK} \exp\left\{ -\left[\frac{RT}{E_S} \ln\left(\frac{C_1^0}{C_1^e} \right) \right]^n \right\} \tag{9.14}$$

where n is a constant parameter. Equation 9.13 is similar to the previously explained Dubinin equation (see Section 3.2) where E_S, which is related to the characteristic energy that must be temperature invariant [40].

Currently, the majority of authors use Langmuir equation, that is, Equation 9.7 instead of Equation 9.14 adapted to solution work. The Freundlich equation, on the other hand, suffer from the shortcoming that the variation of their main parameters with temperature cannot be predicted in a simple fashion, as opposed to Dubinin's equation [40].

Finally, it is necessary to state that any one of the equations exposed in the present section to correlate the relation between the amounts adsorbed q_1^e with the equilibrium concentration in solution C_1^e corresponds to a specific model for adsorption from the solutions. That is, it should be considered as empirical isotherm equations.

9.4 MODEL DESCRIPTION OF ADSORPTION FROM THE LIQUID PHASE ON SOLIDS

Adsorption from a binary liquid mixture can be described as a phase-exchange reaction [3]:

$$N_2^l \text{ (solute in solution)} + N_1^s \text{ (adsorbed solvent)} \Leftrightarrow$$
$$N_2^s \text{ (adsorbed solute)} + N_1^l \text{ (solvent in solution)}$$

On a homogeneous surface, which is characterized by a unique adsorption energy for all the surface, the equilibrium constant is defined by the following equation [2,20,43]:

$$K_{12} = K_0 \exp\left(\frac{\varepsilon_1 - \varepsilon_2}{RT} \right) = \frac{x_1^s x_2^l \gamma_1^l \gamma_2^l}{x_2^s x_1^l \gamma_2^l \gamma_1^l} = \alpha \beta_{12} \tag{9.15}$$

where:

x_i^l and x_i^s are the mole fractions of the ith component in the bulk l, and surface s phases

γ_i^l and γ_i^s are the activity coefficients in the bulk and surface phases, respectively

α is the separation factor

ε_i is the adsorption energy of the ith component

K_0 is the preexponential factor

In addition, the validity of the monolayer character of the adsorption process, the equality of the molecular sizes of solution components, and the constancy of the total number of molecules in the surface phase in the present model description will be considered [43]. In this case, since $x_1^s + x_2^s = 1$, then [3]

$$x_1^s = \frac{K_{12}x_{12}^l\gamma_{12}(x_1^s, x_1^l)}{1 + K_{12}x_{12}^l\gamma_{12}(x_1^s, x_1^l)}$$

where $x_{12}^l = x_1^l/x_2^l$, $\gamma_{12}^l = \gamma_1^l/\gamma_2^l$, and $\gamma_{21}^s = \gamma_2^s/\gamma_1^s$.

However, most adsorbents have heterogeneous surfaces, that is, the heat evolved during adsorption is dependent on the adsorption amount. To be more precise, a heterogeneous surface is characterized by different surface sections, which have different adsorption energies. Consequently, the surface is characterized by a distribution of adsorption energies and each surface section is the place of an independent adsorbed phase [36,44].

Let us then suppose adsorption on a heterogeneous solid while supposing the random distribution of surface sites. Consequently, the overall integral equation for the calculation of the mole fraction of component of the first component over the whole surface phase has the following form [20,43]:

$$x_{1,t}^s = \int_\Delta \frac{K_{12}x_{12}^l\gamma_{12}(x_{1t}^s, x_1^l)}{1 + K_{12}x_{12}^j\gamma_{12}(x_{1t}^s, x_1^l)}F(\varepsilon_{12})d\varepsilon_{12} \tag{9.16}$$

where $x_{12}^l = x_1^l/x_2^l$, $\gamma_{12}^l = \gamma_1^l/\gamma_2^l$, $\gamma_{21}^s = \gamma_2^s/\gamma_1^s$, and $F(\varepsilon_{12})$ are the distribution function of differences of adsorption energies $\varepsilon_1 - \varepsilon_2$ and Δ is the integration region.

The integral Equation 9.16 generates various isotherm equations, one of them is a general expression introduced by Jaroniec and Marczewski [20,45]:

$$x_{1t}^s = \left(\frac{[\overline{K}_{12}x_{12}^l\gamma_{12}(x_1^s, x_1^l)]^n}{[1 + \overline{K}_{12}x_{12}^l\gamma_{12}(x_1^s, x_1^l)]^n}\right)^{m/n} \tag{9.17}$$

The parameters $\overline{K}_{12}$, m, and n characterize the distribution function, $F(\varepsilon_{12})$ and $\overline{K}_{12}$, the characteristic equilibrium constant describes the position of $F(\varepsilon_{12})$ on the axis of energy differences $\varepsilon_{12} = \varepsilon_1 - \varepsilon_2$ and $0 < m < 1$, and $0 < n < 1$ are the heterogeneity parameters characterizing the distribution function.

Equation 9.17 is called the general Langmuir equation [20] for special values of the heterogeneity parameters, m and n. It is reduced to different well-known isotherm equations, such as the Langmuir–Freundlich equaution for $0 < m = n < 1$, the generalized Freundlich isotherm equation, in the case where $n = 1$ and $0 < m < 1$, and the Toth isotherm equation for $m = 1$ and $0 < n < 1$ [20].

The earlier discussion of a model description of adsorption from the liquid phase on solids has justified the use of empirical equations formally similar to the Langmuir, Freundlich, Sips, and Toth isotherms [27–29]. The use of Dubinin-type equations was also phenomenologically justified previously [40]. Besides, in the model described in the present section, the effect of surface heterogeneity, nonideality of the bulk, and the surface phases on adsorption equilibrium was taken into account. However, the complexity of liquid adsorption systems makes it very difficult to separate the effects of solution nonideality and surface heterogeneity [43]. It is observed rather the global nonideality of experimental systems was caused by both adsorbate and adsorbent imperfections [20]. The effect of solid surface heterogeneity and solution nonideality on adsorption equilibria can be discussed in terms of some theoretical models, and the values of surface activity coefficient may also be measured experimentally. Nevertheless, the surface activity coefficients and heterogeneity can be regarded only for assumed models of adsorption process [20].

These difficulties makes, in general, the researchers to use the following approach: first, experimentally determine the adsorption isotherm, applying a methodology that is analogous to those explained in Section 9.2 of the present chapter, and then correlates the dataset with the Langmuir, Freundlich, or Dubinin isotherms [27–29,40]. In the next section, we will see some examples of this line of attack.

9.5 SOME APPLICATIONS OF LIQUID–SOLID ADSORPTION

Waters polluted by organic compounds are often treated by adsorption processes in which activated carbons and other materials act as adsorbents [10,13–23]. Several studies have been made in order to comprehend thoroughly the adsorption mechanisms involved in these processes [10,13–23,25–35,39–43,46–62]. The information obtained in this mode leads to a superior design and to a better effectiveness for the materials used as adsorbents.

9.5.1 ACTIVATED CARBONS

Activated carbons are the most widely used industrial adsorbent for removing contaminants and pollutants from gaseous, aqueous, and nonaqueous streams, due in part, to their uniquely powerful adsorption properties and the ability to readily modify their surface chemistry [48,63]. Carbon is the mainly used adsorbent in the case of liquid–solid adsorption systems. As was also affirmed before, the adsorption capacity of an activated carbon depends on the nature of the adsorbent, the nature of the adsorbate, and the solution conditions, that is, pH, temperature, and ionic strength [10,13]. Regarding the activated carbon surface, the main components are the carbon basal planes, edges and crystal defects, and ash impurities, that is, metal oxides and oxygen surface groups [10,13]. The last components are principally located on the edges of the graphitic basal planes (see Chapter 7, Sections 7.5 and 7.6) [10,13]. Surface functional groups can be classified as acidic, that is, carboxyl, carbonyl, phenolic, hydroxyl, lactone, anhydride, and basic chromene- and pyrone-like structures [61]. Despite being a small fraction of the overall carbon surface, the oxygen groups are however very active, exhibiting a significant influence on the adsorption capacity [62].

Phenol is one of the most important compounds adsorbed by carbon from the liquid phase. Phenol is a fundamental structural part for a diversity of synthetic organic compounds. Then, wastewater originating from many chemical plants and pesticide and dye manufacturing industries contains phenols [46]. In addition, wastewater originating from other industries such as paper and pulp, resin manufacturing, gas and coke manufacturing, tanning, textile, plastic, rubber, pharmaceutical, and petroleum also includes diverse types of phenols [22,46]. In addition, as the phenols generated as a consequence of industrial activity, wastewaters also contain phenols produced as a product of decay of vegetation [46]. In view of the wide prevalence of phenols in different wastewaters and their toxicity to human and animal life even at low concentration, it is essential to remove them before the discharge of wastewater into water bodies [22,46]. Several techniques, such as oxidation with ozone/hydrogen peroxide, biological methods, membrane filtration, ion exchange, electrochemical oxidation, reverse osmosis, photocatalytic degradation, and adsorption have been used for the removal of phenols [46]. However, regardless of the accessibility of the aforementioned methods for the removal of phenols, the adsorption process even now remains the best as it can, in general, to remove all types of phenols, and the effluent treatment is convenient because of the simple design and easy operations [46].

Phenol adsorption and substituted phenols from aqueous solutions on activated carbons is one of the most studied of all liquid-phase applications of carbon adsorbents [22,40,47]. Today, it is known that the adsorption process on carbon materials basically depends on several variables, such as the pH of the solution, the electron-donating or electron-withdrawing properties of the phenolic compound, and the surface area of the adsorbent and its surface chemistry, which is determined by the nature of its oxygen surface functionalities and its surface charge [10]. Phenol and substituted phenol compounds in aqueous solutions, in their uncharged form, are adsorbed on the carbon surface by dispersion forces between the π electrons of aromatic ring and the π electrons of the graphene layers. Nevertheless, the pH of the solution can affect the charge of phenolic compounds and consequently affects the electrostatic interactions between the adsorbent and the adsorbate [13].

Due to the lack of data, in general, an experimental investigation of the adsorption equilibrium has to be carried out in order to establish the calculation basics for engineering [23]. To analyze the adsorption isotherms of phenolic compounds obtained at equilibrium, many authors use the Langmuir or Freundlich equations [21,28]. It is also possible to investigate the adsorption of phenol compounds from aqueous solutions [40] within the framework of Dubinin's theory [40–42]. With the obtained empirical correlations of equilibrium data it is possible to get the calculation basis for the design of liquid-phase adsorptive separation processes.

Active carbons are also applied for the removal of oil and other organic compounds from effluent water in petroleum refining, petrochemicals, metal extraction, detergent, margarine, and soft fat manufacture, mineral extraction, and other industries [10,28, 49–52]. In the food and beverage industries, active carbon is used to remove the color or odor from the produced products. In the chemical and pharmaceutical industry, active carbons are applied for the elimination of impurities to improve product quality [10]. As was previously stated, one of the most significant part of the needed information in the analysis and design of adsorption separation processes are the liquid-phase adsorption isotherm data of various organic compounds in carbon and also in other systems, these data have been obtained and reported over the years [10,13–23,25–35,39–43,47–60].

For example, benzene and toluene are significant materials in the chemical industry, as they are initial materials in the chemical production of a lot of products, and they are frequently used as solvent in a broad diversity of manufacturing processes [28,51,52]. Given that these compounds are classified as flammable and toxic materials, their occurrence in the environment, generally in wastewater at low concentration, is of main concern. For this reason, the subtraction of these compounds from an aqueous waste stream is also necessary [28]. Adsorption onto an activated carbon offers a viable method for the elimination of these organic pollutants from industrial wastewater [28].

9.5.2 PRECIPITATED SILICA

Another important adsorbent is silica (see Sections 7.1 through 7.4) [64–85]. Adsorption phenomena in silica also play a major role in many industrial and technological processes [70–74] as well as in liquid–solid chromatography [75]. In the case of silica adsorption the difficulties in the interpretation of the result arises from a variety of interactions between molecules of liquid components and the previously analyzed complex character of the solid surface, which is usually energetically heterogeneous [20,76–78]. The distribution at the surface of morphological irregularities and various silanol groups determine the surface heterogeneity of silica gel [76–78]. In addition, dimensions of the open space in capillaries of silica gels establish the adsorption of molecules, particularly in the form of clusters or mixed complexes that go through the pore systems of solids, then because of steric reasons, the adsorption process of clusters and complexes are restricted in very narrow pores. In this case, the complexes containing weakly bonded molecules may be destroyed and the adsorption is difficult in comparison to adsorption in macropores [20]. Consequently, it is possible to observe the influence of the geometrical structure on the adsorption selectivity as well as on the global characteristic of surface heterogeneity [70,71].

As one example of adsorption in silica, the competition of liquid components for silica gel surface in the binary liquid mixtures: methanol-benzene and 2-propanol-n-heptane was reported by Goworek et al. [20]. The binary liquid mixture–silica gel adsorption isotherms were measured using the commercial silica gels Si-40 and Si-100 from Merck (Table 9.1 [20]). On the basis of specific surface excess isotherms of the surface layer capacities were calculated. In the present case, $\Delta x_i^l = x_i^0 - x_i^e$ is also the change in the mole fraction of component i, resulting from bringing a specified mass of adsorbent m_a, into contact with a specified amount of solution n^0 [1,2], but it will be reported as the areal reduced excess amount, which is defined as [11]

$$\Gamma_1^e = \frac{n^0}{A}\left(x_1^0 - x_1^e\right)$$

TABLE 9.1
Textural Characterization of Silica Gels

Adsorbent	BET Area, S (m²/g)	Pore Volume (cm³/g)	Average Pore Diameter (nm)
Si-40	814	0.60	3
Si-100	348	1.15	10

where A is the area exposed by the adsorbent. The areal reduced excess adsorption isotherms for methanol (component 1) + benzene (component 2), and 2-propanol (component 1) + n-heptane (component 2) systems on silica gel Si-40 and Si-100 are presented in the form shown in Figure 9.2, and the shape of experimental isotherms is of the type II according to Schay-Nagy classification [9].

9.5.3 ZEOLITES

Zeolites are crystalline materials containing pores and cavities of molecular dimensions (ca. 3–15 Å) creating a nanoscale framework, which can be filled with water or other guest molecules (see Chapter 8, Sections 8.2.1, 8.3.1, and 8.3.2) [17,86–91]. The resulting molecular sieving ability has enabled the creation of new types of selective separation processes. More specifically, hydrophobic zeolites, such as all the silica zeolites or zeolites with very low aluminum content possess a high capacity for adsorbing organic compounds dissolved in water. Some recent studies showed that hydrophobic, dealuminated zeolites, adsorbed organic compounds from water as effectively as activated carbon [15,18,92,93]. The hydrophobicity of zeolitesis is controlled, fundamentally, by changing the Si/Al ratio in the framework by synthesis conditions and postsynthesis modification treatments [91,92,94].

The most studied hydrophobic zeolite in adsorption of organic compounds from water solutions is silicalite-1 [18]. Silicalite-1 is a molecular sieve with MFI structure composed of pure silica. The MFI framework has a 10-MR channel system with elliptical pore with diameters of 5.2 × 5.7 Å [87]. In addition, other zeolites, such as all silica beta zeolite [95], which possesses a three-dimensional 12-membered ring and an interconnected channel system with pore diameters of 7.1 × 7.3 Å [87] have been applied to the removal of methyl tert-butyl ether (MTBE) from water solutions [15]. These diameters are similar to those of dealuminated mordenite, a 12-membered ring zeolite with 6.5 × 7.0 Å pores [87]. A recent study [93] showed that hydrophobic, dealuminated mordenite adsorbed MTBE from water better than an activated carbon. In this study 5 mg of zeolite powders with 25 mL of aqueous solutions containing 100 µg/L of MTBE was equilibrated for 15 min, and the dealuminated mordenite removed 96% of the MTBE [15,93].

Likewise, dealuminated Y (DAY) zeolite and a zeolite with a very low aluminum content have a high capacity for adsorbing organic compounds dissolved in water, for instance, meta-nitrophenol dissolved in water solution (200 mg/L) was efficiently removed by liquid-phase adsorption in DAY zeolite as effectively as with an activated carbon [92].

REFERENCES

1. D.H. Everett (Ed.), IUPAC, Manual on definitions, terminology and symbols in colloid and surface chemistry, *Pure Appl. Chem.*, 58 (1986) 967.
2. C. Berti, P. Ulbig, A. Burdorf, J. Seippel, and S. Schulz, *Langmuir*, 15 (1999) 6035.
3. A.W. Adamson and A.P. Gast, *Physical Chemistry of Surfaces* (6th ed.), John Wiley & Sons, New York, 1997.
4. J. Toth, in *Adsorption: Theory, Modeling and Analysis* (J. Toth, Ed.), Marcel Dekker, New York, 2002, p. 1.

5. A.L. Myers and P.A. Monson, *Langmuir*, 18 (2002) 10261.
6. E.I. Frances, F.A. Siddiqui, D.J. Ahn, C.-H. Chang, and N.-H.L. Wang, *Langmuir*, 11 (1995) 3177.
7. A.L. Myers and J.M. Prausnitz, *AIChE J.*, 11 (1965) 121.
8. A.L. Myers and F. Moser, *Chem. Eng. Sci.*, 32 (1977) 529.
9. G. Schay and L.G. Nagy, *J. Chim. Phys.*, 58 (1961) 149.
10. F. Rodriguez-Reinoso and A. Sepulveda-Escribano, in *Handbook of Surfaces and Interfaces of Materials*, Chapter 9, Vol. 5, (H.S. Nalwa, Ed.), Academic Press, New York, 2001, p. 309.
11. F. Rouquerol, J. Rouquerol, and K. Sing, *Adsorption by Powders and Porous Solids*, Academic Press, New York, 1999.
12. S.J. Gregg and K.S.W. Sing, *Adsorption Surface Area and Porosity*, Academic Press, London, UK, 1982.
13. D. Nevskaia, E. Castillejos-Lopez, V.N. Muñoz, and A. Guerrero-Yuiz, *Environ. Sci. Technol.*, 38 (2004) 5786.
14. L.R. Radovic, C. Moreno-Castilla, and J. Rivera-Utrilla, in *Chemistry and Physics of Carbon* (Vol. 27, L.R. Radovic, Ed.), Marcell Dekker, New York, 2001, p. 227.
15. S. Li, V.A. Tuan, R. Noble, and J. Falcone, *Environ. Sci. Technol.*, 37 (2003) 4007.
16. K. Wagner and S. Schul, *J. Chem. Eng. Data*, 46 (2001) 322.
17. D.W. Ruthven, *Principles of Adsorption and Adsorption Processes*, John Wiley & Sons, New York, 1984.
18. S. Chempath, J.F.M. Denayer, K.M.A. De Meyer, G.V. Baron, and R.Q. Snurr, *Langmuir*, 20 (2004) 150.
19. O.A. Olafadehan and A.A. Susu, *Ind. Eng. Chem. Res.*, 43 (2004) 8107.
20. J. Goworek, A. Derylo-Marczewska, and A. Borowka, *Langmuir*, 15 (1999) 6103.
21. D. Andrieux, J. Jestin, N. Kervarec, R. Pichon, M. Privat, and R. Olier, *Langmuir*, 20 (2004) 10591.
22. L.S. Colella, P.M. Armenante, D. Kafkewitz, S.J. Allen, and V. Balasundaram, *J. Chem. Eng. Data*, 43 (1998) 573.
23. J. Seippel, P. Ulbig, and S. Schulz, *J. Chem. Eng. Data*, 45 (2000) 780.
24. S. Sircar, J. Novosad, and A.L. Myers, *Ind. Eng. Chem. Fundam.*, 11 (1972) 249.
25. D.P. Valenzuela and A.L. Myers, *Adsorption Equilibrium Data Book*, Prentice Hall, Englewood Cliffs, NJ, 1989.
26. M.B. Rao and S. Sircar, *Langmuir*, 15 (1999) 7258.
27. J. Oscik, *Adsorption*, Ellis Horwood, Chichester, UK, 1982.
28. H. Hindarso, S. Ismadji, F. Wicaksana, Mudjijati, and N. Indraswati, *J. Chem. Eng. Data*, 46 (2001) 788.
29. C. Tien, *Adsorption Calculations and Modeling*, Butterworth, Boston, MA, 1994.
30. D.D. Do, *Adsorption Analysis: Equilibria and Kinetics*, Imperial College Press, London, UK, 1998.
31. I. Abe, K. Hayashi, and T. Hirashima, *J. Colloid Interface Sci.*, 94 (1983) 577.
32. J. Avom, J.K. Mbadcam, C. Noubactep, and P. Germain, *Carbon*, 35 (1997) 365.
33. M.A. Khan and Y.I. Khattak, *Carbon*, 30 (1992) 957.
34. H. Teng and C.T. Hsieh, *Ind. Eng. Chem. Res.*, 37 (1998) 3618.
35. R.-S. Juang, F.-C. Wu, and R.-L. Tseng, *J. Chem. Eng. Data*, 41 (1996) 487.
36. W. Rudzinski and D.H. Everett, *Adsorption of Gases in Heterogeneous Surfaces*, Academic Press, London, UK, 1992.
37. Origin®, Scientific Graphing and Analysis Software, OriginLab Corporation, Northampton, MA, 2003.
38. N.R. Draper and H. Smith, *Applied Regression Analysis*, John Wiley & Sons, New York, 1966.
39. M. Jaroniec and A. Derylo, *J. Colloid Interface Sci.*, 84 (1981) 191.

40. F. Stoeckli, M.V. Lopez-Ramon, and C. Moreno-Castilla, *Langmuir*, 17 (2001) 3301.
41. M.M. Dubinin, *Carbon*, 27 (1989) 457.
42. F. Stoeckli, in *Porosity in Carbons* (J. Patrick, Ed.), Arnold, London, UK, 1995.
43. M. Jaroniec and R. Madey, *Physical Adsorption on Heterogeneous Surfaces*, Academic Press, London, UK, 1988.
44. S. Ross and J.P. Olivier, *On Physical Adsorption*, John Wiley & Sons, New York, 1964.
45. M. Jaroniec and A.W. Marczewski, *Monatsh. Chem.*, 15 (1984) 997.
46. A. Jain, V.K. Gupta, S. Jain, and Suhas, *Environ. Sci. Technol.*, 38 (2004) 1195.
47. L.R. Radovic, C. Moreno-Castilla, and J. Rivera-Utrilla, *J. Chem. Phys. Carbon*, 27 (2000) 227.
48. B. Singh, S. Madhusudhanan, V. Dubey, R. Nath, and N.B.S.N. Rao, *Carbon*, 34 (1996) 327.
49. S.H. Lin and F.M. Hsu, *Ind. Eng. Chem. Res.*, 34 (1995) 2110.
50. G. McKay and B.A. Duri, *Chem. Eng. Process*, 24 (1988) 1.
51. D. Chatzopoulos, A. Varma, and R.L. Irvine, *AIChE J.*, 39 (1993) 392027.
52. J. Choma, W. Burakiewitz-Mortka, M. Jaroniec, and R.K. Gilpin, *Langmuir*, 9 (1993) 2555.
53. J.T. Cookson, P.N. Cheremishinoff, and F. Eclerbusch (Eds.) *Carbon Adsorption Handbook*, Ann Arbor Science, Ann Arbor, MI, 1978.
54. I.H. Suffet and M.J. McGuire (Eds.) *Activated Carbon Adsorption of Organics from the Aqueous Phase* (Vols. 1 and 2), Ann Arbor Science, Ann Arbor, MI, 1980.
55. F.L. Slejko, *Adsorption Technology. A Step-by-Step Approach to Process Valuation, and Application*, Marcel Dekker, New York, 1985.
56. S.D. Faust and O.M. Aly, *Adsorption Processes for Water Treatment*, Butterworth Publishers, London, UK, 1987.
57. J.R. Perrich, *Carbon Adsorption for Wastewater Treatment*, CRC Press, Boca Raton, FL, 1981.
58. N.P. Cheremishinoff, *Carbon Adsorption for Pollution Control*, Prentice Hall, Upper Saddle River, NJ, 1993.
59. D.M. Nevskaia, A. Santianes, V. Munoz, and A. Guerrero-Ruiz, *Carbon*, 37 (1999) 1065.
60. D.M. Nevskaia and A. Guerrero-Ruiz, *J. Colloid Interface Sci.*, 234 (2001) 316.
61. H.P. Boehm, *Carbon*, 32 (1994) 759.
62. C. Leon-Leon and L. Radovic, in *Chemistry and Physics of Carbon* (Vol. 24, P. Thrower, Ed.), Marcel Dekker, New York, 1994.
63. J.K. Brennan, T.J. Bandosz, K.T. Thomson, and K.E. Gubbins, *Colloids Surf., A.*, 187–188 (2001) 539.
64. J. Persello, in *Adsorption on Silica Surfaces* (E. Papirer, Ed.), Marcel Dekker, New York, 2000, p. 297.
65. M.A. Hernandez, J.A. Velazquez, M. Asomoza, S. Solis, F. Rojas, V.H. Lara, R. Portillo, and M.A. Salgado, *Energy & Fuels*, 17 (2003) 262.
66. G.M.S. El Shaffey, in *Adsorption on Silica Surfaces* (E. Papirer, Ed.), Marcel Dekker, New York, 2000, p. 35.
67. S.M. Yang, H. Miguez, and G.F. Ozin, *Adv. Funct. Mater.*, 11 (2002) 425.
68. W.W. Porterfield, *Inorganic Chemistry. A Unified Approach*, Academic Press, New York, 1993.
69. H. van Damme, in *Adsorption on Silica Surfaces* (E. Papirer, Ed.), Marcel Dekker, New York, 2000, p. 119.
70. M. Borowko and W. Rzuysko, *Ber. Bunsen-Ges. Phys. Chem.*, 101 (1997) 1050.
71. J. Goworek, A. Nieradka, and A. Dabrowski, *Fluid Phase Equilibr.*, 136 (1997) 333.
72. A. Hamraoui and M. Privat, *J. Chem. Phys.*, 107 (1997) 6936.
73. H. Sellami, A. Hamraoui, M. Privat, and R. Olier., *Langmuir*, 14 (1998) 2402.

74. A. Hamraoui and M. Privat, *J. Colloid Interface Sci.*, 207 (1998) 46.
75. K. Unger, D. Kumar, V. Ehwald, and F. Grossmann, in *Adsorption on Silica* (E. Papirer, Ed.), Marcel Dekker, New York, 2000, p. 565.
76. B.A. Morrow and I.D. Gay, in *Adsorption on Silica Surfaces* (E. Papirer, Ed.), Marcel Dekker, New York, 2000, p. 9.
77. R. Duchateau, *Chem. Rev.*, 102 (2002) 3525.
78. E.F. Vansant, P. van der Voort, and K.C. Vranken, *Stud. Surf. Sci. Catal.*, 93 (1995) 59.
79. T.W. Dijkstra, R. Duchateau, R.A. van Santen, A. Meetsma, and G.P.A. Yap, *J. Am. Chem. Soc.*, 124 (2002) 9856.
80. T. Shimada, K. Aoki, Y. Shinoda, T. Nakamura, N. Tokunaga, S. Inagaki, and T. Hayashi, *J. Amer. Chem. Soc.*, 125 (2003) 4688.
81. A. Anedda, C.M. Carbonaro, F. Clemente, R. Corpino, and P.C. Ricci, *J. Phys. Chem. B*, 107 (2003) 13661.
82. C.J. Brinker and G.W. Scherer, *Sol-Gel Science*, Academic Press, New York, 1990.
83. R. Roque-Malherbe and F. Marquez, *Mater. Sci. Semicond. Process.*, 7(2004) 467.
84. R. Roque-Malherbe and F. Marquez, *Surf. Interface Anal.*, 37 (2005) 393.
85. F. Marquez-Linares and R. Roque-Malherbe, *J. Nanosci. & Nanotech.*, 6 (2006) (In Press).
86. C.S. Cundy and P.A. Cox, *Chem. Rev.*, 103 (2003) 663.
87. C. Baerlocher, W.M. Meier, and D.H. Olson, *Atlas of Zeolite Framework Types*, Elsevier, Amsterdam, the Netherlands, 2001.
88. M.E. Davies, *Nature*, 417 (2002) 813.
89. A. Corma, *Chem. Rev.*, 95 (1995) 559.
90. F. Marquez-Linares and R. Roque-Malherbe, Facets, *IUMRS J.*, 2 (2003) 14.
91. R. Roque-Malherbe, in *Handbook of Surfaces and Interfaces of Materials* (Vol. 5, H.S. Nalwa, Ed.), Academic Press, New York, 2001, p. 495.
92. R. Roque-Malherbe and F. Marquez-Linares, Facets, *IUMRS Journal.*, 3 (2004) 8.
93. M.A. Anderson, *Environ. Sci. Technol.*, 34 (2000) 725.
94. M.L. Occelli and H. Kessler (Eds.), *Synthesis of Porous Materials*, Marcel Dekker, New York, 1997.
95. M.A. Camblor, A. Corma, and S. Valencia, *J. Chem. Soc. Chem. Commun.*, 2365 (1996).

10 Carbon Dioxide Adsorption on Akageneite, Sphere and Particle Packing, and Ordered Amorphous and Microporous Molecular Sieve Silica

10.1 AKAGANEITE

10.1.1 INTRODUCTION

Iron oxides, oxide hydroxides, and hydroxides are common compounds. These are some of the most important materials on earth because of their magnetic [1–4], catalytic [5,6], and adsorption [7,8] properties. In particular, the oxide hydroxide akaganeite is an important constituent of soils, geothermal brine deposits, and corrosion products [9–21]. The common structural building block for all the known oxide hydroxides is the $Fe(O,OH)_6$ octahedron organized in different forms [1]. Particularly, the octahedrons that generate the framework produce a hollandite-type structure by displaying the symmetry of the I2/m (No. 12) space group in which four double chains of octahedrons create tunnels with square cross sections [9]. Further, these tunnels are in part occupied by chloride anions whose charges are compensated by protons or positive ions [11,21]. Moreover, the octahedrons are spatially organized to generate two different octahedral sites for the Fe(III) ions to produce antiferromagnetic behavior owing to the alignment of their magnetic moments [10,13]. However, the small size of the crystallites creates a situation where the direction of the magnetic moment of the crystallites turns out to be unsteady and fluctuates, producing superparamagnetism [14].

In addition, akaganeite, on heating, releases adsorbed water (ca. 350 K). Next, a dehydroxilation process takes place (at ca. 480 K). Finally, an exothermic peak found at ca. 633 K has been associated with both the loss of HCl and the beginning of the formation of hematite and in some cases magnetite [12]. One more important feature of akaganeite is the morphology exhibited by their crystals. In this regard, the recognized crystallite forms for this material are somatoids or spindle shaped and rods or needle-like [15] though additional one-dimensional morphologies, such as tubes and fibers, had been obtained as well [16]. Hence, the essential morphological characteristic of akaganeite is one-dimensionality. This feature is due to the hollandite-type framework, because the channels run through the [010] direction. Accordingly, the crystal is elongated in this direction. Meanwhile, the crystal is narrow along the [100] and [001] crystallographic directions [10]. As a result, the crystal sizes are normally between 100 and 1000 nm*.

Furthermore, as a result of their small crystallite sizes, akaganeites show a high specific surface area (S), that is, this material shows a developed external surface together with the micropores [5]. After thermal dehydration of akaganeite, water molecules are detached. Thereafter, a structure showing an open-channel system that is suitable for the adsorption of small molecules and a relatively high external surface area is produced [7,8]. Moreover, owing to the existence of a developed external surface, akaganeite can adsorb ions as well on this surface [17–20]. In this regard, adsorption of arsenite [17,18] and phosphates [19,20] on the akaganeite outer surface has been investigated, yielding experimental adsorption isotherms, which are fitted with the Freundlich equation in general [19]. Hence, both gas adsorption in the micropores and ion adsorption on the external surface are important properties of akaganeite.

As we state that this book is made from the point of view of materials science, thorough characterization of akageneite phases is investigated in this chapter previous to the study of CO_2 adsorption [10]. Carbon dioxide was selected for the adsorption study, because this molecule is a good probe to study adsorption on microporous materials [22], given that it is an excellent tool for the measurement of the micropore volume [23] and adsorption interactions [24]. In particular, in the present case owing to the magnetic properties of akaganeite and if these magnetic properties could be modified by controlling the electronic structure of the akaganeite through adsorbed molecules, this process could produce effects that might be used in practice. Further, adsorption is a useful process for the separation, storage, and recovery of gases [25]. In particular, the separation, storage, and recovery of carbon dioxide are very important research topics, because it is a reactant in significant industrial processes and a greenhouse gas that contributes to global warming [26].

Hence, the synthesis and structural characterization of two akageneite phases are described in this chapter, together with the investigation of their magnetic and adsorption properties. The research was carried out by measurements and characterizations that consist of powder X-ray diffraction (XRD), energy dispersive X-ray analysis (EDX), thermogravimetric analysis (TGA), Mössbauer spectroscopy (MS), and magnetization experiments, whereas the adsorption investigation was performed using two volumetric adsorption analyzers [10].

10.1.2 SYNTHETIC PROCEDURES

All the consumable chemicals were of analytical grade without additional purification. The water used in the synthesis process was bidistilled, provided the synthesis was performed by the acid hydrolysis of iron chlorides [16,27–35]. In this way, one solution of ferrous chloride and ferric chloride in the ratio of 1:2 was prepared, which is identified as solution A. A second solution of 6 M sodium hydroxide was also prepared, which is identified as solution B. Solution B was mixed under agitation with solution A and was stirred for 60 min at 20°C. Then, this mixture was heated at 90°C for 30 min under stirring conditions. After that, a 0.3 M sodium citrate solution was added. Afterward, the entire resulting mixture was heated at 90°C for 30 min [10]. Then, the precipitated reaction product was washed with acetone and deionized water. Later, it was centrifuged three times for 15 min at 4000 rpm. Finally, the sample was dried in vacuum without heating [13]. The complete synthesis procedure was carried out in nitrogen atmosphere at ambient pressure. In addition, the process was repeated obtaining the same product in both cases, namely the as-synthesized sample, labeled AK-AS.

The first modification of the as-synthesized akaganeite was a thorough washing process with bidistilled water to get the washed sample identified as AK-W. Thereafter, this material was exchanged at temperature during 72 h with a 2 M LiCl aqueous solution at ratio of 1 g of sample per 1 L of solution, obtaining the L-exchanged sample, which is identified as AK-Li.

10.1.3 CHARACTERIZATION

To characterize the tested materials, the phase identification was carried out initially. Thereafter, the structural elucidation was carried out. To complete these tasks, the XRD profiles of the produced samples were gathered with a Bruker D8 Advance system in Bragg–Brentano vertical goniometer configuration. The X-ray radiation source was a ceramic Cu anode tube. In addition, the gathered XRD patterns were refined with the Pawley method to confirm the assigned structure. The computer program used to carry out the Pawley refining processes was the Bruker DIFFRAC*plus* TOPAS™ software package [31]. The XRD profiles of the (a) AK-AS, (b) A-W, and (c) AK-Li samples are exhibited in Figure 10.1. The XRD analysis revealed the presence of akaganeite and sodium chloride in the AK-AS material (Figure 10.1a), whereas only the akaganeite phase was found in the AK-W (Figure 10.1b) and the AK-Li (Figure 10.1c) materials.

The XRD profiles of the AK-W and AK-Li materials were then refined using the Pawley profile fitting method assuming the I2/m space group [13]. The obtained results are exhibited in Table 10.1, together with data that was previously reported in the literature for two different akaganeites [9,13].

The direct comparison of both sets of data yielded a slight dissimilarity in the parameters, possibly due to a different distribution of the chloride cations.

Second, the analysis of the chemical composition of the produced akageneites was performed using energy dispersive X-ray analysis (EDAX) with the help of a spectrometer coupled with the JEOL JSM-6360 electron microscope, provided the electron beam acceleration voltage was 20 kV. To improve the accuracy, the elemental composition of five spots was measured and the average composition was calculated. The analysis reported that the AK-AS samples contain oxygen, sodium, chloride, and iron, whereas the AK-W samples contain oxygen, chloride, and iron. Finally the AK-Li samples contain oxygen, lithium, chloride, and iron [10]. The results were similar to those reported for standard akaganeite phases [13].

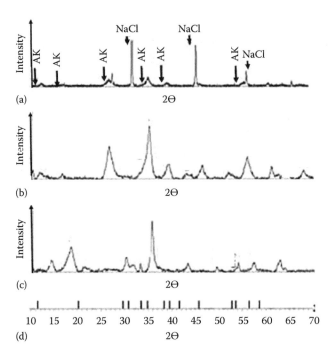

FIGURE 10.1   XRD profiles of (a) AK-AS, (b) the AK-W, and (c) AK-Li; in addition, the peaks corresponding to the akageneite phase are shown (d).

TABLE 10.1
Cell Parameters and Crystallite Size of the Tested Samples Were Refined with the Pawley Method Assuming the I2/m Space Group, Together with the Literature Data

Sample	a (Å)	b (Å)	c (Å)	β (degree)	Φ (nm)
AK-W	10.37(4)	3.08(3)	10.76	90.9(3)	10(1)
AK-Li	9.24(8)	3.24(5)	10.11(1)	92(10)	10(1)
AK [9]	10.5536(7)	3.03449(8)	10.5740(4)	90.086(5)	35

Source: Stahl, K. et al., *J. Corr. Sci.*, 45, 2563e2575, 2003; Post, J.E. et al., *Am. Mineral.*, 88, 782, 2003.

Next, a TGA testing process was carried out by means of a TA, Q-500 equipment [10], provided the temperature was linearly scanned from 25°C to 600°C at a rate of 5°C/min, at a flow of 100 mL/min of pure N_2 [26]. The profiles corresponding to the AK-W and AK-Li samples demonstrated that the effect of heating in akaganeite initially removed water up to 380 K, later dehydroxilation occurred from 443 K to 493 K, and finally released HCl at ca. 543 K [9]. Hence as evidenced in both profiles, the synthesized akaganeites behaved as expected [13]. In addition, these data made possible to conclude that the composition in weight percent (wt. %) of the AK-AS sample was akaganeite—48 wt. %, sodium chloride—wt. 47%, and sodium citrate—wt. 5%, whereas the other samples were composed of a pure akageneite phase [10].

The magnetic properties of the produced materials were tested with Mössbauer spectroscopy (MS) and magnetic measurements. The Mössbauer test was carried out with a SEECo spectrometer that operated in the constant acceleration mode, including a 50 mCi 57Co gamma-ray source at Rh matrix made by Rietverc GmbH. Moreover, the 1024-point raw data were analyzed using a least-squares fitting spectral analysis program (PeakFit, Seasolve Software Inc., Framingham, Massachusetts), provided all the spectra were collected at 300 K [10]. The spectra of the AK-W and AK-Li phases were fitted with two overlapping paramagnetic doublets, using a Lorentzian function to simulate the peaks, to calculate the isomeric effect (δ), quadrupole splitting (Δ), and amplitude A of the synthesized materials. Thereafter, the parameters corresponding to the tested materials are reported in Table 10.2, together with data reported in the literature for a pure akageneite phase [35].

In addition, the magnetization curve (M vs. H) was collected at room temperature (300 K) in the vibrating sample magnetometer (VSM), Lakeshore 7400 Series, provided the maximum magnetic field used was 2.2 T. The curves indicated that both tested materials show antiferromagnetic behavior. This fact means that two spin sublattices are coupled along the [010] direction of the crystalline structure in the framework of these materials [36,37]. Then, given that the Fe(III) cation spins pointed in opposite direction, no net magnetization should be observed. However, surface spins produce an imbalance. Subsequently, a net magnetic moment is conferred to the akaganeite nanoparticles [38,39], producing an anomalous antiferromagnetic behavior [32,40,41]. Now, as iron is located in nonequivalent sites near to the chlorine anion sites, the chlorine content should have an altering effect in the magnetic performance of this material [42]. In addition, this antiferromagnetic behavior could be masked, because the crystallite sizes are very small. Then, the Fe(III) ions experience an average-to-zero effective internal magnetic field producing a superparamagnetic behavior [44], as was observed for both materials [10].

TABLE 10.2
Mössbauer Parameters Corresponding to the Washed AK-W and AK-Li Phases

Sample	δ_1	δ_2	Δ_1	Δ_2	A_1	A_2	References
AK-W	0.373	0.537	0.367	0.945	59	41	[10]
AK-Li	0.377	0.547	0.370	0.956	62	38	[10]
Standard akaganeite	0.372	0.55	0.375	0.940	60	40	[43]

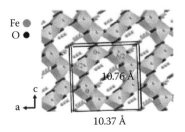

FIGURE 10.2   Akaganeite framework. (From Post, J.E. et al., *Am. Mineral.*, 88, 782, 2003.)

Finally, using the crystallographic information reported for the AK-W and AK-Li phases (Table 10.1), together with the atomic positions and occupancy factors reported by Post et al. [13] and the Wyckoff sites taken from the International Tables for Crystallography [34], a representation (Figure 10.2) of the akaganeite framework was produced along the [010] crystallographic direction with the Bruker DIFFRAC*plus* TOPAS™ software. In relation to the framework, it is necessary to state that the evidenced 4 Å × 4 Å wide pores along the [010] direction is the most important feature for the adsorption applications of this material [36,37].

10.1.4   NITROGEN AND CARBON DIOXIDE ADSORPTION

To measure the akaganeite outer specific surface, the adsorption isotherm of nitrogen was collected at 77 K. In this regard, the adsorption data were fitted to the Brunauer–Emmett–Teller (BET) isotherm equation in linear form to make the calculation (See Chapter 3):

$$y = \frac{x}{n_a(1-x)} = \left(\frac{1}{N_m C}\right) + \left(\frac{C-1}{CN_m}\right)x = b + mx \qquad (10.1)$$

where:
$b = (1/N_m C)$
$m = (C-1/CN_m)$
$y = x/n_a(1-x)$
$x = P/P_0$

in the region: $0.05 < x < 0.4$.

On this ground, N_m was estimated yielding 2.16 mmol/g. Then, the specific surface area was calculated producing $S = 220$ m^2/g. Thereafter, given that the equivalent particle diameter (Φ_{BET}) of the synthesized akaganeite can be computed with the equation, $\Phi_{BET} = 6/S_{BET}\rho$, where ρ is the akaganeite density [32], $\Phi_{BET} = 8$ (1) nm, a value, within the experimental error, equivalent to the Gaussian spherical diameter measured with the XRD data [10,13].

The high-pressure carbon dioxide adsorption isotherms at 263, 273, 300, and 318 K in the pressure range of $0.01 < P < 30$ bar were collected on the tested akaganeites (Figure 10.3).

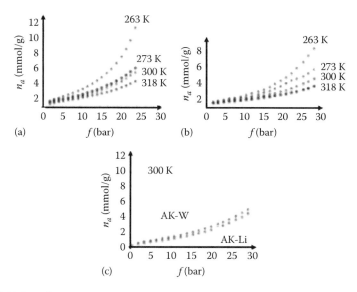

FIGURE 10.3   High-pressure carbon dioxide adsorption on the tested akaganeites.

Now, the experimental high-pressure carbon dioxide adsorption data at different temperatures on both akaganeite samples degassed at 473 K were fitted with the linear form of the Dubinin–Radushkevitch (DR) equation [5]:

$$y = \ln(n_a) = \ln(N_a) - \left(\frac{RT}{E}\right)^n \ln\left(\frac{P_0}{P}\right)^n = b - mx \qquad (10.2)$$

where:

$y = \ln(n_a)$
$b = \ln(N_a)$
$m = (RT/E)^n$
$x = \ln(P_0/P)^n$

then, the fitting process allowed the calculation of the best fitting parameters of Equation 10.2, that is, N_m, E, for $n = 2$, along with the regression coefficient, and the standard errors. In Table 10.3, parameters calculated with the D–E equation are reported, namely the maximum amount adsorbed, N_m, along with the micropore volume that is calculated using the Gurtvich rule, that is, $W_{MP}^{CO_2} = V_{CO_2}^L N_m$ ($V_{CO_2}^L = 48.3$ cm^3/mol, molar volume of carbon dioxide) along with the characteristic energy of adsorption, E, and the isosteric heat of adsorption that is calculated using the following equation, q_{iso} $(\theta = 0.37) = 1.16E$ [5,10].

The previously reported data shows that the carbon dioxide molecule freely passed through the tested material pores, whereas nitrogen was only adsorbed on the outer surface. Afterward, as the diameters of the carbon dioxide and nitrogen molecules are 3.30 and 3.64 Å, respectively [42], we concluded that the tunnels existing in the akaganeite framework [13] have a pore diameter, $D = 3.5 \pm 0.1$ Å [10].

TABLE 10.3

Parameters Calculated by Fitting the Carbon Dioxide Adsorption Data to the DR Adsorption Isotherm Equation

Sample	N_m (mmol/g)	W (cm³/g)	E (kJ/mol)	q_{iso} (kJ/mol)
AK-W	1.13 (5)	0.055 (2)	25	27
AK-Li	1.08	0.053	23	29

Next, the isosteric heat of adsorption is reported in Figure 10.4, which is calculated according to a relation similar to the Clausius–Clapeyron equation [5,41]:

$$q_{iso} \approx RT^2 \left[\frac{d \ln P}{dT} \right]_{n_a} \approx RT_1T_2 \left[\frac{\ln P_2 - \ln P_1}{T_2 - T_1} \right]_{n_a} \tag{10.3}$$

where P_1 and P_2 are the equilibrium adsorbate pressures at n_a = constant, for the temperatures T_1 and T_2.

The results of these calculations (Figure 10.4), namely a plot of the isosteric heat of adsorption versus the recovery, $\theta = n_a/N_m$, where n_a is the amount adsorbed and N_m is the maximum amount adsorbed in the pores, are consistent with the fact that the carbon dioxide molecule becomes subjected to the field gradient quadrupole together with dispersion, repulsion, and acid–base interactions. This fact can be corroborated by observing the figure reported for the isosteric heat of adsorption, that is, q_{iso} = 27–29 kJ/mol. The value is typical when dispersion and quadrupolar interactions are the fundamental adsorption fields that are present [5,41,42]. As discussed in Chapter 2, when a molecule is adsorbed, it is subjected to the following interactions: dispersion, repulsion, polarization, field dipole, field gradient quadrupole along with adsorbate–adsorbate interactions and in some cases the acid–base reactions [43,44]. In our case, in view of the fact that carbon dioxide has a noticeable quadrupolar moment (Q_{CO2} = 4.3 × 10⁻⁴² C·m²) and slight acidity [45], it should strongly interact within the akaganeite micropores, particularly at low coverage because of the small size of the akaganeite channels.

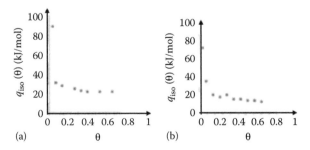

FIGURE 10.4 Isosteric heat of adsorption for carbon dioxide adsorption on AK-W: (a) and AK-Li (b).

Moreover, these data fairly well coincide with the isosteric heat of adsorption that is calculated using the *D–R* adsorption isotherm equation (Table 10.3).

Now, one assumption is made to make a thermodynamic analysis of the reported data. Specifically, the adsorbent plus the adsorbed phase is considered as a solid solution, that is, the adsorbate–adsorbent system (*aA*). Thereafter, the fundamental thermodynamic equation, within the frame of the solution thermodynamic approach, is given by [47–50]

$$dU_{aA} = TdS_{aA} - PdV_{aA} + \mu_a dn^a + \mu_A dm$$

where μ_a (in J/mol) and μ_A (in J/Kg) are the chemical potentials of the adsorbed and solid phases in the solution, and n^a and m are the amount of moles of adsorbate and the adsorbent mass in the system aA. Hence [47],

$$U_{aA} = TS_{aA} - PV_{aA} + \mu_a n^a + \mu_A m \tag{10.4}$$

On this ground, it is possible to write the internal energy per unit mass of adsorbent as follows: $\bar{U}_{aA} = T\bar{S}_{aA} - P\bar{V}_{aA} + \mu_a n_a + \mu_A$, where $n_a = n^a/m$. Now, the internal energy for the unit mass of empty adsorbent is given as follows: $\bar{U}_A = T\bar{S}_A - P\bar{V}_A + \mu_A^0$. Therefore [48],

$$U_a = \bar{U}_{aA} - \bar{U}_A = TS_a - PV_a + \mu_a n_a + \Phi \tag{10.5}$$

where:
$$S_a = \bar{S}_{aA} - \bar{S}_A$$
$$V_a = \bar{V}_{aA} - \bar{V}_A$$
$$\Phi = \mu_A - \mu_A^0$$

whereas, the fundamental thermodynamic equation for the adsorbed phase is given by

$$dU_a = TdS_a - PdV_a + \mu_a dn_a + d\Phi \tag{10.6}$$

Therefore, as the grand potential is defined as $\Omega = U - TS - \mu n$, $\Omega_a = U_a - TS_a - \mu_a n_a - \Phi$. Thereafter [50],

$$d\Omega_a = dU_a - TdS_a - S_a dT - d\mu_a n_a - \mu_a dn - d\Phi \tag{10.7}$$

Consequently, substituting Equation 10.6 in Equation 10.7 we get:

$$d\Omega_a = -PdV_a - S_a dT - n_a d\mu_a \tag{10.8}$$

Now, in equilibrium: $d\mu_a = d\mu_g = RT \ln f$, where f is the fugacity of the adsorbate. Thereafter,

$$d\Omega_a = -PdV_a - S_a dT - n_a RT \ln f = d\Phi \tag{10.9}$$

Consequently, $\Omega_a = \Phi$. Therefore, for $T = \text{const}$:

$$\Phi = RT \int_0^f n_a d(\ln f) - V_a P \qquad (10.10)$$

Now, in the frame of the osmotic theory of adsorption [39], $\Phi = \Pi W$. Hence,

$$\Pi W = RT \int_0^f n_a d(\ln f) \quad V_a P \qquad (10.11)$$

where W is the micropore volume of the adsorbent, whereas $\Pi = P_0 - P$ is the osmotic pressure in which Π is a pressure that can compress the adsorbate and can produce the same effect created by an adsorption field, that is, an amount, n_a, of molecules adsorbed in the micropores [5]. But, in our case, the experimental pressure was not very high. Thus, the approximate relation is as follows:

$$\Pi W = RT \int_0^P n_a d(\ln P) \qquad (10.12)$$

where, W, n_a, and P are the micropore volume, the magnitude of adsorption, and the equilibrium pressure, respectively, and $\Pi = P_0 - P$, the osmotic pressure, can be applied [50].

The physical meaning of Π can be explained by a model, where it is assumed that the adsorption field within the adsorption space is zero, explicitly we have a mere geometric volume. In this case, a pressure P_0 can compress the adsorbate in this geometric volume and can produce the same effect created by an adsorption field, that is, an amount n_a of molecules compressed in this volume [41].

Now, we plotted the pore volume (W) versus the osmotic pressure (Π) that is calculated according to Equation 10.11 (Figure 10.5a and b). These data show an increase in the osmotic pressure with the micropore volume. Effect takes place because carbon dioxide molecule has an ellipsoidal form, 5.4 Å long with a diameter of 3.4 Å [42]. Hence, the adsorbate–adsorbent interactions are greatly dependent on the pore geometry, specially, in the case that is studied here when the pore width of the adsorbent is very near to the size of the adsorbed molecules [43].

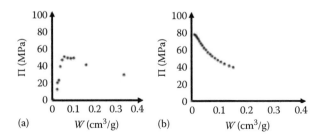

FIGURE 10.5 Osmotic pressure (Π) versus pore volume (W) plot.

Thus, the cooperative interaction of the carbon dioxide molecules with the framework and between them produces an increase in the osmotic pressure, Π, which rises with the equilibrium adsorption pressure of the CO_2 in the gaseous phase, causing the onset of the adsorbent deformation in some cases, which is in fact accompanied by an abrupt increase in the micropore volume [46, 51, 52].

10.2   SPHERE AND PARTICLE PACKING AND ORDERED AMORPHOUS AND MICROPOROUS MOLECULAR SIEVE SILICA

10.2.1   INTRODUCTION

Adsorption is not only a very important industrial process but also a powerful methodology for the characterization of the surface of materials. Here, both aspects will be emphasized [5,41,53–57]. In this regard, the interaction of carbon dioxide [26] with the surface of a group of well-characterized sphere and particle packing [58,59] and ordered amorphous [60,61] and crystalline microporous molecular sieve silica [5] is discussed here, with the help of different research methods along with an appropriate theoretical methodology [62–65].

Porous silicas are very important materials in different fields because they are profusely applied as adsorbents, catalyst supports, for membrane preparation, and for other large surface and porosity-related uses [66]. The molecular properties of silica are strongly affected by the nature of their surface sites. In sol-gel synthesized porous silica surfaces, that is, those of interest here, the unsaturated surface valences are satisfied by surface hydroxyl functionalities [67] that, depending on the calcination temperature, exists as (a) vicinal (hydrogen-bonded silanols), (b) geminal (two hydroxyl groups attached to the same silicon atom), or (c) isolated (no hydrogen bonds possible) silanol sites [5,41].

Adsorption on nanoporous materials is an excellent method for the separation and recovery of gases [25,69–71]. In this regard, the separation and recovery of carbon dioxide are, in particular, very important research problems because it is a greenhouse gas that contributes to global warming and is a reactant in significant industrial processes, for example, in the Solvay process for the production of $NaHCO_3$ [69]. Hence, independent of the fact that carbonaceous adsorbents, molecular sieves, alumina, or silica gel can be applied for carbon dioxide separation by applying the pressure swing adsorption (PSA) process, activated alumina is currently considered as the most suitable sorbent for removing carbon dioxide from air in a PSA process [72]. Nevertheless, modified silica materials, with a high specific surface area, are currently widely studied as carbon dioxide adsorbents. Particularly, the impregnation of amines in silica is a flourishing field of research [73,74]. However, these amines are not stable at high temperature, given that these organic substances, on the surface, begin to decompose at about 200°C [33]. Consequently, another form of modification of silica, by loading oxides in the silica surface, is in development because carbon dioxide will strongly interact with the formed Lewis centers [75].

Another important application of carbon dioxide adsorption as a characterization tool is the measurement of the micropore volume and the pore size distribution (PSD) of nanoporous materials [76–79]. Usually, these measurements are carried out with the help of nitrogen adsorption isotherms measured at 77 K [41,53,55]. However, at such a low temperature, the transport kinetics of the nitrogen molecules through the micropores is very slow in pores smaller than 0.7 nm [76]. However, in the case of carbon dioxide adsorption, at 273 K, the higher temperature of adsorption applied causes the carbon dioxide molecules to have a bigger kinetic energy. Subsequently, they are capable of going into the microporosity for pores of sizes less than 0.7 nm [23]. Therefore, an experimental solution for this difficulty is the use of carbon dioxide adsorption at 273 K instead of nitrogen at 77 K, to apply the Dubinin adsorption isotherm to describe carbon dioxide adsorption and to calculate the correspondent adsorption parameters [5,47,80,81].

Another significant application of carbon dioxide adsorption as a means for the characterization of the surface chemistry of adsorbents is the study of the infrared spectra of adsorbed carbon dioxide because it is a small weakly interacting probe molecule that is very useful for the study of the acid and basic properties of solid surfaces [62–65].

In this section, we discuss the study of carbon dioxide adsorption interactions in well-characterized amorphous silica: specifically, three sphere packing (SP) [60,61,82,83] and three extremely high specific surface area particle packing (PP) silica xerogels [59], being these materials produced with the help of a modification of the Stober–Fink–Bohn (SFB) methodology [68], that is, the change of ammonia by an amine as catalyst [26].

In addition, a hexagonal silica mesoporous molecular sieve (MMS) was tested for comparison, namely MCM-41 [84] which is ordered but not crystalline because of the lack of precise atomic positioning in the pores' wall structure, along with one unidirectional 12-membered ring (MR) channels silica molecular sieve (MS), that is, SSZ-24, which is the siliceous counterpart of AlPO4-5, showing the AFI framework [85] and two synthetic molecular sieves (MSs) with a high Si/Al relation, that is, two ZSM-5 and the extremely high Si/Al relation molecular sieve named dealuminated Y (DAY), provided the ZSM-5 is a 10MR molecular sieve showing the MFI framework type and the DAY is a 12MR molecular sieve exhibiting the FAU framework type [86].

10.2.2 SYNTHESIS

All the consumed chemicals were of analytical grade without additional purification, namely water was bidistilled and the reagents were of analytical grade. The SP (Stöber silica) and PP silicas were synthesized with the help of the SFB method [68] if the source materials for the synthesis: tetraethyl orthosilicate (TEOS), double-distilled water (DDW), methanol (MeOH), isopropanol, and ammonium hydroxide (40 wt. % NH_4OH in water) were provided. However, to prepare the PP silica, a modification to the SFB method was introduced, concretely ammonia was substituted by an amine as catalyst. Moreover, two silica aerogel samples were synthesized to get materials by the supercritical drying procedure as follows: gels were flushed three times with liquid CO_2 at 100 bar and 300 K during 3 h for each flush. Thereafter, the temperature was increased to 313 K and the samples were flushed at the same pressure with gaseous CO_2 at 100 bar during 6 h to get the AER-309 aerogel and 24 h in the case of the AER-909 aerogel [26]. The supercritical drying process was carried out in the Helix supercritical dryer that is developed and supplied by Applied Separations Inc., Allentown, PA. The MMS–MCM-41, that is, the hexagonal phase of the family of MMSs, was used as a standard in the present study because it is a well-studied silica surface [84]. Concretely, the MCM-41 used in the present study was synthesized using the methods described in the literature [84], whereas the molecular sieve (MS), MS–SSZ-24, was provided by R. Lobos. In addition, the tested MS–DAY, DAY-20F (Si/Al = 20), was produced by Degussa AG, Germany [87], whereas the MS–ZSM-5, both 3020 and 5020, were provided by the PQ Corporation Malvern [26].

10.2.3 Characterization

The scanning electron microscopy (SEM) study was performed with a JEOL CF 35 electron microscope in secondary electron mode at an accelerating voltage of 25 kV to image the surface of the studied samples. These samples were attached to the sample holder with an adhesive tape and then coated under vacuum by cathode sputtering with a 30–40 nm gold film before observation. The surface morphology was revealed from SEM images, and the average grain size was estimated, albeit qualitatively. The corresponding SEM images of the SP silicas SP-80 and SP-FM3a-GEO are shown in Figure 10.6, where the spherical form of the Stöber silica along with a micrograph of the particle packing silica PP-68bs1e is evidently shown in which it is obviously revealed that the spherical geometry was broken [26].

Next, as previously explained, the nitrogen adsorption specific surface area measurement and carbon dioxide pore volume along with the PSD investigation of the produced samples were made in a Quantachrome Autosorb-1 automatic physisorption analyzer. In this regard, with the help of the adsorption isotherm of N_2 at 77 K in samples degassed at 573 K during 7 h in high vacuum (10^{-6} Torr), the specific surface area, S_{BET}, was determined by applying the BET method (see Chapter 3), whereas the Saito–Foley (SF) method [53] was used at the same time for the calculation of the micropore size distribution (MPSD) as the density functional theory (DFT) method [54,55] was employed for the calculation of the PSD in the mesopore region (see Chapter 4). In this regard, the calculated parameters [5] are reported in Table 10.4, where S_{BET} is the BET specific surface area; W_{Mic} is the micropore volume; and W_{DFT} is the DFT pore volume that includes the mesopore region. Meanwhile d_{Mic} and d_{Mes} are the SF and DFT pore width mode, respectively. Finally Φ is the particle size.

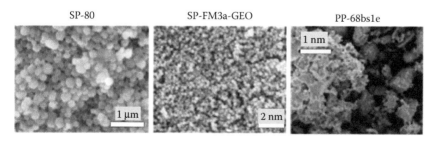

FIGURE 10.6   SEM images of the sphere packing silicas SP-80, SP-FM3a-GEO, and PP-68bs1e.

TABLE 10.4

Morphological Data of the Different Tested Silica

Sample	S_{BET} (m²/g)	W_{Mic} (cm³/g)	W_{DFT} (cm³/g)	d_{Mic} (nm)	d_{Mes} (nm)	Φ (nm)	References
SP-80	400	0.16	0.49	0.57	3.90	250	[58,59]
SP-69	300	0.06	0.52	0.59	8.10	350	[58,59]
SP-FM3a						250	[58,59]
MCM-41	800		1.69		3.50		[26]
PP-70bs2	1600	0.18	3.00	0.55	6.50		[60,61]
PP-74bs5	1200	0.14	1.70	0.57	6.50		[60,61]
PP-68bs1	1500	0.25	2.40	0.55	8.10		[60,61]
PP-79bs1	1400	0.16	2.70	0.57	12.10		[60,61]
AER-309	500	0.16	4.20	0.58	5.00		[60,61]
AER-909	1000	0.40	2.00	0.57	7.3		[60,61]
DAY		0.30			0.60		[87]
ZSM(3020)		0.11			0.54		[5]
ZSM(5020)		0.12			0.54		[5]

The previously reported morphologic data of the SP and PP silica samples indicate that these adsorbents have a complex pore structure, including micropores and mesopores, where this morphology is a result of the particular internal structure of the SP and PP silicas, which are composed of the secondary particles, evidenced by SEM where these agglomerates are composed of primary particles. That is, the observed morphology is explained by the agglomeration of the primary particles, which is revealed with the help of the SAXS study [26], to form the secondary particles because void spaces are created between the primary particles in the agglomeration process. This morphology will be reflected in the adsorption of carbon dioxide as will be explained later.

The thermal gravimetric investigation of the tested silicas [26] shows that the release of loosely linked water after dehydration takes place at relatively low temperatures, that is, below 200°C. Dehydroxylation takes place at temperatures higher than 200°C, where the surface silanol groups begin their condensation to form water molecules [88]. Finally, only isolated silanols are present on the surface at 700°C–800°C [89].

10.2.4 CARBON DIOXIDE ADSORPTION ON THE TESTED SILICA

The adsorption isotherms of CO_2 at 273 K in samples degassed at 573 K during 3 h in high vacuum (10^{-6} Torr) were obtained in an upgraded Quantachrome Autosorb-1 automatic sorption analyzer. As the vapor pressure of carbon dioxide at 273 K is $P_0 = 26,141$ Torr [77], the adsorption process takes place in the following relative pressure range: $0.00003 < P/P_0 < 0.03$, where the adsorption process is well described by the DR adsorption isotherm equation, namely Equation 10.2 [47,80,81].

The fitting process of this equation to the experimental adsorption of carbon dioxide in the tested silica allowed us to calculate the best fitting parameters, that is, N_m, E, together with W, for $n = 2$.

The parameters calculated with the D–R adsorption isotherm equation allow us to evaluate not only the micropore volume of the sample, as was previously explained, but also the adsorption heat released during adsorption, data that allow the assessment of the adsorbate–adsorbent interaction. In this regard, we will now show that it is possible to calculate q_{iso} in an original way by using only one isotherm as follows [81]:

$$q_{iso} = -\Delta G(\Theta) + EF(T,\Theta) \tag{10.13}$$

where $\theta = n_a/N_m$. Then,

$$\Delta G(\Theta) = RT \ln\left(\frac{P}{P_0}\right) \tag{10.14}$$

Along with:

$$F(T,\Theta) = (\alpha T/2)\left[\ln(1/\Theta)^{\frac{1}{n}-1}\right] \tag{10.15}$$

It is also possible now to assert that as $E = \Delta G(1/e)$, where $\theta = 1/e$ in which $e \approx 2.71828183$ is the base of the Nmpierian logarithm system. Now, with the help of Equations 10.13 through 10.15, for $\theta = 1/e \approx 0.37$, it is possible to get the following equation:

$$q_{iso}(0.37) = -\Delta G(0.37) + EF(T,0.37) = [1 - F(T,0.37)]E$$

To calculate $F(T,0.37)$, we need trustworthy experimental calorimetric data reported in the literature. In this regard, the experimental heat of adsorption data that will be applied to compute $F(T,0.37)$ is $q_{iso} = 22$ kJ/mol for the adsorption of carbon dioxide at 298 K, in the range of $0.1 < n_a < 0.7$ mmol/g in MCM-41 [90]. Accordingly, it is possible to estimate that $F(T,0.37) \approx 1.16$. Finally, the equation used to calculate the isosteric heat of adsorption reported in Table 10.5 is

$$q_{iso}(0.37) = 1.16E \tag{10.16}$$

The reported data indicate that the tested SP and PP silicas show similar values for the isosteric heat of adsorption. This value is also identical to the experimental isosteric heat of adsorption. Concretely, $q_{iso} = 28$ kJ/mol measured for the adsorption of carbon dioxide at 296–306 K in the range of $0.1 < n_a < 1.5$ mmol/g in silicalite [91]. The similarity between our data and the values reported in the literature for the isosteric heat of adsorption of carbon dioxide on silicalite is due to the similitude between silicalite and the tested silica in the micropore range. That is, the studied silica samples,

TABLE 10.5

Parameters Calculated by the Fitting of the Dubinin Equation to the Experimental Carbon Dioxide Adsorption on the Tested Silica at 273 K

Sample	N_m (mmol/g)	W (cm³/g)	E (kJ/mol)	q_{iso} ($\Theta = 0.37$) (kJ/mol)
SP-69	4.31	0.21	24	28
SP-80	2.67	0.13	24	28
SP-FM3a	4.72	0.23	24	28
PP-75bs1	4.72	0.23	24	27
PP-68bs1	3.48	0.17	24	28
MMS	4.89	0.24	19	22
ZSM-5	3.48	0.17	33	38
DAY	6.77	0.33	17	20

in the micropore region, exhibit pore sizes in the range of 0.56 nm $< d_{Mic} <$ 0.59 nm, and the 10 MR channels of the MFI framework of silicalite display pore openings of 0.51×0.55 nm² and 0.53×0.56 nm². Consequently, as the pore size is the main parameter in the determination of the adsorption heat in porous materials with a similar surface chemistry [42], our conclusion reasonably results.

The previously described adsorption data indicate that the interaction of the carbon dioxide molecule with silica is not strong. That is, we are in the presence of a physical adsorption process, as is normally accepted for the adsorption of carbon dioxide on silica. It is accepted that this physical adsorption process binds the carbon dioxide molecule inside the silica surface micropores by the influence of the dispersive forces and the attraction of the quadrupole interaction [91–93]. We know that, in general, if a molecule contacts the surface of a solid adsorbent, it becomes subjected to diverse interaction fields, such as the dispersion energy (φ_D) repulsion energy (φ_R), polarization energy (φ_P), field dipole energy ($\varphi_{E\mu}$), field gradient quadrupole energy (φ_{EQ}), and some specific interactions, such as the acid–base interaction (φ_{AB}) and the adsorbate–adsorbate interaction energy (φ_{AA}) [5,41,94]. Dispersion and repulsion are the fundamental forces present during physical adsorption in all adsorbents. In the case of molecules, such as H_2, Ar, CH_4, N_2, and O_2, these are the only forces present, given that the dipole moments of these molecules is zero. The quadrupole moment is very low or absent, and the polarization effect will be only noticeable in the case of adsorbents with high electric fields [5,25,56,57]. Dispersion and repulsion interactions are present in all adsorption gas–solid systems. Therefore, they are nonspecific interactions [95]. The electrostatic interactions between the adsorbed molecule and the adsorbent framework, that is, φ_P, $\varphi_{E\mu}$, and φ_{EQ}, depend on the structure and composition of the adsorbed molecule and the adsorbent itself (Chapter 2). That is, the interaction between the adsorbent and molecules with, for example, a noticeable quadrupolar moment, such as the carbon dioxide molecule gives rise to specific interactions, where the combination of the dispersive and electrostatic.

REFERENCES

1. R.M. Cornell and U. Schwertmann, *The Iron Oxides: Structure, Properties, Reactions, Occurrence and Uses*, VCH, New York, 1996.
2. N. Pinney, J.D. Kubicki, D.S. Middlemiss, C.P. Grey, and D. Morgan, *Chem. Mater.*, 21 (2009) 5727e5742.
3. R. Zboril, L. Machala, and D. Petridis, *Chem. Mater.*, 14 (2002) 969e982.
4. L.H. Reddy, J.L. Arias, J. Nicolas, and P. Couvreur, *Chem. Rev.*, 112 (2012) 5818e5878.
5. R. Roque-Malherbe, *Physical Chemistry of Materials: Energy and Environmental Applications*, CRC Press, Boca Raton, FL, 2009.
6. L. Machala, J. Tucek, and R. Zboril, *Chem. Mater.*, 23 (2011) 3255e3272.
7. A. Navrotsky, L. Mazeina, and J. Majzlan, *Science*, 319 (2008) 1635e1638.
8. L. Mazeina, S. Deore, A. Navrotsky, *Chem. Mater.*, 18 (2006) 1830e1838.
9. K. Stahl, K. Nielsen, J. Jiang, B. Lebech, J.C. Hanson, P. Norby, and J. Lanschot, *J. Corr. Sci.*, 45 (2003) 2563e2575.
10. R. Roque-Malherbe, F. Lugo, C. Rivera, R. Polanco, P. Fierro, and O.N.C. Uwakweh, *Curr. Appl. Phys.*, 15 (2015) 571.
11. J. Kim and C.P. Grey, *Chem. Mater.*, 22 (2010) 5453e5462.
12. S. Goni-Elizalde and M.E. Garcia-Clavel, *Thermochim. Acta*, 129 (1988) 325e334.
13. J.E. Post, P.J. Heaney, P.J.R.B. von Dreele, and R.B.J.C. Hanson, *Am. Mineral.*, 88 (2003) 782.
14. S. Bedanta and W. Kleemann, *J. Phys. D: Appl. Phys.*, 42 (2009) 28, 013001.
15. A.L. Mackay, *Mineral Mag.*, 33 (1962) 270.
16. N.K. Chaudhari and Y.-S. Yu, *J. Phys. Chem. C*, 112 (2008) 19957.
17. E.A. Deliyanni, L. Nalbandian, and K.A. Matis, *J. Colloid Interface Sci.*, 302 (2006) 458.
18. G.A. Waychunas and B.A. Rea, *Geochim. Cosmochim. Acta*, 57 (1993) 2251.
19. R. Chitrakar, S. Tezuka, A. Sonoda, K. Sakane, K. Ooi, and T. Hirotsu, *J. Colloid Interface Sci.*, 298 (2006) 602.
20. E.A. Deliyanni, E.N. Peleka, and N.K. Lazaridis, *Sep. Purific. Technol.*, 52 (2007) 478.
21. J. Kim, U.G. Nielsen, and C.P. Grey, *J. Am. Chem. Soc.*, 130 (2008) 1285.
22. R. Roque-Malherbe, O.N.C. Uwakweh, C. Lozano, R. Polanco, A.H. Maldonado, P. Fierro, F. Lugo, and J.N. Primera-Pedrozo, *J. Phys. Chem. C*, 115 (2011) 15555.
23. D. Cazorla-Amorós, J. Alcañiz-Monge, M.A. de la Casa-Lillo, and A. Linares-Solano, *Langmuir*, 14 (1998) 4589.
24. J.A. Dunne, M. Rao, S. Sircar, R.J. Gorte, and A.L. Myers, *Langmuir*, 12 (1996) 5896.
25. D.M. Ruthven, *Principles of Adsorption and Adsorption Processes*, John Wiley & Sons, New York, 1984.
26. R. Roque-Malherbe, R. Polanco-Estrella, and F. Marquez, *J. Phys. Chem. C*, 114 (2010) 17773.
27. W. Wang, J.Y. Howe, and B. Gu, *J. Phys. Chem. C*, 125 (2008) 9203.
28. F. Jones, M.I. Odgen, A. Oliveira, G.M. Parkinson, and W.R. Richardson, *Cryst. Eng. Comm.*, 5 (2005) 159.
29. S. Bashir, R.W. McCabe, C. Boxall, M.S. Leaver, and D. Mobbs, *J. Nanopart. Res.*, 11 (2009) 701.
30. J.C. Acuna and F.E. Echevarria, *Sci. Tech.* 13 (2007) 993.
31. L. Fuentes-Cobas, Synchrotron radiation diffraction and scattering in ferroelectrics, in *Multifunctional Polycrystalline Ferroelectric Materials* (L. Pardo and J. Ricote Eds.), Springer, Berlin, Germany, 2011, p. 217.
32. C. Luna, M. Ilyn, V. Vega, V.M. Prida, J. Gonzalez, and R. Mendoza-Resendez, *J. Phys. Chem. C*, 118 (2014) 21128.
33. J.H. Scheckman, P.H. McMurryr, and S.E. Pratsinis, *Langmuir*, 25 (2009) 8248.

34. T. Hahn (Ed.), *International Tables for Crystallography* (Vol. A, 5th ed.), The International Union of Crystallography, Springer, Dordrecht, the Netherlands, 2005.
35. G.J. Long and F. Grandjean (Eds.), *Mossbauer Spectrometry Applied to Magnetism and Materials Science*, Springer, Berlin, Germany, 2013.
36. J.E. Post and V.F. Buchwald, *Am. Mineral.*, 76 (1991) 272.
37. J.M.D. Coey, M. Venkatesan, H. Xu, Introduction to magnetic oxides, in: *Functional Metal Oxides: New Science and Novel Applications* (S.B. Ogale, T.V. Venkatesan, and M.G. Blamire (Eds.), Wiley-VCH Verlag, New York, 2013, pp. 1–49.
38. S. Mørup, M.F. Hansen, Superparamagnetic particles, in: *Handbook of Magnetism and Advanced Magnetic Materials*, (vol. 4, H. Kronmüller and S.P.S. Parkin Eds.), John Wiley & Sons, New York, 2007, pp. 2159.
39. C.A. Barrero, K.E. Garcia, A.L. Morales, S. Kodjikian, and J.M. Greneche, *J. Phys. Condens Mat*, 18 (2006) 6827.
40. F. Gonzales-Lucena, PhD Dissertation, University of Ottawa, Canada, 2010.
41. R.T. Yang, *Adsorbents: Fundamentals, and Applications*, John Wiley & Sons, New York, 2003.
42. R. Roque-Malherbe and F. Diaz-Castro, *J. Mol. Catal. A*, 280 (2008) 194.
43. O. Deutschmann, H. Knozinger, K. Kochloefl, and T. Turek, *Heterogeneous Catalysis and Solid Catalysts*, Ullmann's Encyclopedia of Industrial Chemistry, Wiley-VCH Verlag GmbH & Co. KGaA, Weinheim, Germany, 2009, pp. 1–110.
44. H.-J. Freund and M.W. Roberts, *Surf. Sci. Rep.*, 25 (1996) 255.
45. A.D. Buckingham, *Proc. R. Soc. Lond. Ser. A*, 273 (1963) 275.
46. X. Song and J.-F. Boily, *J. Phys. Chem. C*, 116 (2012) 2303.
47. M.M. Dubinin, *Prog. Surf. Membr. Sci.*, 9 (1975) 1.
48. B.P. Bering, M.M. Dubinin, and V.V. Serpinskii, *J. Colloid Interface Sci.*, 38 (1972) 185.
49. R. Roque-Malherbe, *Micropor. Mesopor. Mat.*, 41 (2000) 227.
50. V.P. Bering and V.V. Serpinskii, *Bull. Acad. Sci. USSR*, 24 (1974) 2342.
51. F.-X. Coudert, *Phys. Chem. Chem. Phys.*, 12 (2010) 10904.
52. A. Boutin, F.-X. Coudert, M.-A. Springuel-Huet, A.V. Neimark, G. Ferey, and A.H. Fuchs, *J. Phys. Chem. C*, 114 (2010) 22237.
53. A. Saito and H.C. Foley, *AIChE J.*, 37 (1991) 429.
54. R. Evans, in *Fundamentals of Inhomogeneous Fluids* (D. Henderson, Ed.), Marcel Dekker, New York, 1992, p. 85.
55. Y. Rosenfeld, M. Schmidt, H. Lowen, and P. Tarazona, *Phys. Rev. E.*, 55 (1997) 4245.
56. S.J. Gregg and K.S.W. Sing, *Adsorption Surface Area and Porosity*, Academic Press, London, UK, 1982.
57. W.A. Steele, *The Interaction of Gases with Solid Surfaces*, Pergamon Press, Oxford, UK, 1974.
58. F. Marquez and R. Roque-Malherbe, *J. Nanosci. Nanotech.*, 6 (2006) 1114.
59. R. Roque-Malherbe, F. Marquez, W. del Valle, and M. Thommes, *J. Nanosci. Nanotech.*, 8 (2008) 5993.
60. R. Roque-Malherbe and F. Marquez, *Surf. Interface Anal.*, 37 (2005) 393.
61. R. Roque-Malherbe and F. Marquez, *Mat. Sci. Semicon. Proc.*, 7 (2004) 467.
62. H. Knozinger and S.J. Huber, *Chem. Soc. Faraday Trans.*, 94 (1998) 2047.
63. E. Garrone, B. Fubini, B. Bonelli, B. Onida, and C. Otero-Arean, *Phys. Chem. Chem. Phys.*, 1 (1999) 513.
64. B, Bonelli, B. Civalleri, B. Fubini, P. Ugliengo, C. Otero-Arean, and E. Garrone, *J. Phys. Chem. B*, 104 (2000) 10978.
65. F.X. Llabres-Xamena and A. Zecchina, *Phys. Chem. Chem. Phys.*, 4 (2002) 1978.
66. G.M. El Shafey, in *Adsorption on Silica Surfaces* (E. Papirer, Ed.), Marcel Dekker, New York, 2000, p. 34.
67. C. J. Brinker and G. W. Scherer, *Sol-Gel Science*, Academic Press, New York, 1990.

68. W. Stöber, A. Fink, and E. Bohn, *J. Colloid Interface Sci.*, 26 (1968) 62.
69. Z. Yong, V. Mata, and A.E. Rodrigues, *J. Chem. Eng. Data*, 45 (2000) 1093.
70. C. Knöfel, C. Martin, V. Hornebecq, and P.L. Llewellyn, *J. Phys. Chem. C*, 113 (2009) 21726.
71. H.Y. Huang, R.T. Yang, D. Chinn, and C.L. Munson, *Ind. Eng. Chem. Res.*, 42 (2003) 2427.
72. S. Deng, in *Encyclopedia of Chemical Processing*, Taylor & Francis Group, Boca Raton, FL, 2006, p. 2825.
73. Y. Belmabkhout, G. De Weireld, and A. Sayari, *Langmuir*, 25 (2009) 13275.
74. Z. Bacsik, R. Atluri, A.E. Garcia-Bennett, and N. Hedin, *Langmuir*, 26 (2010) 10013.
75. R. Bal, B.B. Tope, T.K. Das, S.G. Hegde, and S. Sivasanker, *J. Catal.*, 204 (2001) 358.
76. D. Cazorla-Amoros, J. Alcañiz-Monge, and A. Linares-Solano, *Langmuir*, 12 (1996) 2820.
77. Powder Tech Note 35, Quantachrome Instruments, Boynton Beach, FL.
78. S. Lowell and J.E. Shields, *Powder Surface Area and Porosity*, Chapman & Hall, London, UK, 1991.
79. F. Rodriguez-Reinoso and A. Linares-Solano, in *Chemistry and Physics of Carbon* (Vol. 21, P.A. Thrower, Ed), Marcel Dekker, New York, 1988.
80. M.M. Dubinin, *American Chemical Society Symposium Series* (Vol. 40), American Chemical Society, Washington, DC, 1977, p. 1.
81. B.P. Bering, M.M. Dubinin, and V.V. Serpinskii, *J. Colloid Interface Sci.*, 38 (1972) 185.
82. R.K. Iler, *The Chemistry of Silica*, John Wiley & Sons, New York, 1979.
83. A.C. Pierre and G.M. Pajonk, *Chem. Rev.*, 102 (2002) 4243.
84. C.T. Kresge, M.E. Leonowicz, W.J. Roth, J.C. Vartuli, and J.S. Beck, *Nature*, 359 (1992) 710.
85. R.F. Lobo, M. Pan, I. Chan, H.X. Li, R.S. Medrud, S.I. Zones, P.A. Crozier, and M.E. Davis, *Science*, 262 (1993) 1543.
86. C. Baerlocher, W.M. Meier, and D.M. Olson, *Atlas of Zeolite Framework Types* (6th revised ed.) Elsevier, Amsterdam, the Netherlands, 2007.
87. C. Muñiz, F. Diaz, and R. Roque-Malherbe, *J. Colloid Interface Sci.*, 329 (2009) 11.
88. R. Mueller, H.K. Kammler, K. Wegner, and S.E. Pratsinis, *Langmuir*, 19 (2003) 160.
89. P.K. Jal, S. Patel, and B.K. Mishra, *Talanta*, 62 (2004) 1005.
90. Y. He and N.A. Seaton, *Langmuir*, 22 (2006) 1150.
91. J.A. Dunne, R. Mariwala, M. Rao, S. Sircar, R.J. Gorte, and A.L. Myers *Langmuir*, 12 (1996) 5888.
92. R. Bai, J. Deng, and R.T. Yang, *Langmir*, 19 (2003) 2776.
93. R.M. Barrer and R. Gibbons, *Trans. Faraday Soc.*, 61 (1965) 948.
94. A. Corma, *Chem. Rev.*, 95 (1995) 559.
95. R.M. Barrer, *Zeolites and Clay Minerals as Sorbents and Molecular SieVes*, Academic Press, London, UK, 1978.

11 Study of Carbon Dioxide Adsorption on Nitroprussides and Prussian Blue Analogs

11.1 NITROPRUSSIDES

11.1.1 INTRODUCTION

Pentacyanonitrosylferrates, commonly named nitroprussides (labeled here as NPs) are a group of metal cyanides, consisting of microporous frameworks that are assembled from $[Fe(CN)_5NO]^{2-}$ units bridged though M^{2+} cations by means of the CN^- ligands [1–13]. In other words, they are transition metal cyanides displaying frameworks built with transition metals bridged through the linear cyanide ion, that is, it is shaped by an octahedral ferrous center surrounded by five cyanide-bound ligands along with one linear nitric oxide ligand displaying porous frameworks that show interesting magnetic [1,13], adsorption [11,12], and other properties that make these coordination polymers an interesting class of materials. Likewise, NPs exhibit a marked polymorphic nature [11]. Then, minor changes in the synthesis process of these materials could lead to materials with fairly different properties.

In the tridimensional porous structure of the NPs, O atoms at the end of the NO ligands stay free, and the M^{2+} cation coordinates to one or more water molecules at the same time as the rest of the water molecules are hydrogen-bonded to the coordinated water molecules [7–9]. In some of these materials, the M^{2+} cations are spin sites bridged by a spacer, that is, $[Fe(CN)_5NO]^{2-}$. Then, super-exchange interactions coupling the spin sites [1] generating magnetism are possible. Concretely, Fe–, Co–, Ni–, and Cu–NPs follow the modified Curie–Weiss law, a characteristic behavior of paramagnetic materials, and at low temperatures show ferromagnetic order [13,14].

In addition, in porous NPs after thermal dehydration, both the coordinated and the hydrogen-bonded or zeolitic water molecules are detached leading to materials with an open channel system suitable for the adsorption of small molecules [12,15–20]. In this regard, a good probe for the measurement of the micropore volume [20] and adsorption heats [21] is the carbon dioxide molecule [18]. Furthermore, because it is a small and weakly interacting adsorbate, their IR spectra are very informative [3,4,16,22,23]. As a final point, adsorption is an excellent process for the separation, storage, and recovery of gases [17], in particular separation, storage, and recovery of carbon dioxide are very important research topics, since it is a reactant in significant industrial processes and a greenhouse gas that contributes to global warming [18].

In this regard, since NPs could have a great significance in the sequestration of gases on their porous frameworks, particularly adsorbing atmospheric gaseous pollutants produced by factories, power plants, cars, planes, and homes, which have a large effect in our environment, their study is significant [24]. Particularly CO_2 occupies a special position, since their emissions cause global warming because it influences the balance of the received and emitted energy leading to the increase in the Earth temperature, stimulating sea-level elevation, the increase in the oceans acidity, and the strengthening of tropical storms between other harmful effects [25].

11.1.2 SYNTHESIS

All the consumed chemicals were of analytical grade without additional purification. The water used in the synthesis process was bidistilled. The produced NPs were synthesized by adding 0.025 moles of solid $Na_2[Fe(CN)_5NO]_3$ $2H_2O$ to a solution containing 0.025 moles of the corresponding Ni(II), Zn(II), and Cd(II) salt, that is, $Ni(NO_3)_2$, $Zn(NO_3)_2$, and $Cd(NO_3)_2$ in 250 mL of water under constant stirring [14]. The formed precipitates were then separated, successively washed with distilled water, and dried at 70°C during 24 h. The final concentrations of the reagents were 0.1 M. In previous reports, 0.01 M solutions were normally used and the $Na_2[Fe(CN)_5NO]$ was added in water solution [4].

11.1.3 CHARACTERIZATION

X-ray diffraction (XRD) tests were carried out with a Bruker D8 Advance system in a Bragg–Brentano vertical goniometer configuration. The angular measurements were made with a θ/2θ of 0.0001 reproducibility, applying steps of 0.01° from 5° to 80° to get XRD profiles that could be accurately resolved by least squares methods. The X-ray radiation source was a ceramic XRD Cu anode tube type KFL C 2K of 2.2 kW with a long fine focus. A variable computer-controlled motor-driven divergence slit with Soller slits were included to allow and keep the irradiated area on the sample surface constant. A Ni filter was placed before the detector to eliminate CuK_α radiation. A LynxEye one-dimensional detector was employed, being based on a Bruker AXS compound silicon strip technology that increases the measured intensities without sacrificing the resolution and peak shape. This together with the use of small scanning step resulted in high-quality XRD profiles suitable for mathematical treatment.

The gathering of the *in situ* XRD profiles of the dehydrated NP samples at room temperature was performed using an Anton-Paar HTK-1200N stage, which was designed to be used in the range from room temperature up to 1200°C. To be dehydrated the samples were treated at 100°C for 2 h under a N_2 (Praxair, 99.99%) flowing at a rate of 50 mL/min, provided the sample is mounted on an alumina sample holder with temperature sensor located just below the sample. The same chamber was applied to collect the *in situ* XRD profiles of the NPs during carbon dioxide adsorption at 298 K and 1 atm. For the adsorption experiment, CO_2 (Praxair, 99.99%) was

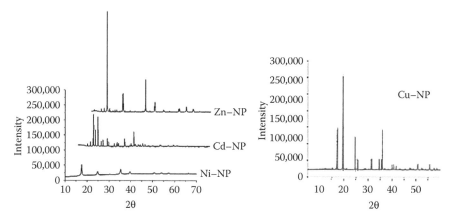

FIGURE 11.1   XRD profiles of the Zn–, Cd–, Ni–, and Cu–nitroprussides.

allowed into the chamber at a flow rate of 50 mL/min. In this sense, in Figure 11.1 the XRD profiles of the four tested NPs are reported [3,4].

Next, the thermogravimetric analysis (TGA) testing process was carried out with a TA Q-500 instrument. Samples were placed onto a ceramic sample holder suspended from an analytical balance. The sample and holder were heated according to a predetermined thermal cycle: the temperature was linearly scanned from 25°C to 300°C at a heating rate of 5°C/min under a pure N_2 flow of 100 mL/min. The instrument software automatically controlled the data collection, temperature control, heating rate, and gas flow. The TGA data was collected as a wt. % versus T (°C) profile, where wt. % $= (M_t/M_0) \times 100$ and is the percent ratio of the sample mass during the thermal treatment M_t, and the initial mass of the sample M_0. In Figure 11.2 the XRD profiles of the Zn, Cd, Ni, and Cu–NPs are reported [3,4].

Diffuse reflectance infrared Fourier transform spectrometry (DRIFTS) were gathered using a Thermo Scientific Nicolet iS10 FTIR spectrometer. The data of the hydrated and dehydrated samples were collected at a resolution of 4 cm^{-1} employing 100 scans per sample. A background with pure KBr provided by Nicolet that applied the same conditions was always gathered before the sample collection. Both the hydrated and dehydrated sample spectra were obtained at room temperature under N_2 (Praxair, 99.99%) flowing at a rate of 50 cc/min. Sample dehydration was performed at 100°C for 2 h under a pure N_2 flow of 50 mL/min in the IR high temperature cell. For gathering of the DRIFTS spectra for the NPs during carbon dioxide adsorption, the background was carefully measured using the dehydrated sample at room temperature. In Figure 11.2 the DRIFTS spectra of the Zn–, Cd–, Ni–, and Cu–NPs are reported [3,4].

In Figure 11.3 the vibrations observed in the range between 1300 and 2700 cm^{-1} are shown, which are related to the v(CN) stretching vibration at around 1610 cm^{-1}, a 1950 cm^{-1} band corresponding to v(NO) stretching vibration, and a 2200 cm^{-1} band due to δ(Fe–CN) vibration [26], which are all distinctive of NPs [27]. It is necessary to state that all these vibrations are maintained after dehydration with minor shifts

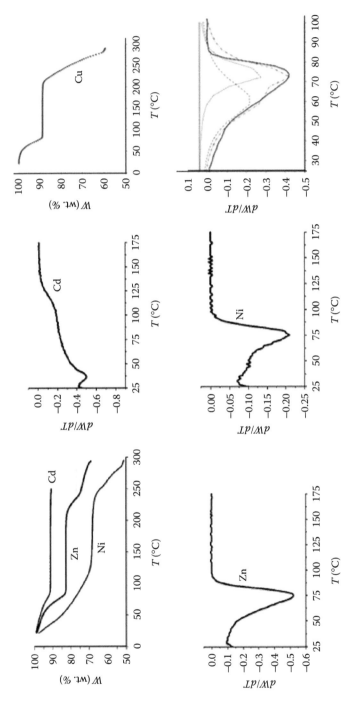

FIGURE 11.2 TGA profiles of the Zn–, Cd–, Ni–, and Cu–NPs together with their derivative.

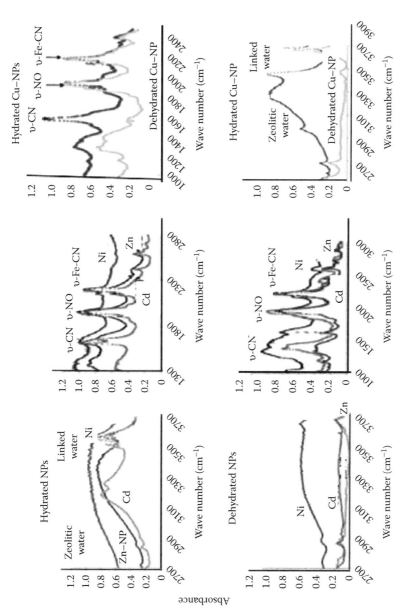

FIGURE 11.3 DRIFTS spectra of the hydrated and dehydrated Zn–, Cd–, Ni–, and Cu–NPs.

excluding the $v(CN)$ stretching vibration corresponding to the unlinked cyanide, which disappears in the Cd–NP. This band, possibly, vanishes since all the CNs are linked in the dehydrated state [12] as is the case for Cu–NP [14].

Alternatively, the vibrations related to the water molecules contained in the NP porous framework are usually present in the 3000–3800 cm^{-1} range, being characterized by one broad band in the 3250–3400 cm^{-1} range corresponding to hydrogen bonded or zeolitic water and narrow peaks at around 3650 cm^{-1} related to linked water [7]. Then, both bands disappear after dehydration. It is necessary to affirm now that, in general, the DRIFTS results correlate well with the TGA data.

Afterward, to complete the characterization of the tested NPs, a Mössbauer (excluding the Cu–NP) along with an investigation of their magnetization curves was made. In this regard, the room-temperature spectra of the synthesized NPs were gathered by applying a SEECo spectrometer operating at constant acceleration mode with a 50 mCi ^{57}Co γ-ray source in a Rh matrix made by Rietverc GmbH, being the velocity scale chosen to ensure a thorough scan of the materials to reveal all the features associated with the sample material, and the calibration of the velocity scale was made with reference to Fe metal. Next, to make the analysis of the spectra, the 1024 point raw data were folded and analyzed using the software WMOSS, a public domain Mössbauer spectral analysis program available at www.SEECo.us, which was formerly WEBRES company. Finally, the Mössbauer spectra were fitted with Lorentzians functions characterized by three parameters, that is, position, width, and intensity with the help of the software Peakfit. Moreover, the magnetic measurements were made with a vibrating sample magnetometer (VSM) to the get the magnetization curves.

The recorded Mössbauer spectra consisted only of one quadrupole doublet, a fact undoubtedly demonstrated by the fitting of the experimental spectra [3]. Hence, the spectra were well-resolved doublets, provided the fitting exercise was carried out without constraining the parameters, that is, the full width at half maximum (FWHM), δ is the isomer shift, Δ is the quadrupole splitting, and h_+/h_- is the doublet asymmetric parameter, that is, the difference between the doublet peak heights. The reproducibility of the data implied high confidence on the results shown in Table 11.1 [3].

The reported δ and Δ Mössbauer parameters were almost unaffected by the M^{2+} outer cation, that is, Ni^{2+}, Zn^{2+}, and Cd^{2+}. Meanwhile, the value reported for the isomer shift is characteristic of Fe^{2+} cations in a low spin (L.S.) electronic configuration [17,28], given that Fe is surrounded by strongly bonded ligands. Therefore, a strong

TABLE 11.1

Results of the Mössbauer Spectra-Fitting Process

Sample	FWHM	δ (mm/s)	Δ	h_+/h_-
Zn–NP	0.234	1.864	0.661	0.995
Ni–Np	0.253	1.892	0.664	0.971
Cd–NP	0.240	1.892	0.664	0.973

crystal field that produces the LS configuration in octahedral coordination is generated [29]. Besides, the value measured for the quadrupole splitting is considerable
due to a significant charge asymmetry produced in the region surrounding the Fe^{2+}
cations by the NO group [12].

The magnetization curves (M vs. H) of the Zn–, Cd–, and Ni–NPs were measured
out at room temperature (298 K) in the VSM, Lakeshore 7400 Series, being the maximum magnetic field applied 2.2 T, that is, 20 Kilo Oersted. The powder sample was
weighted, located in the sample holder, and then applied to the ramp from −2.2 to
+2.2 T and thereafter in backward direction.

Magnetochemistry [30–32] is the branch of chemical physics, which studies the
magnetic properties of materials. In this sense, a parameter that allows an assessment of these properties is the effective magnetic moment μ_{eff}, measured in Bohr
magnetrons μ_B, since it is independent of temperature and the external field strength.
Concretely, the assessment of the magnetic properties is generally made with the
help of the magnetization curves. In our case, the results of the measurements with
the VSM are reported in Figure 11.4. The obtained data allowed the determination
of the effective magnetic moment (μ_{eff}) for Ni–NP yielding $\mu_{eff} = 2.9\ \mu_B$, a value that
is to some extent larger than the spin-only value 2.83 μ_B [30,31], given that the electronic configuration of Ni^{2+} is [Ar] $3d^8$, it is located in octahedral sites and exhibit a
high-spin electronic configuration. Therefore, the magnetic measurements are consistent with high-spin Ni^{2+} located in octahedral sites. These data allow the determination of the effective magnetic moment (μ_{eff}) for Ni–NP yielding $\mu_{eff} = 2.9\ \mu_B$, a
value that is to some extent larger than the spin-only value 2.83 μ_B [32]. Given that
the electronic configuration of Ni^{2+} is [Ar] $3d^8$, it is located in octahedral sites, and
exhibits a high-spin electronic configuration [30]. Therefore, the magnetic measurements are consistent with high-spin Ni^{2+} located in octahedral sites. Besides, Zn^{2+}
and Cd^{2+} are also located in octahedral coordination surrounded by CN and H_2O

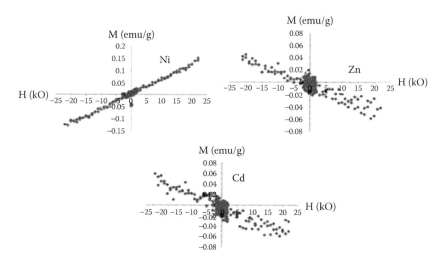

FIGURE 11.4   Magnetization curves of the Zn–, Cd–, and Ni–NPs.

in the Zn– and Cd–NPs. Nevertheless, since these cations show the following electronic configurations: Zn^{2+} [Ar] $3d^{10}$ and Cd^{2+} [Kr] $4d^{10}$, they do not have unpaired electrons. Consequently, the measured magnetization curves show a diamagnetic behavior [3].

Afterward, to investigate the structural effects and the interactions of the CO_2 molecule with the Ni–, Zn–, Cd–, and Cu–NP frameworks during adsorption, DRIFTS of adsorbed molecules, high-pressure volumetric adsorption of carbon dioxide, and *in situ* XRD investigations of CO_2 adsorption experiments were performed. With this means, a novel quantitative research of the carbon dioxide interactions within the adsorption space of Ni–, Zn–, and Cd–NP [3] and Cu–NPs [4] were performed. In this regard, the volume of the available adsorption space, the intensity of the adsorption field within the adsorption space, the specific interactions of the carbon dioxide molecule with the NP frameworks, and the influence of the adsorption process in the structure of the NPs were studied.

11.1.4 Carbon Dioxide Adsorption

To analyze the CO_2 adsorption data the DR adsorption isotherm equation was applied, since it is usually applied to mathematically model the equilibrium adsorption process in microporous crystalline materials, such as zeolites, metal organic frameworks (MOFs), and other porous coordination polymers as NPs [33], giving that a vast amount of data indicate that the adsorption process in the micropore range is very well described by it [46–49]. In this regard, as have been previously discussed, this adsorption isotherm equation can be represented in a log–log scale as follows:

$$y = \ln(n_a) = \ln(N_m) - \left(\frac{RT}{E}\right)^n \ln\left(\frac{P_0}{P}\right)^n = b - mx \qquad (11.1)$$

in which $y = \ln(n_a)$, $b = \ln(N_a)$, $m = (RT/E)^n$, and $x = \ln(P_0/P)^n$, n_a is the amount adsorbed, P_0/P is the inverse of the relative pressure, E is a parameter named as the characteristic energy of adsorption, and N_m is the maximum amount adsorbed in the micropore volume. Thereafter, the fitting process allowed the calculation of the best-fitting parameters of Equation 11.1, that is, N_m and E for $n = 2$ along with the regression coefficient and the standard errors. In Table 11.2, parameters calculated with the D–E equation are reported, that is, N_m along with the micropore volume calculated using the Gurtvich rule, that is, $W_{MP}^{CO_2} = V_{CO_2}^L N_m$ ($V_{CO_2}^L = 48.3$ cm³/mol, molar volume of carbon dioxide), E, and the isosteric heat of adsorption calculated using the following equation: q_{iso} ($\Theta = 0.37$) $= 1.16E$ [3,4,24,25].

The parameters calculated with the DR adsorption isotherm equation not only allow us to evaluate the micropore volume of the sample but also the magnitude of the adsorption heat, which is related to the adsorption interaction between the adsorbate and the adsorbent, being used for this assessment of the characteristic energy of adsorption E. In this regard, we know that, in general, if a molecule contacts the surface of a solid adsorbent, it becomes subjected to diverse interaction fields, such as the dispersion energy, repulsion energy, polarization energy, field dipole energy,

TABLE 11.2

Parameters Calculated by Fitting the Experimental CO_2 Adsorption Data with the Dubinin–Radushkevich Isotherm Equation

Sample	T (K)	N_m (mmol/g)	W (cm³/g)	E (kJ/mol)	q_{iso} (kJ/mol)
Ni–NP	273	5.649	0.275	9	11
Ni–NP	298	5.895	0.287	9	11
Zn–NP	273	4.971	0.242	12	14
Zn–NP	298	5.156	0.251	12	14
Cd–NP	273	2.609	0.127	21	25
Cd–NP	298	2.937	0.143	22	26
Cu–NP	273	2.609	0.154	25	30
Cu–NP	298	2.937	0.181	25	30

field gradient quadrupole energy ϕ_{EQ}, and some specific interactions, such as the acid–base interaction and the adsorbate–adsorbate interaction energy [17].

Dispersion and repulsion are the fundamental forces present during physical adsorption in all adsorbents. To be precise, dispersion and repulsion interactions are present in all adsorption gas–solid systems since they are nonspecific interactions. However, the electrostatic interactions between the adsorbed molecule and the adsorbent framework that depends on the structure and composition of the adsorbed molecule and the adsorbent itself is very important in our case [15]. In the case of a molecule such as the carbon dioxide molecule, which shows a noticeable quadrupolar moment [34], it gives rise to specific interactions where the electrostatic attractive interactions are normally stronger than the dispersion ones.

The quantitative evaluation of the interaction between carbon dioxide and the tested NPs reported in Table 11.2 was carried out with the help of the isosteric heat of adsorption q_{iso}, calculated using only one isotherm as follows: $q_{iso}(0.37) = 1.16E$. Hence, the contribution of the oxygen-covered region of the adsorption space was estimated with the help of the previously measured isosteric heat adsorption of CO_2 on the microporous molecular sieve DAY, that is, dealuminated Y zeolite, which is $q_{iso}(0.37) \approx 20$ kJ/mol. Hence, the contribution of the specific interactions of the CO_2 molecule with the M^{2+} cations to the isosteric heat of adsorption should be $q_{iso}(0.37) \approx 10$ kJ/mol [4,22].

The results reported in Table 11.2 indicate that the isosteric heat of adsorption for $\theta = 0.37$ depends on the size of the adsorption space. Thereafter, the bigger value is those reported for the Cu– and Cd–NPs, where the carbon dioxide molecules are more confined than in the rest of NPs because of the size of their adsorption space, as the pore size of the channels formed in frameworks of the Cu– and Cd–NPs is close to the kinetic diameter of the CO_2 molecule. This fact causes an effect named as confinement or surface interaction potential overlapping. This fact together with the high amount of CO_2 adsorption on Ni– and Zn–NPs at high pressures indicates that Ni– and Zn–NPs are excellent for CO_2 storage and Cu– and Cd–NPs are good for gas cleaning, since they could selectively adsorb water along with possible contaminants. This effect favors relatively strong attraction and repulsion interactions between the framework of the adsorbent and the adsorbed molecule, and also between the adsorbate molecules that affect the framework of the Cd–NP.

11.2 PRUSSIAN BLUE ANALOGS

11.2.1 Introduction

Transition metal cyanides also known as Prussia blue analogs (PBAs) exhibit structures built with transition metals attached through the linear cyanide chain, that is, $-M-C \equiv N-M-N \equiv C-M-$ [35–51], provided these compounds are very useful owing to their magnetic [42], adsorption [35,36], and other properties [52].

The structural framework of PBAs is associated to the perovskite structure, even though the metal centers are evidently linked by the cyanide bridges instead of oxide ions [43]. Subsequently, since the basic structural element is straight, PBAs should crystallize in the cubic system more precisely in the $Fm3m$ space group and to less extent in the $Pm3m$. Although for some metals, specific distortions produce the crystallization in the $I4mm$ and $I\bar{4}m2$ [44].

To understand the adsorption properties of PBAs it is necessary to remark now that the presence of $[M'(X)(CN_6)]^{\mu-}$ vacancies produces a channel network [42] shaped by cavities, where the exchangeable metal A is located on these voids to generate charge balance [44] also acting as charge centers capable of producing electrostatic interactions with the adsorbed molecules, specifically in the case of carbon dioxide, given that this molecule has a quadrupole moment that strongly interacts with the electric field gradients within the cavity [17].

In the specific case of PBAs the generalized formula can be expressed as follows: $AM[Fe(X)(CN_6)]nH_2O$, where A is generally an alkaline metal, M is a transition metal, $Fe = Fe(X) X = II$ or III, whereas n H$_2$O are coordinated together with loosely bound water molecules filling the $[Fe(X)(CN)_6]^{\mu-}$ vacancies that forms the microporous framework [45]. Moreover, carbon dioxide adsorption, given that this molecule has a quadrupole moment, it strongly interacts with the electric field gradients within the cavity [15–17].

During the past years, PBAs have been widely studied [42–51], even though their high-pressure adsorption properties have not been largely considered. Moreover, soft porous crystals, that is, adsorbents that display structural flexibility also have been extensively studied [52–54]. However, this property has not been reported for Cu(II) and Zn(II) hexacyanoferrate(II). Therefore, the key questions addressed here were the structure elucidation together with the investigation of the adsorption space and the framework expansion effect of Cu(II) hexacyanoferrate(III) polymorph and Zn(II) hexacyanoferrate(II) labeled Cu–PBA-I and Zn–HII. Moreover, given that zeolite-like hexacyanometallates undergo different framework transformations produced by external stimuli, such as mechanical stress, adsorption, or temperature [52], when, as in our case, the interaction of the guest molecules with the framework is the effect producing the structural transformations, then atypical adsorption–desorption isotherm patterns are observed [53], particularly gate opening effects characterized by a big hysteresis loop between the adsorption and desorption branch of the isotherm could be found [54].

In addition, CO_2 is a good probe molecule to test framework transformations during adsorption owing to the relatively high value of their quadrupolar moment ($Q_{CO_2} = -4.3 \times 10^{-42}$ Cm^2 [34]). Consequently, when it is adsorbed at low coverage, it becomes subjected to the dispersion, repulsion, and field gradient quadrupole interaction energies [16]. Moreover, carbon dioxide molecule has an ellipsoidal form 5.4 Å long and with a diameter of 3.4 Å [55,56] allows it to penetrate in the majority of channels and cavities, making it a very good test molecule [20].

In the research discussed here, the structural analysis was performed with a broad set of characterization methods. In addition, a low- and high-pressure carbon dioxide adsorption investigation was performed. Concretely, the structural characterizations were made with XRD, scanning electron microscopy (SEM), energy dispersive X-ray analysis (EDAX), DRIFTS, and TGA, simultaneously with the investigation of the adsorption of their low- (up to 1 bar) and high-pressure (up to 30 bar) carbon dioxide adsorption properties to be found if these materials show structural flexibility [35,36].

Hence, the main questions examined were synthesis along with the structural characterization of Cu(II) and Zn(II) hexacyanoferrate(II) (labeled Cu–PBA and Zn–HII) together with the investigation of carbon dioxide adsorption in the low-pressure (up to 1 bar) and high-pressure (up to 30 bar) ranges to investigate the structural transformations induced by adsorption in the framework of the tested PBAs. To perform the corresponding experimental work the structural characterizations were done using XRD and a morphological study of the synthesized hexacyanoferrates was made with a SEM. Furthermore, the elemental composition was measured by means of the EDAX facility included in the SEM. Thereafter, the state of water in the studied material was investigated using TGA. Finally, low- and high-pressure carbon dioxide adsorption was studied to closely examine the adsorption space of the produced hexacyanoferrate.

11.2.2 SYNTHESIS

All chemicals used were of analytical grade and water was bidistilled. The synthesis performed was carried out by following the recipes reported in literature [8,11,13] as follows: 0.025 moles of solid potassium hexacyanoferrate(II) and solutions containing 0.025 moles of $Zn(NO_3)_2$ for 250 mL of water were mixed at 70°C under constant stirring. Next, the formed precipitate was filtered, washed with distilled water, and dried at 343 K for 24 h to get the sample labeled as Cu–PBA-I and Zn–HII. Meanwhile, the synthesis of the Cu(II) hexacyanoferrate(III) was made using a modified version of the standard one, that is, it took place by mixing 0.025 moles of solid potassium hexacyanoferrate(III) and solutions containing 0.025 moles of $Cu(NO_3)_2$ in 250 mL of water at 70°C under constant stirring. Next, the formed precipitate was filtered, washed with distilled water, and dried at 343 K for 24 h to get the sample labeled as Cu–PBA-I.

11.2.3 CHARACTERIZATION

11.2.3.1 Methods

The SEM images were obtained at 20 kV (JEOL, JSM-6360) for sample grains placed on a carbon tape, at the same time as elemental chemical analysis was performed using the EDAX included in the SEM.

XRD profiles were collected (Bruker D8 Advance) using a Bragg–Brentano vertical goniometer configuration. The angular measurements were made using steps of 0.01°. The X-ray radiation source was a ceramic XRD Cu anode tube-type KFL C 2K. In addition, a Soller slit was included and a Ni filter was placed before the detector. In addition, a LynxEye™ one-dimensional detector was used for X-ray detection.

Finally, the curve-fitting processes were performed with the analysis and peak separation software PeakFits (Seasolve Software Inc., Framingham, Massachusetts) based on a least square procedure. The program made possible the calculation of the best-fitting parameters in the case of curve fitting. For peak separation the software made the calculation of the band parameters possible, that is, intensity, peak position, and half-bandwidth using the spectrum-fitting Lorentzian functions.

The XRD profiles were collected and resolved into separate Bragg components using the Pawley powder pattern decomposition method by means of the Bruker AXS TOPASs powder XRD analysis software.

11.2.3.2 Structure Determination

The SEM images are reported in Figure 11.5 along with the elemental chemical analysis, which are reported in Table 11.3 [35,36].

Cu–PBA-I

Zn–HII

FIGURE 11.5 SEM images of the tested PBAs.

TABLE 11.3
Chemical Composition of the Tested PBAs

Sample	C_K (wt. %)	C_{Fe} (wt. %)	C_{Cu} (wt. %)
Cu–PBA-I	0.2	1.2	1.5
Z–HII	0.4	1.0	2.1

TABLE 11.4
Cell Parameter Calculated with the Help of the Pawley Method

Sample	a (Å)	b (Å)	V (Å³)	S.P.
Cu–PBA-I	9.87	10.52	927	$I\bar{4}m2$
Z–HII	10.23		1071	$Fm\bar{3}m$

The XRD profiles were fitted using the Pawley method [57] assuming the $I\bar{4}m2$ space group (SG) for the as-synthesized Cu–PBA and the $Fm\bar{3}m$ space for the as-synthesized Zn–HII sample, being the calculation of the cell parameter as the result of the refining process, which is shown in Table 11.4.

Now, using the previously reported crystallographic information for the Cu–PBA-I and Zn–HII (Table 11.4) together with the atomic positions, Wyckoff sites and occupancy factor for the $I\bar{4}m2$ and $Fm\bar{3}m$ space groups reported in the International Tables for Crystallography [58] was generated with the software PowderCell-2.4, the unit cell representation (Figure 11.6) corresponding to the synthesized PBAs.

Hence, we can conclude that the synthesized PBAs, that is, the Cu–PBA-I and the Zn–HII materials are different polymorphs than those reported in the literature [28,42,46].

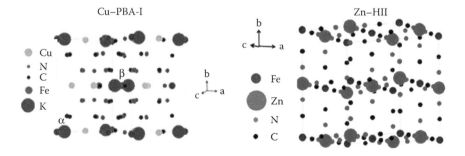

FIGURE 11.6   Frameworks of the Cu–PBA-I and Zn–HII PBAs.

11.3 CARBON DIOXIDE ADSORPTION

Carbon dioxide (Praxair, 99.99%) adsorption was investigated at 300 K and pressure up to 1 bar (LP) (Quantachrome AS-1) and at 300 K and pressure up to 30 bar (HP) (Quantachrome iSorbHP-100) on samples degassed at 423 K for 3 h in high vacuum (10?6 Torr). The backfilling process was performed using helium (Praxair, 99.99%). Concretely, to analyze the CO_2 adsorption data, the DR adsorption isotherm Equation 11.1 was used. Thereafter, the fitting process allowed the calculation of the best-fitting parameters of Equation 11.1, that is, N_m, E for $n = 2$ along with the regression coefficient and the standard errors. In Table 11.5 parameters calculated with the D–E equation are reported, that is, N_m along with the micropore volume calculated using the Gurtvich rule, that is, $W_{MP}^{CO_2} = V_{CO_2}^L N_m$ ($V_{CO_2}^L = 48.3 \, cm^3/mol$, molar volume of carbon dioxide), E, and the isosteric heat of adsorption calculated by using the following equation: q_{iso} ($\Theta = 0.37$) $= 1.16E$ [3,4,24,25].

Subsequently, the carbon dioxide high-pressure adsorption isotherms are reported in Figure 11.7. In this sense, the high-pressure CO_2 adsorption data indicated that as

---

TABLE 11.5
Parameters Calculated by Fitting the Experimental CO_2 Adsorption Data with the Dubinin–Radushkevich Isotherm Equation

Sample	T (K)	N_m (mmol/g)	W (cm³/g)	E (kJ/mol)	q_{iso} (kJ/mol)
Cu–PBA-1	300	1.84	0.090	16	19
Zn–HII	300	5.19	0.49	57	66

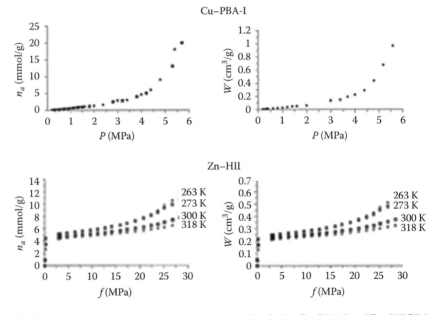

FIGURE 11.7   High pressure carbon dioxide adsorption in the Cu–PBA-I and Zn–HII PBAs.

a consequence of this process, the adsorption space was not saturated, instead the amount adsorbed continuously increased, because the adsorbent in the course of the adsorption processes experienced framework deformation, that is, adsorption can produce the expansion of the adsorbent structure [52–54].

To work out the framework expansion during the adsorption process, the adsorbed phase was considered as a solid solution, that is, during adsorption the adsorbate–adsorbent system (*aA*) was formed. Thereafter, using the thermodynamic approach described in Section 10.1.4 [32,34], the analysis was made with the help of the Π (osmotic pressure) versus *P* (equilibrium adsorption pressure) plot shown in Figure 11.8 [35,36], being shown in this plot that due to the cooperative interaction of the carbon dioxide molecules with the framework an increase in the osmotic pressure Π with the equilibrium adsorption pressure of the CO_2 in the gaseous phase is produced, this process causes the start of the CU–PBA-I and Zn–HII frameworks deformation fact accompanied by an increase of the micropore volume as shown in Figure 11.7.

The physical meaning of Π can be explained by a model, where it is assumed that the adsorption field within the adsorption space is zero. Explicitly, we have a mere geometric volume, in this case a pressure P_0 can compress the adsorbate in this geometric volume and can produce the same effect created by an adsorption field, that is, an amount n_a of molecules compressed in this volume [16]. Hence, these effects take place because carbon dioxide molecule has an ellipsoidal form 5.4 Å long and with a diameter of 3.4 Å [54,55]. Hence, the adsorbate–adsorbent interactions are greatly dependent on the pore geometry, especially in the case where the pore width of the adsorbent is very near to the size of the adsorbed molecules [52]. Thus, the cooperative interaction of the carbon dioxide molecules with the framework and between them produces the osmotic pressure Π, which increases with the equilibrium adsorption pressure of the CO_2 in the gaseous phase causing the onset of the adsorbent deformation fact accompanied by an increase in the micropore volume. This effect was possible owing to the well-known flexibility of the cyanide chain linkers $-M-C \equiv N - M - N \equiv C - M-$ that shapes the framework [39–41].

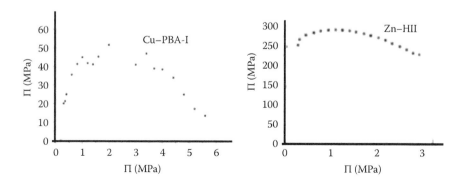

FIGURE 11.8   Π versus *P* plot for the Cu–PBA-I and the Zn–HII PBAs.

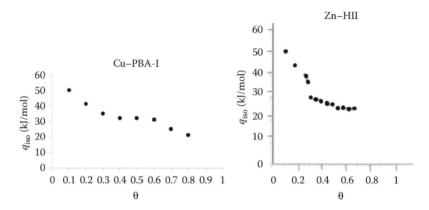

FIGURE 11.9 Plot of the isosteric heat of adsorption (q_{iso}) versus the recovery θ plot for the Cu–PBA-I and the Zn–HII PBAs.

Finally, the isosteric heat of adsorption q_{iso} was calculated using a relation similar to the Clausius–Clapeyron equation [15]:

$$q_{iso} \approx RT^2 \left[\frac{d \ln P}{dT} \right]_{n_a} \approx RT_1T_2 \left[\frac{\ln P_2 - \ln P_1}{T_2 - T_1} \right]_{n_a} \tag{11.2}$$

where P_1 and P_2 are the equilibrium adsorbate pressure at a constant magnitude of adsorption, that is, n_a = constant for the temperatures T_1 and T_2. In Figure 11.9 the isosteric heat of adsorption (q_{iso}) for both PBAs versus the recovery (θ) are plotted, where $\theta = n_a/N_m$, where n_a and N_m are the magnitude of adsorption and the maximum magnitude of adsorption, respectively.

In this case the results are consistent with the fact that the carbon dioxide molecule becomes subjected to the field gradient quadrupole together with dispersion and repulsion interactions, being in the case analyzed here, the whole interaction provided by the quadrupolar field in cooperation with the dispersion interactions [59]. Concretely, as was previously stated when the CO_2 molecule contacts the surface of both solid adsorbents, it becomes subjected to the dispersion energy, repulsion energy, polarization energy, field dipole energy, field gradient quadrupole energy, and possibly to some specific interactions, such as the acid–base interaction along with the adsorbate–adsorbate interaction energy [16]. Dispersion and repulsion are the two fundamental forces present during physical adsorption in all adsorbents. However, the electrostatic interactions between the adsorbed molecule and the adsorbent framework that depends on the structure and composition of the adsorbed molecule and the adsorbent itself are very important in our case [17]. In this regard, we have previously discussed that the carbon dioxide molecule, which shows a noticeable quadrupolar moment [34], a fact which gives rise to specific interactions where the electrostatic attractive interactions are normally stronger than the dispersion ones [35–37], specifically as it was yet specified that the contribution of the oxygen covered region of the adsorption space is $q_{iso} \approx$ 20–21 kJ/mol [3,4]. Hence, the contribution of the specific interactions of the CO_2 molecule with the Cu^{2+} and Zn^{2+} cations to the isosteric heat of adsorption range between 20 and 30 kJ/mol is shown by the results reported in Figure 11.9 [26,57].

Moreover, since the isosteric heat of adsorption depends on the size of the adsorption space, thereafter the adsorption spaces of the Cu–PBA-I and the Zn–HII PBAs are similar in size, giving that there is no significant differences in the reported q_{iso} versus θ plots for both materials. In addition, this plot indicates that the adsorption process is heterogeneous [60], fundamentally because at low coverage the interaction with the surface dominate the process, whereas at higher coverage the adsorbate–adsorbate interaction becomes predominant [36].

All the previously discussed CO_2 adsorption properties of the tested PBAs and specifically together the high amount of CO_2 adsorption at high pressures implies that both materials are excellent for CO_2 storage together with gas cleaning.

REFERENCES

1. Z.-Z. Gu, O. Sato, T. Iyoda, H. Hashimoto, and A. Fujishima, *Phys. Chem.*, 100 (1996) 18289.
2. L.A. Gentil, E.J. Baran, and P.J. Aymonino, *Inorg. Chim. Acta*, 20 (1976) 251.
3. R. Roque-Malherbe, O.N.C. Uwakweh, C. Lozano, R. Polanco, A. Hernandez-Maldonado, P. Fierro, F. Lugo, and J.N. Primera-Pedrozo, *J. Phys. Chem. C*, 115 (2011) 15555–15569.
4. R. Roque-Malherbe, C. Lozano, R. Polanco, F. Marquez, F. Lugo, A. Hernandez-Maldonado, and J. Primera-Pedroso, *J. Solid State Chem.*, 184 (2011) 1236–1244.
5. L.A. Gentil, P.J. Baran, and. P.J. Aymonino, *Inorg. Chim. Acta*, 20 (1976) 251.
6. D.F. Mullica, E. Sappenfield, D.B. Tippin, D.H. Leschnitzer, *Inorg. Chim. Acta*, 164 (1989) 99–103.
7. A. Benavente, J.A. de Moran, and P.J. Aymonino, *J. Chem. Crystallogr.*, 27 (1997) 343.
8. E. Reguera, A. Dago, A. Gomez, and J.F. Bertran, *Polyhedron*, 15 (1996) 3139.
9. K.E. Funck, M.G. Hilfiger, C.P. Berlinguette, M. Shatruk, W. Wernsdorfer, and K.R. Dunbar, *Inorg. Chem.*, 48 (2009) 3438.
10. E. Reguera, A. Dago, A. Gomez, and L.M.D. Cranswick, *Polyhedron*, 20 (2001) 165.
11. J. Balmaseda, E. Reguera, A. Gomez, J. Roque, C.M. Vazquez, and M. Autie, *J. Phys. Chem. B*, 107 (2003) 107.
12. J.T. Culp, C. Matranga, M. Smith, E.W. Bittner, and B. Bockrath, *J. Phys. Chem. B*, 110 (2006) 8325.
13. L. Reguera, J. Balmaseda, C.P. Krap, and E. Reguera, *J. Phys. Chem. C*, 112 (2008) 10490.
14. M. Zentkova, M. Mihalik, I. Toth, Z. Mitroova, A. Zentko, M. Sendek, J. Kovac, M.M. Lukacova, M. Marysko, and M. Miglierini, *J. Magn. Magn. Mater.*, 272–276 (2004) E753.
15. F. Rouquerol, J. Rouquerol, and K.S.W. Sing, *Adsorption by Powder Porous Solids*, Academic Press, New York, 1999.
16. R. Roque-Malherbe, *Adsorption and Diffusion in Nanoporous Materials*, CRC Press, Boca Raton, FL, 2007.
17. R. Roque-Malherbe, *Physical Chemistry of Materials: Energy and Environmental Applications*, CRC Press, Boca Raton, FL, 2010.
18. R. Roque-Malherbe, R. Polanco-Estrella, and F. Marquez, *J. Phys. Chem. C*, 114 (2010) 17773.

19. R. Roque-Malherbe, R. Wendelbo, A. Mifsud, and A. Corma, *J. Phys. Chem.*, 99 (1995) 14064.

20. D. Cazorla-Amoros, J. Alcaniz-Monge, M.A. de la Casa-Lillo, and A. Linares-Solano, *Langmuir*, 14 (1998) 4589.

21. M.M. Dubinin, *Prog. Surf. Membr. Sci.*, 9 (1975) 1.

22. J.A. Dunne, M. Rao, S. Sircar, R.J. Gorte, and A.I. Myers, *Langmuir*, 12 (1996) 5896.

23. H. Knozinger and S. Huber, *J. Chem. Soc. Faraday Trans.*, 94 (1998) 2047.

24. F.X. Llabres-i-Xamena and X. Zecchina, *Phys. Chem. Chem. Phys.*, 4 (2002) 1978.

25. F. Lugo, MSc Thesis, School of Science, University of Turabo, 2012.

26. F. Lugo, PhD Dissetation, School of Science, University of Turabo, 2015.

27. G.S. Pawley, *J. Appl. Crystallogr.*, 14 (1984) 357.

28. K. Nakamoto, *Infrared and Raman Spectra of Inorganic and Coordination Compounds: Part A: Theory and Applications in Inorganic Chemistry*, John Wiley & Sons, New York, 1997.

29. P. Aymonino, *J. Pure Appl. Chem.*, 60 (1988) 1257.

30. R.V. Parish, in *Mössbauer Spectroscopy* (D.P.E. Dickson and F.J. Berry, Eds.), Cambridge University Press, Cambridge, UK, 1986, p. 17.

31. G.J. Long, in *Mössbauer Spectroscopy* (D.P.E. Dickson and F.J. Berry, Eds.), Cambridge University Press, Cambridge, UK, 1986, p 70.

32. R.L. Carlin, *Magnetochemistry*, Springer-Verlag, Berlin, Germany, 1986.

33. O. Kahn, *Molecular Magnetism*, VCH, Weinheim, Germany, 1993.

34. A.F. Orchard, *Magnetochemistry*, Oxford University Press, Oxford, UK, 2003.

35. A.D. Buckingham, *Proc. Royal Soc. Series A.*, 273 (1963) 275.

36. R. Roque-Malherbe, F. Lugo, and R. Polanco, *Appl. Surf. Sci.*, 385 (2016) 360–367.

37. R. Roque-Malherbe, E. Carballo, R. Polanco, F. Lugo, and C. Lozano, *J. Phys. Chem. Solids*, 86 (2015) 65–73.

38. J. Balmaseda, E. Reguera, A. Gomez, J. Roque, C. Vazquez, and M. Autie, *J. Phys. Chem. B*, 107 (2003) 11360.

39. M. Verdaguer and G. Girolami, in: *Magnetism: Molecules to Materials V* (J.S. Miller and M. Drillon Eds.), Wiley-VCH Verlag GmbH & Co. KGaA, Weinheim, Germanny, 2004.

40. S. Natesakhawat, J.T. Culp, C. Matranga, and B. Bockrath, *J. Phys. Chem. C*, 111 (2007) 1055.

41. J. Jimenez-Gallegos, J. Rodriguez-Hernandez, H. Yee-Madeira, and E. Reguera, *J. Phys. Chem. C*, 114 (2010) 5043.

42. M. Okubo, D. Asakura, Y. Mizuno, J.-D. Kim, T. Mizokawa, T. Kudo, and I. Honma, *J. Phys. Chem. Lett.*, 1 (2010) 2063.

43. M. Avila, J. Rodriguez-Hernandez, A. Lemus-Santana, and E. Reguera, *J. Phys. Chem. Solids*, 72 (2011) 988.

44. A. Ludi and H.U. Güdel, *Struct. Bond.*, 14 (1973) 1–21.

45. C. Loos-Neskovic, S. Ayrault, V. Badillo, B. Jimenez, E. Garnier, M. Fedoroff, D.J. Jones, and B. Merinov, *J. Solid State Chem.*, 2004 (1817) 177.

46. N.R. de Taconi, K. Rajeshwar, and R.O. Lezna, *Chem. Mater.*, 15 (2003) 3046.

47. R. Martinez-Garcia, E. Reguera, J. Rodriguez, and R. Balmaseda, *J. Powder Diffr.*, 19 (2004) 284.

48. C. Loos-Neskovic and M. Fedoroff, *React. Polym. Ion Exch. Sorbents*, 7 (1988) 173.

49. M. Avila, L. Reguera, J. Rodriguez-Hernandez, J. Balmaseda, and E. Reguera, *J. Solid State Chem.*, 181 (2008) 2899.

50. F. Adekola, M. Fedoroff, S. Ayrault, C. Loos-Neskovic, E. Garnier, and L.-T. Yu, *J. Solid State Chem.*, 132 (1996) 399.

51. S. Adak, L.L. Daemen, M. Hartl, D. Williams, J. Summerhill, and H. Nakotte, *J. Solid State Chem.*, 184 (2011) 2854.

52. J. Lehto and S. Haukka, *Thermochim. Acta*, 160 (1990) 343.
53. (a) V.A. Tvardoskiy, *Sorbent Deformation*, Elsevier, Amsterdam, the Netherlands, 2006. (b) F.-X. Coudert, *Phys. Chem. Chem. Phys.*, 12 (2010) 10904.
54. T.K. Maji and S. Kitagawa, *Pure Appl. Chem.*, 79 (2007) 2155.
55. P. Bhatt, N. Thakur, M.D. Mukadam, S.S. Meene, and S.M. Yusuf, *J. Phys. Chem. C*, 117 (2013) 2676.
56. W.N. Haynes (chief editor), *Handbook of Physics and Chemistry* (97th ed), CRC Press, Boca Raton, FL, (2016–2017).
57. H. Omi, T. Ueda, K. Miyakubo, and T. Eguchi, *Appl. Surf. Sci.*, 252 (2005) 660.
58. T. Hahn (Ed.), *International Tables for Crystallography* (Vol. A, 5th ed.), The International Union of Crystallography, Springer, Dordrecht, the Netherlands, 2009.
59. P.L. Llewellyn, S. Bourrelly, C. Serre, A. Vimont, M. Daturi, L. Hamon, M. De Weireld et al., *Langmuir*, 24 (2008) 7245.
60. W. Rudzinski and D.H. Everett, *Adsorption of Gases in Heterogeneous Surfaces*, Academic Press, London, UK, 1992.

12 Metal and Covalent Organic Frameworks

12.1 METAL ORGANIC FRAMEWORKS

12.1.1 INTRODUCTION

Metal organic frameworks (MOFs) are materials consisting of metal nodes and organic spacers showing permanent porosity [1–3]. The methodology for the synthesis of these zeolite-like compounds was created in the first years of the 1990s [4,5]. After that, an exponential growth of new MOFs has been steadily produced to be used in catalysis [6], gas adsorption [2], separations [3], and other applications. Currently, there exists several porous and large specific surface area materials such as zeolites [7], single-walled carbon nanotubes [8], silica [9], nitroprussides [10], akaganeites [11], Prussian blue analogs (PBAs) [12,13], and covalent organic frameworks (COFs) [14,15]. However, the synthesis of a moderately stable flexible framework MOF is an interesting task. In this regard, the succinate linker could be a good choice, because it is a strong and flexible linker [16] that could result in a material with a wide adsorption space [17].

On the other hand, CO_2 was selected to study the adsorption space [18] and the surface chemistry [19] of the synthesized material, giving that it is a good test molecule, owing to the relatively high quadrupole moment presented by this molecule [20] together with the fact that this molecule freely passes through the tested material pores, because the kinetic diameter of the carbon dioxide molecule is $\sigma_{CO_2} = 3.30$ Å [21].

Furthermore, the adsorption process is able to produce the deformation of the adsorbent, that is, the solid adsorbent is not inert during the course of the adsorption processes. Thus, it can experience expansion, contraction, and swelling [22]. Incidentally, a group of materials known as soft porous crystals have been investigated on account of their structural flexibility, since it undergoes framework expansion induced by an external stimuli [23].

Thereafter, the main purpose of this chapter is the study of the synthesis, elucidation of the structure, study of the low- and high-pressure carbon dioxide adsorption along with the investigation of the framework expansion during the adsorption process together with a test of the thermal stability of the produced Cu–succinic MOF, hereby labeled as Cu–Su–MOF.

To meet the objectives related to the elucidation, thermal stability test and structure elucidation, the synthesized material was investigated with scanning electron microscopy (SEM)-energy dispersive X-ray analysis (EDAX) [9], diffuse reflectance infrared Fourier transform spectrometry (DRIFTS) [24,25] together with magnetic [11] and X-ray diffraction (PXRD) [12] measurements along with physical adsorption studies [13]. Finally, to clearly understand the framework expansion effect during high-pressure adsorption, an approach based on the osmotic theory of adsorption [26,27] to measure some parameters characterizing this process was applied [28–30].

12.1.2 SYNTHESIS

The chemicals used in the synthesis were all of analytical grade without additional purification, whereas water was bidistilled. The synthesis procedure was 1 mmol of copper nitrate ($Cu(NO_3)_2$) and 1 mmol of succinic acid ($C_4H_6O_4$) were dissolved in 50 mL of dimethylacetamide (DMA, $CH_3CON(CH_3)_2$). Afterward, to the previous solution 29 mL of H_2O_2 (5.0%) was added. Next, the mixture was cooled by adding 24.1 mL of a 70.3 mM triethylamine ([TEA], $N(CH_2CH_3)_3$) solution in DMA. Later, the flask was sealed and placed in a hood for 24 h at a temperature of 70°C. Finally, the product was washed with distilled water and dried at 70°C.

Specifically, the synthesis was made in DMA as solvent, the source for the MOF organic spacer was $C_4H_6O_4$, whereas the metal source was anhydrous $Cu(NO_3)_2$. In addition, hydrogen peroxide (H_2O_2) was added as a source of atomic oxygen to increase the reaction rate, whereas triethylamine ([TEA], $N(CH_2CH_3)_3$) was used during the synthesis to remove hydrogen from the dicarboxylic acid.

12.1.3 CHARACTERIZATION METHODS

First, the SEM study was carried out with a JEOL JSM-6360 microscope whose electron beam acceleration voltage was 20 kV. The tested sample grains were homogeneously placed on a carbon tape. Hence, images of the tested materials were acquired [13]. In addition, the elemental chemical analysis of the as-synthesized and washed samples was performed using an EDAX spectrometer coupled into the microscope.

Second, DRIFTS were gathered using a Thermo Scientific Nicolet iS10 FTIR spectrometer. The IR spectra are normally reported as band intensity versus wavenumber [in cm^{-1}], where the intensity is expressed as transmittance, $T = (I_0/I)$, where I_0 is the intensity of the incident radiation and I is the intensity of the transmitted radiation, or as absorbance, $A = (1/T)$ [25]. Otherwise, during the Raman process, photons are not absorbed or emitted, as in one IR process, they are shifted in frequency by an amount corresponding to the energy of the vibration transition [26], corresponding to the absorption in the IR effect (Stoke processes). Concretely, the scattered photons are shifted to inferior frequencies, that is, the molecules remove energy from the exciting photons. The data were collected at a resolution of 4 cm^{-1} employing 100 scans per sample. The hydrated sample spectra were obtained at room temperature under N_2 flow (Praxair, 99.99%) at a rate of 50 cc/min. To get the spectra of the dehydrated samples, it was heated up to 100°C under a flow of N_2 (Praxair, 99.99%) at a rate of 50 cc/min for 2 h. Thereafter, spectra of the dehydrated materials were obtained at room temperature under N_2 (Praxair, 99.99%) flow at a rate of 50 cc/min. Moreover, DRIFTS spectra of carbon dioxide adsorbed in the MOF framework were obtained using the dehydrated sample as a background at room temperature.

After that, CO_2 (Praxair, 99.99%) flow at a rate of 50 cc/min for 3 min was passed through the dehydrated samples. Then, the sample was purged under N_2 (Praxair, 99.99%) flow at a rate of 50 cc/min for 1 min [31–33]. Hence, the spectrum of the carbon dioxide molecule adsorbed on the PBA microporous framework was then obtained at room temperature under N_2 flow [34,35].

The micro-Raman spectra were collected with a Jobin–Yvon T64000 Raman spectrophotometer, consisting of a double pre-monochromator coupled to a mono-chromator/spectrograph with 1800 grooves/mm grating. A Leica microscope with an 80X objective was applied to focus the 514.5 nm radiation of an Ar^+ laser [25].

Meanwhile, the magnetization curves (M versus H) were collected at room temperature (298 K) in the vibrating sample magnetometer (VSM), Lakeshore 7400 Series. The powder sample was weighted, located on the sample holder, and subsequently used the ramp from −2.2 to 2.2 T and backward directions [36,37].

Next, the XRD profiles were gathered with a Bruker D8 Advance system in Bragg–Brentano vertical goniometer configuration. The 2θ angular measurements were made by applying steps of 0.01°. The X-ray radiation source was a ceramic Cu anode tube ($\lambda = 1.542$ Å). Variable Soller slits were included and a Ni filter was placed before the detector. In addition, a LynxEye™ one-dimensional detector was used to produce large counting that resulted to high-quality XRD profiles, which can be accurately resolved by least square methods. To confirm the proposed structure, the gathered XRD patterns were refined with the Pawley and Rietveld methods. In this regard, the computer software used in the refining processes was the Bruker DIFFRAC*plus* TOPAS™ software package, where the emission profile was shaped by Lorentzians.

Finally, carbon dioxide (Praxair, 99.99%) adsorption was investigated at 273 K and 300 K in the low-pressure (LP) range (pressure up to 1 bar) on samples degassed at 373 K for 4 h in high vacuum (10^{-6} Torr) in a Quantachrome AS-1 automatic sorption analyzer [10]. To measure the carbon dioxide (Praxair, 99.99 %) adsorption at temperatures of 273 K, 300 K, and 318 K on samples degassed at 373 K for 4 h in high vacuum (10^{-6} Torr) in the high-pressure range (pressure up to 10 bar) a Quantachrome iSorbHP-100 was used. The backfilling process was performed using helium (Praxair, 99.99%) in both cases [11].

Finally, the curve-fitting processes were performed with the analysis and peak separation software PeakFit® (Seasolve Software Inc., Framingham, Massachusetts) based on a least square procedure using the method developed by Levenberg and Marquardt, making the calculation of the best-fitting parameters possible.

12.1.4 Structure Elucidation

12.1.4.1 Scanning Electron Microscopy–Energy Dispersive X-Ray Analysis Study

The as-synthesized Cu–MOF micrograph is reported in Figure 12.1. It displays 1–6 μm wide and long and 1–1.5 μm high coffin-shaped crystals composed of C, O, N, and Cu.

FIGURE 12.1   SEM image of the as-synthesized Cu–Su–MOF.

12.1.4.2   Evaluation of the Tested Material with Infrared Spectroscopy

The spectra were collected *in situ* in a high-temperature IR cell under pure N_2 flow, at different temperatures: 298 K, 373 K, 448 K, and 523 K (Figure 12.2a and b). In the spectra are observed, on one hand, the characteristic absorption bands of the -COO- groups, at 1450 and 1650 cm^{-1} (Figure 12.2a) [31], whereas a broad band (3200–3700 cm^{-1}) corresponding to the stretching vibrations produced by adsorbed

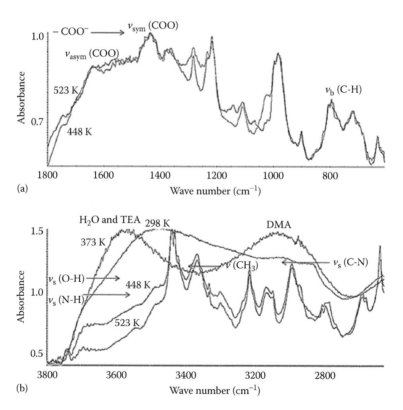

FIGURE 12.2   DRIFTS spectra of the Cu–Su–MOF as-synthesized (298 K) and degassed at 373 K, 448 K, and 523 K in the range between 1800–3800 cm^{-1} (a) and degassed at 448 K and 523 K in the range between 600–1800 cm^{-1} (b).

H_2O and TEA molecules were observed in the spectrum reported in Figure 12.2b. Moreover, at 447 K and 523 K, the broad band disappeared. Thereafter, H_2O and TEA were released after degassing at 447 K and 523 K, whereas since the bands in the range 2700–3300 cm^{-1} produced by the vibrations corresponding to DMA also vanished, this compound was also discharged.

12.1.4.3 Diffuse Reflectance Infrared Spectrometry Study of Adsorbed Carbon Dioxide and Raman Investigation of the Tested Cu–Su–MOF

The spectrum of carbon dioxide adsorbed at 760 Torr and 298 K on the Cu–Su–MOF is reported in Figure 12.3.

In Figure 12.3 a band is observed at ca. 2339 cm^{-1}. This peak is the result of the adsorption of carbon dioxide by dispersive and electrostatic forces within the adsorption space produced by the Cu–Su–MOF framework, in which adsorption produces the confinement of the carbon dioxide molecule. For this reason, the frequency shifts from 2349 cm^{-1} to 2339 cm^{-1}. Similarly, the band at ca. 2321 cm^{-1} is assigned to a combination band [31]. These assignments are consistent with the symmetry shown by the free carbon dioxide molecule, that is, the $D\infty h$ group symmetry, which exhibits four fundamental vibration modes, that is, the symmetric stretching, v_1 at 1338 cm^{-1}, the doubly degenerate bending vibration, v_{2a} and v_{2b} (667 cm^{-1}), and the asymmetric stretching vibration v_3 is located at 2349 cm^{-1} [32], where v_2 and v_3 modes are infrared active, whereas the v_1 is only Raman active in the free molecule, whereas the asymmetric stretching vibration, v_3, corresponds to carbon dioxide that is physically adsorbed [33]. In addition, since carbon dioxide is amphoteric, it interacts with acid and basic surface sites. Hence, another band must be normally observed at ca. 2362 cm^{-1} [31]. In the present case, this band should correspond to adsorption of carbon dioxide on an electron-accepting Lewis acid site forming the following adducts: $Cu^{2+} ... O = C = O$ [34,35], given that this band was not observed. Hence, their absence indicated that the Cu^{2+} ions are saturated by the oxalate oxygen atoms.

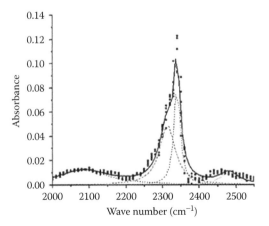

FIGURE 12.3 DRIFTS spectrum of adsorbed CO_2 on the degassed Cu–Su–MOF.

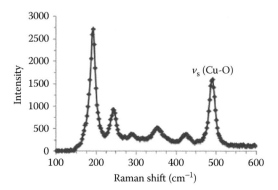

FIGURE 12.4 Raman spectrum of the as-synthesized Cu–Su–MOF.

Meanwhile, the room temperature Raman spectrum of the as-synthesized Cu–Su–MOF (Figure 12.5) show a band located on ca. 490 cm^{-1} distinctive of the stretching vibration mode of CuO$_6$ octahedra, corresponding to Cu^{2+} ions are linked to the succinate in a bidentate mode [15,36,37], that is, as the only spacer present was the succinic acid, hence it can be concluded that it was synthesized as a Cu–succinic material (Figure 12.4).

12.1.4.4 Magnetic Properties of the Synthesized Material

In Figure 12.5, the magnetization curves of the as-synthesized Cu–Su–MO and Ni–nitroprusside (Ni–NP), a standard for the calibration of the vibrating sample magnetometer is shown.

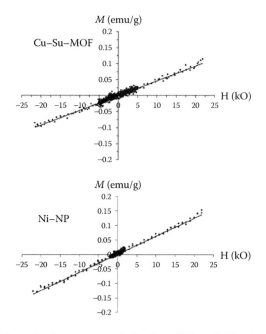

FIGURE 12.5 Magnetization curves of Cu–Su–MOF and Ni–nitroprusside used as standard.

Hence, the effective magnetic moment reported in literature for the Ni–NP is $\mu_{eff} = 2.9$ μB, (μB, Bohr magnetron) [36]. Subsequently, the effective magnetic moment for the Cu–Su–MOF yielded $\mu_{eff} = 2.15$ μB, when compared with the Ni–NP sample used as a standard (Figure 12.5). Accordingly, given that Cu^{2+} in octahedral coordination exhibits magnetic moments in the following range: $1.7 < \mu_{eff} < 2.2$ [37]. Hence, we conclude that the metallic node in the MOF framework is Cu^{2+} in the octahedral coordination as was also demonstrated by the Raman study.

12.1.4.5 Crystallographic Analysis

The powder X-ray diffraction (PXRD) profiles of the as-synthesized material at 298 K and degassed at 373 K, 448 K, 473 K, 523 K, and 573 K were collected *in situ* in a high temperature chamber under N_2 flow as shown in Figure 12.6. These PXRD profiles were fitted with the Pawley method using the Bruker DIFFRAC*plus* TOPAS™ software, assuming (based on the crystal shapes observed in Figure 12.1), the monoclinic *Pm* space group lattice to represent the framework of the Cu–Su–MOF. Consequently, the refining process made the calculation of the cell parameters at different temperatures possible (Table 12.1).

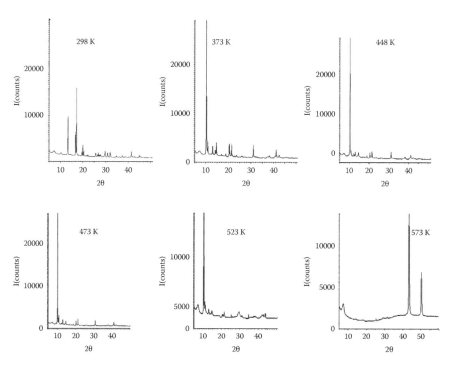

FIGURE 12.6 XRD profiles of the as-synthesized (298 K) and degassed at 373 K, 448 K, 473 K, and 523 K Cu–Su–MOF.

TABLE 12.1

Cell Parameters of the Degassed at 373 K, 448 K, 473 K, and 523 K Cu–Su–MOF Calculated by Fitting the XRD Profiles Using the Pawley Method

Sample	a (Å)	b (Å)	c (Å)	β	V_c (Å³)
373 K	8.171(1)	8.647(1)	9.284(1)	86.02(2)	654.1(1)
448 K	8.139(1)	8.645(1)	9.437(1)	87.66(2)	663.2(1)
473 K	8.143(1)	8.637(1)	9.416(1)	87.51(2)	661.3(1)
523 K	8.130(3)	8.651(2)	9.521(2)	88.14(4)	669.8(5)

12.1.4.6 Geometry of the Adsorption Space of the Tested Metal-Organic Frameworks

The micropore volume was measured using one adsorption isotherm equation, formally alike to the Langmuir isotherm but describing the volume filling rather than surface recovery [27]:

$$n_a = \frac{N_a KP}{1 + KP} \tag{12.1}$$

In which, $K_1 = K_0 \exp\left([(E_0^g - E_0^a)]/RT\right)$ is a constant for $T = \text{const}$, E_0^g is the reference energy state for the gas molecule, E_0^a is the reference energy state for the adsorbed molecule in the homogeneous adsorption field inside the cavity or channel, whereas n_a is the amount adsorbed and N_a is the maximum amount adsorbed. The linear for Equation 12.1 is [28]

$$y = P = N_a \left(\frac{P}{n_a}\right) - \frac{1}{K(T)} = mx = b \tag{12.2}$$

Then the experimental carbon dioxide adsorption data at 298 K on Cu–Su–MOF samples degassed at 373 K, 448 K, and 473 K were fitted with Equation 12.2. Hence, the linear fitting made the determination of the maximum amount adsorbed possible, along with the micropore volume that was calculated using the Gurtvich rule, that is, $W_{MP}^{CO_2} = V_{CO_2}^L N_a$, in which $V_{CO_2}^L = 48.3 \text{ cm}^3/\text{mol}$ is the molar volume of carbon dioxide (Table 12.2).

12.1.4.7 Elucidation of the Framework Structure

The succinate anion as other dicarboxylic acids has many potential donor sites to produce different coordination modes, including bidentate chelate, monodentate chelate, monodentate–bidentate, bis-monodentate, and bis-bidentate bringing [15]. Hence, using the data provided by the XRD profile measured at 373 K, in conjunction

TABLE 12.2
Parameters Calculated with the Langmuir-Type Isotherm Equation for Low-Pressure CO_2 Adsorption at 298 K on the Cu–Su–MOF Degassed at 373 K, 448 K, and 473 K

Temperature (K)	N_m (mmol/g)	W (cm³/g)
373	1.99	0.096
448	2.17	0.105
473	2.73	0.131

with the DRIFTS, Raman, and adsorption results, the knowledge related to the linking options of Cu^{2+} as node and the succinic acid as spacer [36], along with the unit cell parameter reported for the material (Table 12.1), the atomic positions, Wyckoff sites together with the occupancy factors registered in the International Tables for Crystallography [38] was proposed, a monoclinic *Pm* space group unit cell representation for the structure of the Cu–Su–MOF framework. Next, using the software Powder-Cell [39], the framework representation of the Cu–Su–MOF along the three axes was generated (Figure 12.7) with the computation of the chemical composition of a unit cell of the material framework: $CuC_{12}O_{12}H_{18}$.

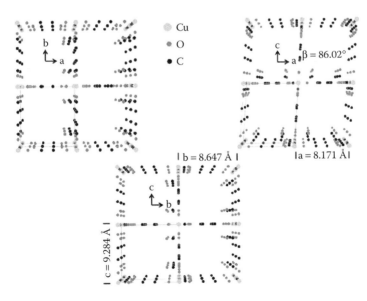

FIGURE 12.7 Cu–Su–MOF framework along [100], [001], and [010] crystallographic directions.

12.1.5 CARBON DIOXIDE ADSORPTION

The high-pressure carbon dioxide adsorption (Ad) and desorption (Des) isotherms at 273 K and 298 K in the pressure range: $0.01 < P < 10$ MPa on the tested Cu–Su–MOF degassed at 448 K were collected (Figure 12.8a).

Afterward, to compute the parameters relevant for the characterization of carbon dioxide adsorption, these isotherms were fitted to the linear form of the Dubinin–Radushkevitch (D–R) adsorption isotherm equation [26]:

$$\ln(n_a) = \ln(N_a) - \left(\frac{RT}{E}\right)^2 \left[\ln\left(\frac{P_0}{P}\right)\right]^2 \tag{12.3}$$

Where, n_a is the amount adsorbed, P/P_0 is the relative pressure, that is, P is the adsorption equilibrium pressure and P_0 is the vapor pressure of carbon dioxide at 298 K, E is the characteristic energy of adsorption, and N_a is the maximum adsorption amount. The linear fitting of the isotherm data by Equation 12.3 (Figure 12.9b) made the determination of the E, N_m possible. Along with W the micropore volume was calculated using the Gurtvich rule (Table 12.3).

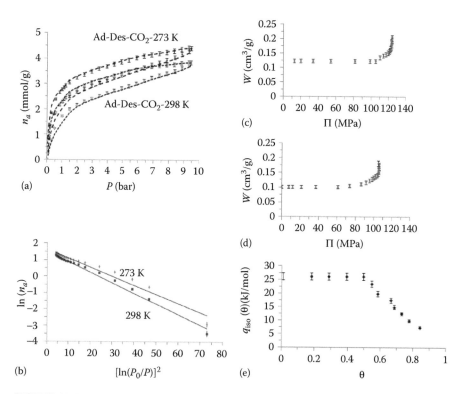

FIGURE 12.8 High-pressure CO_2 adsorption–desorption isotherms (a), D–R plots of the high-pressure adsorption data (b), Micropore volume (W) versus osmotic pressure (Π) at 273 K (c), and 298 K (d) and isosteric heat of adsorption versus θ plot (e).

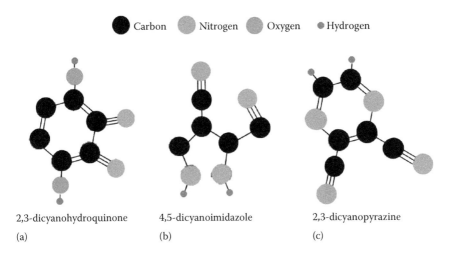

FIGURE 12.9 The three monomers used in the COFs synthesis.

TABLE 12.3

Parameters of the D–R Equation and Isosteric Heat of Adsorption at
$\theta = n_a/N_m = 0.37$

Temperature (K)	N_m (mmol/g)	E (kJ/mol)	W (cm³/g)	q_{iso} (0.37) (kJ/mol)
273	4.44	24	0.217	28
298	4.85	25	0.255	29

These results allowed us to recognize the presence of the sorbent deformation effect [22], giving that the micropore volume calculated with the high-pressure data is bigger than the micropore volume determined with low pressure isotherms (Tables 12.2 and 12.3).

Now, the thermodynamic analysis of the reported data is made using the previously explained methodology (Chapter 10). In this regard, we plotted the pore volume (W) versus the osmotic pressure (Π) in Figure 12.9c and d. This plot shows an initial state where the volume did not notably change, even though, at higher pressure a rapid increase in the micropore volume is evidenced, a fact pointing to the deformation of the adsorbent during the adsorption process.

The effects shown in Figure 12.8c and d take place because carbon dioxide molecule has an ellipsoidal form, 5.4 Å long and with a diameter of 3.4 Å [21]. Hence, the adsorbate–adsorbent interactions are greatly dependent on the pore geometry, specially, in the case when the pore width of the adsorbent is very near to the size of the adsorbed molecules [40,41]. Thus, the cooperative interaction of the carbon dioxide molecules with the framework and between them produces the osmotic pressure, Π, which increases with the equilibrium adsorption pressure of the CO_2 in the gaseous phase, causing the onset of the adsorbent (Cu–Su–MOF) deformation at

$\Pi \approx 100$ MPa by 273 K and $\Pi \approx 90$ MPa on 298 K, fact accompanied by an abrupt increase of the micropore volume. This effect was possible owing to the well-known flexibility of the succinate linkers that shapes the framework.

Finally, the isosteric heat of adsorption q_{iso} was calculated using a relation similar to the Clausius–Clapeyron equation [29]:

$$q_{iso} \approx RT^2 \left[\frac{d \ln P}{dT} \right]_{n_a} \approx RT_1 T_2 \left[\frac{\ln P_2 - \ln P_1}{T_2 - T_1} \right]_{n_a} \qquad (12.4)$$

Where, P_1 and P_2 are the equilibrium adsorbate pressure at $n_a = $ constant for the temperatures T_1 and T_2. The results of these calculations (Figure 12.9e) are consistent with the fact that the carbon dioxide molecule becomes subjected to the field gradient quadrupole together with dispersion, repulsion, and acid–basic interactions. In the case analyzed here, the whole interaction was provided by the quadrupolar field in cooperation with the dispersion interactions due to the absence of the adducts Cu^{2+} ... $O = C = O$ with the consequent lack of the acid–basic interaction [29].

This fact can be corroborated observing the figure reported for the isosteric heat of adsorption, that is, $q_{iso} = 26.1$ kJ/mol, a value typical when dispersion and quadrupolar interactions are the only adsorption fields present [19]. Moreover, the constant value for the isosteric heat of adsorption observed in the q_{iso} versus Θ graphic is produced by a compensation of the heterogeneity of the adsorption field and the sorbate–sorbate interactions [30]. Finally, for $\theta > 0.5$, the measured isosteric heat of adsorption is abruptly decreased due to the deformation of the framework.

12.2 COVALENT ORGANIC FRAMEWORKS

12.2.1 INTRODUCTION

As was previously commented MOFs, zeolites, silica, nitropussides, akaganeites, together with PBAs are organometallic or inorganic in nature-adsorbent materials showing permanent porosity [1–13]. Meanwhile, COFs are extended crystalline organic polymeric structures, where the building blocks are organic molecules topologically linked by covalent bonds into extensive lattices producing porous materials [14,15]. During the past years, all these materials have been produced to be used in separations, catalysis, gas adsorption, and other applications [1–15].

In the reported research COFs synthesized using a dynamic polymerization method were investigated, consistent of a reversible ionothermal trimerization of nitriles [42,43] that allowed us to get polymeric adsorbents exhibiting high performance, regular porosity, and high surface area [44,45], where the building blocks are 2,3-dicyanohydroquinone to produce the COF-labeled LH004, 4,5-dicyanoimidazole for the synthesis of the COF identified a LH006 and finally 2,3-dicyanopyrrzine to get the COF-termed LH008 (see Figure 12.10).

Now, it is necessary to state that COFs are normally porous. Hence, gas adsorption is a possible use of these materials [28,42]. Moreover, adsorption is not only an application, it is also a characterization tool, which allows the investigation of their surface chemistry together with the geometry of their adsorption space [8–14].

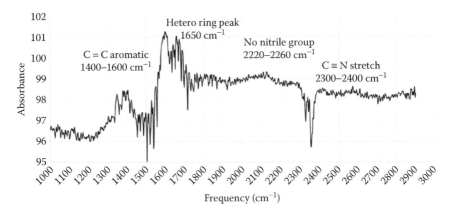

FIGURE 12.10 IR spectrum of the COF-labeled LH004.

Incidentally, in the present research to make the adsorption research was chosen was nitrogen as a test molecule, because it relatively strongly interacts with dispersion, ϕ_D, repulsion, ϕ_R, and field gradient quadrupole, ϕ_{EQ} potential fields. Hence, it is relatively strongly adsorbed at low temperature, that is, 77 K [19].

Thereafter, the main purposes of the research reported in the present paper were: the synthesis, elucidation of the structure of three COFs labeled: LH004, LH006, and LH008. To meet the objectives of the reported investigation, the three COFs were synthesized using a modification of the dynamic polymerization method [42,43]. Thereafter, to elucidate the structure, the materials were investigated with infrared [24] and Raman spectrometry [25], PXRD [46–49], and nitrogen physical adsorption up to 760 Torr in pressure and at a temperature of 77 K [26–30].

12.2.2 Synthetic Procedure

The chemicals used in the synthesis were all analytical grade without additional purification, whereas water was bidistilled. The monomers were 2,3-dicyanohydroquinone, 4,5-dicyanoimidazole, and 2,3-dicyanopyrrzine to produce the LH004, LH006, and LH008 COHs, respectively, whereas the synthesis procedure used was a modification of previous methodologies using the trimerization reaction as the main process [34], that is, a mixture of 1:1 ratio of the concrete monomer/desiccated $ZnCl_2$ was placed in the reactor, then dried in vacuum at 100°C for 12 h in a tubular furnace. Hence, after completion of the drying cycle, the furnace was set to reach 400°C and kept at this temperature for 40 h. Next, after cooling the reactor to room temperature it was opened to get a solid monolith, which was ground thoroughly.

The powder obtained following the previously described procedure was washed in hot water (90°C) with vigorous stirring for 12 h, exchanging the washing solution. Thereafter, the powder was further washed at 90°C for another 12 h with vigorous stirring. Finally, the solid was washed successively with water, dimethyl sulfoxide (DMSO), and distilled acetone was then dried in vacuum at 150°C for 12 h [35].

12.2.3 CHARACTERIZATION

At first, an IR Fourier transform [49] spectrum of the sample LH004 was gathered using the same methodology previously explained. Similarly, the Raman spectra of the COFs LH006 and LH008 were collected also following the procedure previously applied [24].

Thereafter, the IR spectrum of the COF-labeled LH004 is reported in Figure 12.10 [50].

The Raman spectra of the COFs named as LH006 and LH008 are shown in Figure 12.11 [50].

In the analysis of the IR and Raman spectra (Figures 12.10 and 12.11) the principle is applied, which states that the combination of atomic displacements during vibrations gives rise to only two types of vibrations, that is, stretching and bending [51].

In this regard, in the reported IR (Figure 12.10) and Raman (Figure 12.11) spectra were found a band ca. 1400–1600 cm^{-1}, which is assigned to the C=C stretching in the aromatic rings, whereas the peaks at ca. 1550 and 1650 cm^{-1} are related to the hetero ring, that is, the triazine ring (Figure 12.10) [49]. Moreover, no nitrile band was found at ca. 2210 and 2260 cm^{-1} [24]. It is necessary to indicate that the

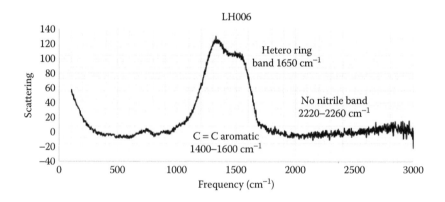

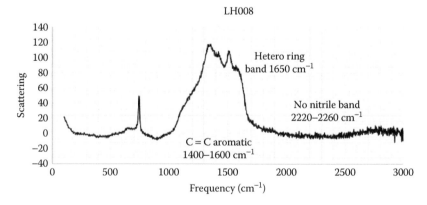

FIGURE 12.11   Raman spectra of the COFs-labeled LH006 and LH008.

differences observed in the intensity of the Raman spectra can be correlated with the different levels of polymerization depending on the hosts.

Consequently, it is possible to state that the IR and Raman spectra allows us to sustain that the polymerized monomers, that is, 2,3-dicyanohifroquinone, 4,5-dicyanoimidalzole, and 2,3-dicyanopyrazine (Figure 12.3) are the building blocks corresponding to the COFs-labeled LH004, LH006, and LH008, respectively.

Next, the XRD profiles were gathered with a Bruker D8 Advance system in Bragg–Brentano vertical goniometer configuration. The 2θ angular measurements were made by applying steps of $0.01°$. The X-ray radiation source was a ceramic Cu anode tube. Variable Soller slits were included and a Ni filter was placed before the detector [26]. In addition, a LynxEye™ one-dimensional detector was used to produce large counting that resulted in high-quality XRD profiles, which can be accurately resolved by least square methods. To confirm the proposed structure, the gathered XRD patterns were refined with the Pawley and Rietveld methods. In this regard, the computer program used to perform the refining processes was the Bruker DIFFRAC*plus* TOPAS™ software package, where the emission profile was shaped by Lorentzians [12].

XRD is a very powerful method for structure elucidation, given that X-rays have wavelengths between 0.15 and 10 nm, that is, dimensions comparable to the distances between planes in crystals [46–48]. By definition, diffraction ≡ scattering + interference. Then, using the Bragg law $2d_{hkl} \sin\theta = \lambda$ (d_{hkl} is the distance between the lattice planes [hkl], θ is the angle between the plane where the sample lies and is the X-ray wavelength) it is possible to calculate the interplanar distances (Table 12.1).

Next, applying the IR and Raman spectra the XRD (Figure 12.12) profiles, the knowledge related to the linking options of the three monomers (Figure 12.10), and

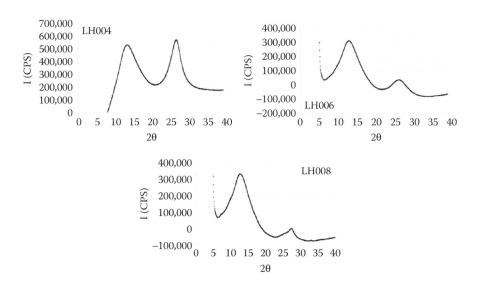

FIGURE 12.12 XRD profiles of the tested COFs.

TABLE 12.4

Interplanar Distance Measured during the XRD Test of the Three COFs

Sample	d_1 (Å)	d_2 (Å)
LH004	6.816	3.934
LH006	6.007	3.520
LH008	6.946	3.253

the International Tables for Crystallography [38], was proposed a tetragonal *P4mmm* space group unit cell representation.

Then, the cell parameters were calculated using $\left(1/d_{hkl}^2\right) = \left[\left((h^2+k^2)/a^2\right) + \left(l^2/c^2\right)\right]^{-1/2}$, that is, the relation valid for the tetragonal system (Table 12.4).

Next, in Figure 12.13 are the proposed frameworks, simulated with the ChemDoodle software [52] and represented included within the tetragonal *P4mmm* space group unit cell.

Finally, with the help of the XRD profile we can calculate the grain size of the powdered COFs applying the Scherrer equation:

$$\delta = \frac{K}{\text{FWHM}}\left(\frac{\lambda}{\cos\theta}\right)$$

where:

δ is the grain size

$K \approx 0.9$ is a shape factor

λ (*K* alpha Cu) = 1,5418 Å is the X-ray wavelength

FWHM = $\Delta(2\Theta)$ = 0.104 radian is the line width at half maximum intensity

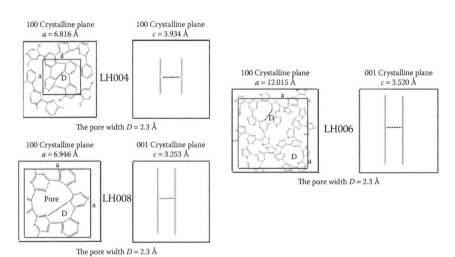

FIGURE 12.13 Frameworks of the tested COFs.

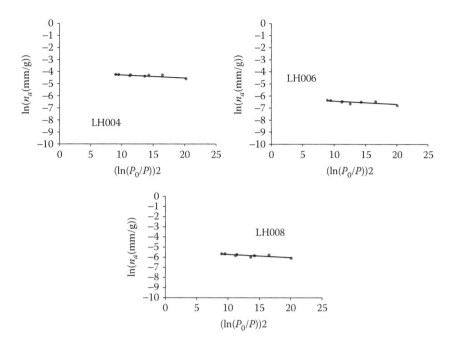

FIGURE 12.14 D–R plots for the adsorption of N_2 at 77 K in the tested COFs.

Then substituting these values in the Scherrer equation we get $\delta = 56$ nm. Thereafter, we can conclude that the three COFs have a very convenient particle size for the use of these materials in dynamic adsorption flow reactors, given that the transference area is very high because of the small amount of the grain size.

Finally, nitrogen (Praxair, 99.99%) adsorption was investigated at 77 K in the pressure range from 0 up to 1 bar on samples degassed at 373 K for 4 h in high vacuum (10^{-6} Torr) in a Quantachrome AS-1 automatic sorption analyzer [18]. Then, in Figure 12.14 are shown the D–R plots corresponding to the adsorption isotherm of N_2, at 77 K, in the tested COFs. In this regard, the fitting of the D–R equation for $n = 2$ allowed us to calculate N_m. E, along with W, the micropore volume, calculated with the following expression: $W = b\,N_a$, where $b = 0.0289$ mol/cc, is the molar volume of liquid nitrogen at 77 K (Table 12.5).

TABLE 12.5
Parameter for the Fitting of D–R Equation to the Nitrogen at 77 K Adsorption

Sample	N_m (mmol/g)	W (cm³/g)	E (kJ/mol)
LH004	0.019	0.00055	14.5
LH006	0.020	0.00056	14.5
LH008	0.005	0.00014	14.4

The N_2 at 77 K adsorption data clearly show that nitrogen do not penetrate into the adsorption space of the tested COFs, provided this phenomenon is a consequence of the fact that the kinetic diameter of nitrogen σ (N_2) = 3.00 Å [18] is bigger than the pore size of the produced COFs, that is, D = 2.3 Å [50].

REFERENCES

1. T.P. Maji and S. Kitagawa, *Pure App. Chem.*, 79 (2007) 2155.
2. D.J. Tranchemontangne, J.L. Mendoza-Cortes, M. O'Keeffe, and O.M. Yaghi, *Chem. Soc. Rev.*, 38 (2009) 1257.
3. P. Horcajada, R. Gref, T. Baati, P.K. Allan, G. Maurin, P. Couvreur, G. Ferey, R.E. Morris, and C. Serre, *Chem. Rev.*, 112 (2012) 1232.
4. K. Sumida, D.L. Rogow, J.A. Mason, T.M. McDonald, E.D. Bloch, Z.R. Herm, T.-H. Bae, and J.R. Long, *Chem. Rev.*, 112 (2012) 724.
5. H.-C. Zhou and S. Kitakawa, *Chem. Soc. Rev.*, 43 (2010) 5415.
6. A. Corma, H. García, and F.X. Llabrés i Xamena, *Chem. Rev.*, 110 (2010) 4606.
7. R. Roque-Malherbe, A. Costa, C. Rivera, F. Lugo, and R. Polanco, *J. Mat. Sci. Eng.*, 3 (2013) 263.
8. F. Marquez, O. Uwakweh, N. Lopez, E. Chavez, R. Polanco, C. Morant, J. M. Sanz et al., *J. Solid State Chem.*, 184 (2011) 655.
9. R. Roque-Malherbe, R. Polanco, and F. Marquez, *J. Phys. Chem. C*, 114 (2010) 17773.
10. R. Roque-Malherbe, O.N.C. Uwakweh, C. Lozano, R. Polanco, A. Hernandez-Maldonado, P. Fierro, F. Lugo, and J.N. Primera-Pedrozo, *J. Phys. Chem. C*, 115 (2011) 15555.
11. R. Roque-Malherbe, F. Lugo, C. Rivera, R. Polanco, P. Fierro, and O.N.C. Uwakweh, *Curr. Appl. Phys.*, 15 (2015) 571.
12. R. Roque-Malherbe, F. Lugo, and R. Polanco, *App. Surf. Sci.*, 385 (2016) 360.
13. R. Roque-Malherbe, E. Carballo, R. Polanco, F. Lugo, and C. Lozano, *J. Phys. Chem. Solids*, 86 (2015) 65.
14. S-Y. Ping and W. Wang, *Chem. Soc. Rev.*, 42 (2013) 548.
15. N. Huang, P. Wang, and P. Jiang, *Nature Rev. Mat.*, 1 (2016) 16068.
16. F.A. Cotton, G. Wilkinson, and C.A. Murillo, *Advanced Inorganic Chemistry* (6th ed.), Wiley-Intersciences, New York, 1999.
17. T. Duangthongyou, S. Jirakulpattana, C. Phakawatchai, M. Kurmoo, and S. Siripaisarnpipat, *Polyhedron*, 29 (2010) 1156.
18. R. Roque-Malherbe, *Adsorption and Diffusion of Gases in Nanoporous Materials*, CRC Press-Taylor & Francis Group, Boca Raton, FL, 2007.
19. R. Roque-Malherbe, *The Physical Chemistry of Materials: Applications in Pollution Abatement and Sustainable Energy*, CRC Press, Boca Raton, FL, 2009.
20. A.D. Buckingham, *Proc. R. Soc. Lond. Ser. A*, 273 (1963) 275.
21. W.N. Haynes (chief editor), *Handbook of Physics and Chemistry* (97th ed.), CRC Press, Boca Raton, FL, 2016.
22. V.A. Tvardoskiy, *Sorbent Deformation*, Elsevier, Amsterdam, the Netherlands, 2006.
23. F.-X. Coudert, *Phys. Chem. Chem. Phys.*, 12 (2010) 10904.
24. W.E. Smith and G. Dent, *Modern Raman Spectroscopy. A Practical Approach*, John Wiley & Sons, New York, 2005.
25. K. Nakamoto, *Infrared and Raman Spectra of Inorganic and Coordination Compounds: Part A: Theory and Applications in Inorganic Chemistry*, John Wiley & Sons, New York, 1997.
26. M.M. Dubinin, *Prog. Surface. Membrane Sci.*, 9 (1975) 1.
27. B.P. Bering, M.M. Dubinin, and V.V. Serpinskii, *J. Collids Interface Sci.*, 38 (1972) 185.

28. R. Roque-Malherbe, *Mic. Mes. Mat.*, 41 (2000) 227.
29. F. Rouquerol, J. Rouquerol, and K. Sing, *Adsorption by Powder Porous Solids*, Academic Press, New York, 1999.
30. W. Rudzinski and D.H. Everett, *Adsorption of Gases in Heterogeneous Surfaces*, Academic Press, London, UK, 1992.
31. B. Bonelli, B. Onida, B. Fubini, C. Otero-Arean, and E. Garrone, *Langmuir*, 16 (2000) 4976.
32. A.L. Goodman, L.M. Campus, and K.T. Schroeder, *Energy Fuels*, 19 (2005) 471.
33. F.X. Llabrés i Xamena and A. Zecchina, *Phys. Chem. Chem. Phys.*, 4 (2002) 1978.
34. P.L. Llewellyn, S. Bourrelly, C. Serre, A. Vimont, M. Daturi, L. Hamon, M. De Weireld et al., *Langmuir*, 24 (2008) 7245.
35. C. Prestipino, L. Regli, J.G. Vitillo, F. Bonino, A. Damin, L. Lamberti, A. Zecchina, P.L. Solari, K.O. Kongshaug, and S. Bordiga, *Chem. Mater.*, 18 (2006) 1337.
36. M. Zentkova, M. Mihalika, I. Toth, I. Mitroovaa, A. Zentkoa, M. Sendeka, J. Kovac, M. Lukakpwa, M. Maryskoc, and M. Miglierinib, *J. Magn. Magn. Mater*, 272–276 (2004) e753.
37. B.N. Figgi and J. Lewis, The Magnetic properties of transition metal complexes, in *Coordination Chemistry* (J. Lewis and R.S. Wilkins, Eds.), Wiley, New York, 1960, p. 400.
38. T. Hahn (Ed.), *International Tables for Crystallography,* Volume A, (5th ed.), The International Union of Crystallography, Springer, Dordrecht, the Netherlands, 2009.
39. W. Kraus and W.G. Nolze, PowderCell-2.4, Federal institute for materials research and testing, in *Rudower Chaussee*, 5 Berlin, Germany, 2000.
40. E.G. Derouane, in *Zeolites Microporous Solids: Synthesis, Structure and Reactivity* (E.G. Derouane, F. Lemos, C. Naccache, and F. Ramos-Riveiro, Eds.), Kluwer Academic Publishers, Dordrecht, the Netherlands, 1992. p. 511.
41. R. Roque-Malherbe and F. Diaz-Castro, *J. Mol. Catal. A*, 280 (2008) 194.
42. E.P. Scriven and C.A. Ramsden, (Eds.), *Advances in Heterocyclic Chemistry*, Vol. 116, (5th ed.), Elsevier, Amsterdam, the Netherlands, 2015.
43. P. Kuhn, A. Thomas, and M. Antonietti, *Macromolecules*, 42 (2009) 319.
44. S. Ren, M.J. Bojdys, R. Dawson, A. Laybourn, Y.Z. Khimyak, D.J. Adams, and A.I. Cooper, *Adv. Mater.*, 24 (2012) 2357.
45. P. Katekomol, J. Roeser, M. Bojdys, J. Weber, and A. Thomas, *Chem. Mater.*, 25 (2013) 1542.
46. R.C. Reynolds, in *Modern Powder Diffraction,* Vol. 20, (D.L. Bish and J.E. Post, Eds.), Reviews in Mineralogy, *The Mineralogical Society of America*, BookCrafters, Chelsea, MI, 1989, p. 1.
47. R. Jenkins and R.L. Snyder, *Introduction to X-Ray Powder Diffractometry*, John Wiley & Sons, New York, 1996.
48. C. Kittel, *Introduction to Solid State Physics* (8th ed.), John Wiley & Sons, New York, 2004.
49. L.J. Bellamy, *The Infrared Spectra of Complex Molecules* (3rd ed.), Chapman & Hall, London, Vol. I, 1975; Vol. II, 1980.
50. L. Hernandez-Colon, A. Emiliano-Paulino, and R. Roque-Malherbe, *J. Mol. Struct.* (submitted, 2017).
51. D.A. Skoog, F.J. Holler, and T.A. Nieman, *Principles of Instrumental Analysis* (5th ed.), Saunders College Publishing, Philadelphia, PA, 1998.
52. ChemDoodle 8, iChemLabs LLC, Technical Wien University, Institute for Applied Synthetic Chemistry.

Index

Note: Page numbers followed by f and t refer to figures and tables respectively.